Student Study Art Notebook

to accompany

Genetics
From Genes to Genomes

Second Edition

Leland H. Hartwell
Fred Hutchinson Cancer Research Center

Leroy Hood
The Institute for Systems Biology

Michael L. Goldberg
Cornell University

Ann E. Reynolds
University of Washington

Lee M. Silver
Princeton University

Ruth C. Veres

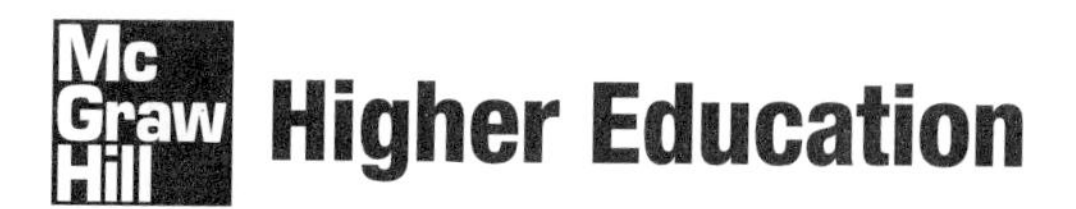

Mc Graw Hill **Higher Education**

Boston   Burr Ridge, IL   Dubuque, IA   Madison, WI   New York   San Francisco   St. Louis
Bangkok   Bogotá   Caracas   Kuala Lumpur   Lisbon   London   Madrid   Mexico City
Milan   Montreal   New Delhi   Santiago   Seoul   Singapore   Sydney   Taipei   Toronto

The McGraw·Hill Companies

Student Study Art Notebook to accompany
GENETICS: FROM GENES TO GENOMES, SECOND EDITION
LELAND H. HARTWELL, LEROY HOOD, MICHAEL L. GOLDBERG, ANN E. REYNOLDS,
LEE M. SILVER, AND RUTH C. VERES

Published by McGraw-Hill Higher Education, an imprint of The McGraw-Hill Companies, Inc.,
1221 Avenue of the Americas, New York, NY 10020. Copyright © 2004 by The McGraw-Hill
Companies, Inc. All rights reserved.

 This book is printed on recycled, acid-free paper containing
10% postconsumer waste.
RECYCLED

1 2 3 4 5 6 7 8 9 0 QPD/QPD 0 9 8 7 6 5 4 3

ISBN 0-07-297796-5

www.mhhe.com

DIRECTORY OF NOTEBOOK FIGURES
TO ACCOMPANY
HARTWELL/HOOD/GOLDBERG/REYNOLDS/ SILVER/VERES
GENETICS: FROM GENES TO GENOMES, 2/e

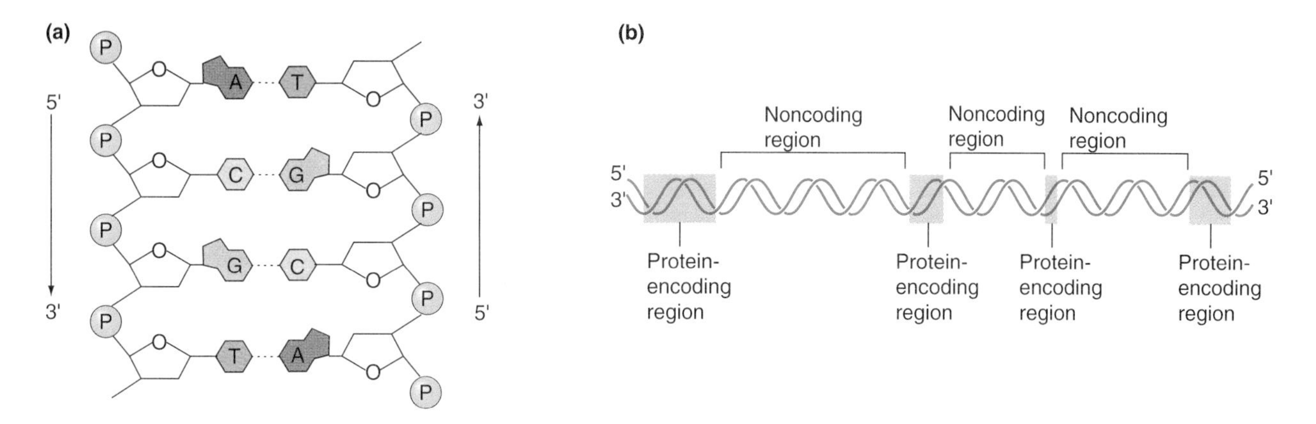

(a)

(b)

Noncoding region   Noncoding region   Noncoding region

Protein-encoding region   Protein-encoding region   Protein-encoding region   Protein-encoding region

Complementary base pairs are a key feature of the DNA molecule
Figure 1.2

(a)

Chymotrypsin — ANTPORLQQASLPLLSNTNCKK‑‑YWGTKIKDAMICAGAS‑GVS
(149)   (189)

Elastase ———— GQLAQTLQQAYLPTVDYAICSSSSYWGSTVKNSMVCAGGDGVRS

(190)   (245)
SCMGDSGGPLVCKKNGAWTLVGIVSWGSS‑TCSTS‑TPGVYARVTALVNWVQQTLAAN

GCQGDSGGPLHCLVNGQYAVHGVTSFVSRLGCNVTRKPTVFTRVSAYISWINNVIASN

A = Ala = alanine	G = Gly = glycine
C = Cys = cysteine	H = His = histidine
D = Asp = aspartic acid	I = Ile = isoleucine
E = Glu = glutamic acid	K = Lys = lysine
F = Phe = phenylalanine	L = Leu = leucine

M = Met = methionine	S = Ser = serine
N = Asn = asparagine	T = Thr = threonine
P = Pro = proline	V = Val = valine
Q = Gln = glutamine	W = Trp = tryptophan
R = Arg = arginine	Y = Tyr = tyrosine

(b)

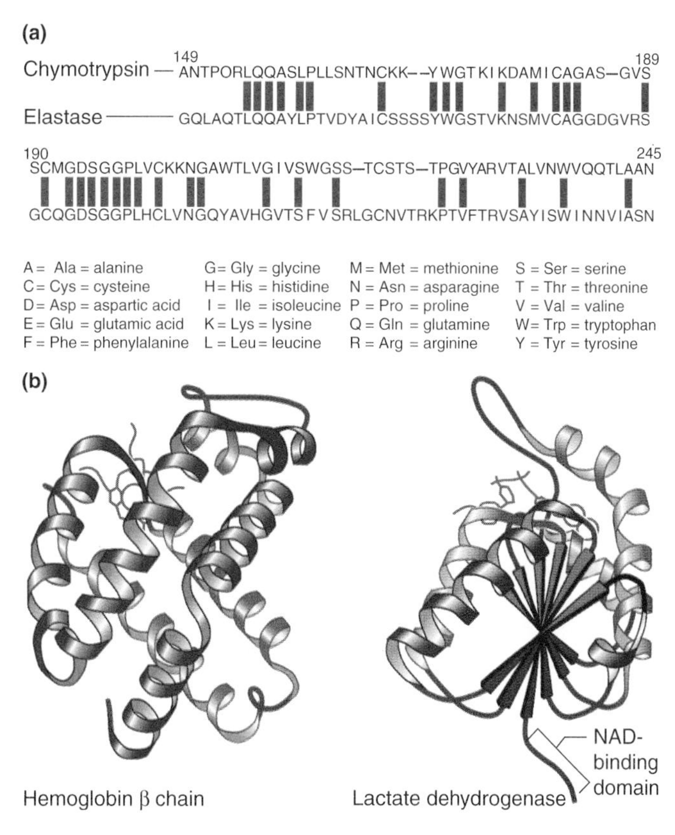

Hemoglobin β chain   Lactate dehydrogenase   NAD-binding domain

Proteins are polymers of amino acids that fold in three dimensions
Figure 1.5

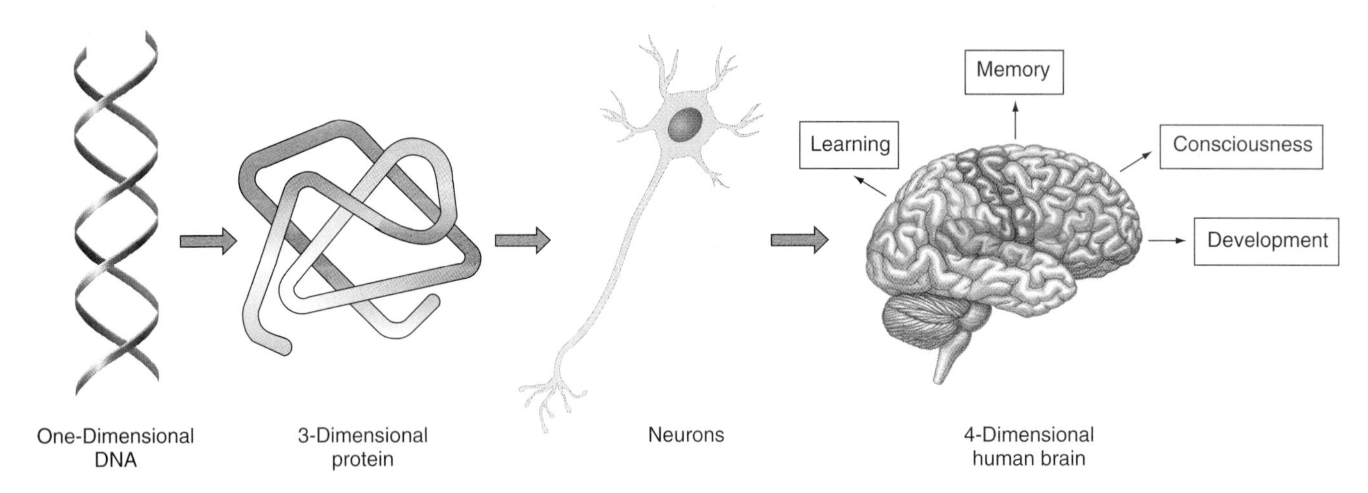

Diagram of the conversion of biological information from a one- to a three- and finally a four-dimensional state.
Figure 1.6

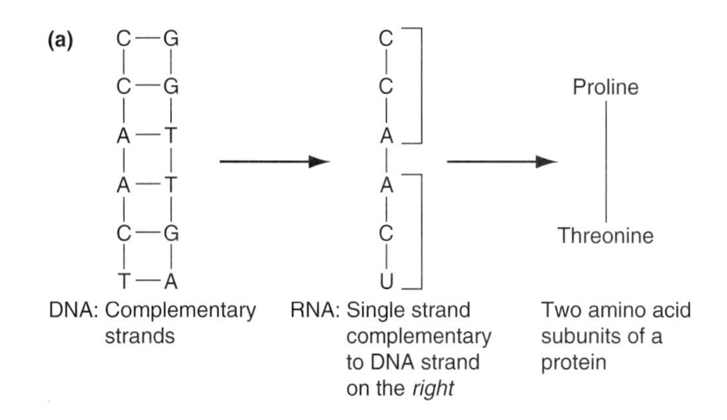

(a)

DNA: Complementary strands

RNA: Single strand complementary to DNA strand on the *right*

Two amino acid subunits of a protein

RNA is an intermediary in the conversion of DNA information into protein via the genetic code
Figure 1.7

(b)

		Second letter			
		U	C	A	G	
First letter	U	UUU ⎫ Phe UUC ⎭ UUA ⎫ Leu UUG ⎭	UCU ⎫ UCC ⎪ Ser UCA ⎪ UCG ⎭	UAU ⎫ Tyr UAC ⎭ UAA Stop UAG Stop	UGU ⎫ Cys UGC ⎭ UGA Stop UGG Trp	U C A G
	C	CUU ⎫ CUC ⎪ Leu CUA ⎪ CUG ⎭	CCU ⎫ CCC ⎪ Pro CCA ⎪ CCG ⎭	CAU ⎫ His CAC ⎭ CAA ⎫ Gln CAG ⎭	CGU ⎫ CGC ⎪ Arg CGA ⎪ CGG ⎭	U C A G
	A	AUU ⎫ AUC ⎪ Ile AUA ⎭ AUG Met	ACU ⎫ ACC ⎪ Thr ACA ⎪ ACG ⎭	AAU ⎫ Asn AAC ⎭ AAA ⎫ Lys AAG ⎭	AGU ⎫ Ser AGC ⎭ AGA ⎫ Arg AGG ⎭	U C A G
	G	GUU ⎫ GUC ⎪ Val GUA ⎪ GUG ⎭	GCU ⎫ GCC ⎪ Ala GCA ⎪ GCG ⎭	GAU ⎫ Asp GAC ⎭ GAA ⎫ Glu GAG ⎭	GGU ⎫ GGC ⎪ Gly GGA ⎪ GGG ⎭	U C A G

Third letter

Comparisons of gene products in different species provide evidence for the relatedness of living organisms
Figure 1.8

S. cerevisiae	---PGSAKKGATLFKTRCQQCHTIEEGGPNKV
A. thaliana	---GDAKKGANLFKTRCAQCHTLKAGEGNKI
C. elegans	--AGDYEKGKKVYKQRCLQCHVVDS-TATKT
D. melanogaster	---AGDVEKGKKLFVQRCAQCHTVEAGGKHKV
M. musculus	---MGDVEKGKKIFVQKCAQCHTVEKGGKHKT
H. sapiens	---MGDVEKGKKIFIMKCSQCHTVEKGGKHKT
	* * ** .. * * *** *

S. cerevisiae	GPNLHGIFGRHSGQVKGYSYTDANINKNVKW
A. thaliana	GPELHGLFGRKTGSVAGYSYTDANKQKGIEW
C. elegans	GPTLHGVIGRTSGTVSGFDYSAANKNKGVVW
D. melanogaster	GPNLHGLIGRKTGQAAGFAYTDANKAKGITW
M. musculus	GPNLHGLFGRKTGQAAGFSYTDANKNKGITW
H. sapiens	GPNLHGLFGRKTGQAPGYSYTAANKNKGIIW
	** *** ** * * * ** . *

S. cerevisiae	DEDSMSEYLTNPKKYIPGTKMAFAGLKKEKDR
A. thaliana	KDDTLFEYLENPKKYIPGTKMAFGGLKKPKDR
C. elegans	TKETLFEYLLNPKKYIPGTKMVFAGLKKADER
D. melanogaster	NEDTLFEYLENPKKYIPGTKMIFAGLKKPNER
M. musculus	GEDTLMEYLENPKKYIPGTKMIFAGIKKKGER
H. sapiens	GEDTLMEYLENPKKYIPGTKMIFVGIKKEER
	... *** ******** * * ** *

S. cerevisiae	NDLITYMTKAAK---
A. thaliana	NDLITFLEEETK---
C. elegans	ADLIKYIEVESA---
D. melanogaster	GDLIAYLKSATK---
M. musculus	ADLIAYLKKATN---
H. sapiens	ADLIAYLKKATN---
	*** .. *

 * Indicates identical and . indicates similar

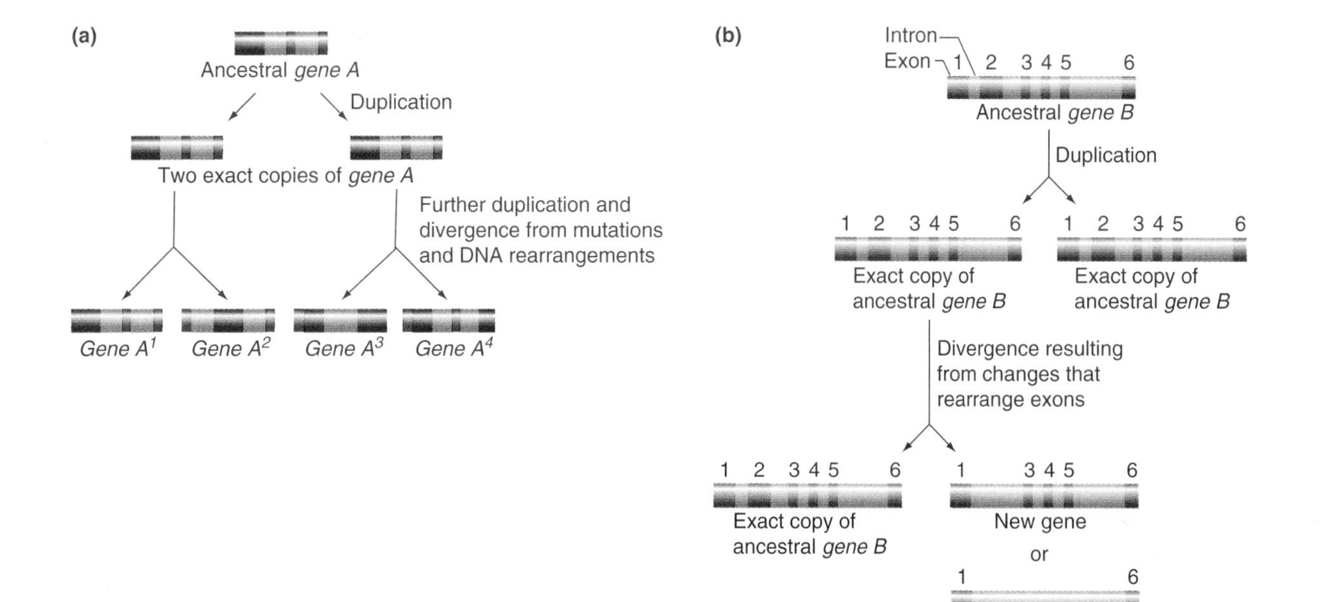

(a)

Ancestral *gene A*

Duplication

Two exact copies of *gene A*

Further duplication and divergence from mutations and DNA rearrangements

Gene A¹   Gene A²   Gene A³   Gene A⁴

(b)

Intron

Exon 1 2 3 4 5   6

Ancestral *gene B*

Duplication

1 2 3 4 5   6

Exact copy of ancestral *gene B*

1 2 3 4 5   6

Exact copy of ancestral *gene B*

Divergence resulting from changes that rearrange exons

1 2 3 4 5   6

Exact copy of ancestral *gene B*

1   3 4 5   6

New gene

or

1   6

Another new gene

How genes arise by duplication and divergence
Figure 1.10

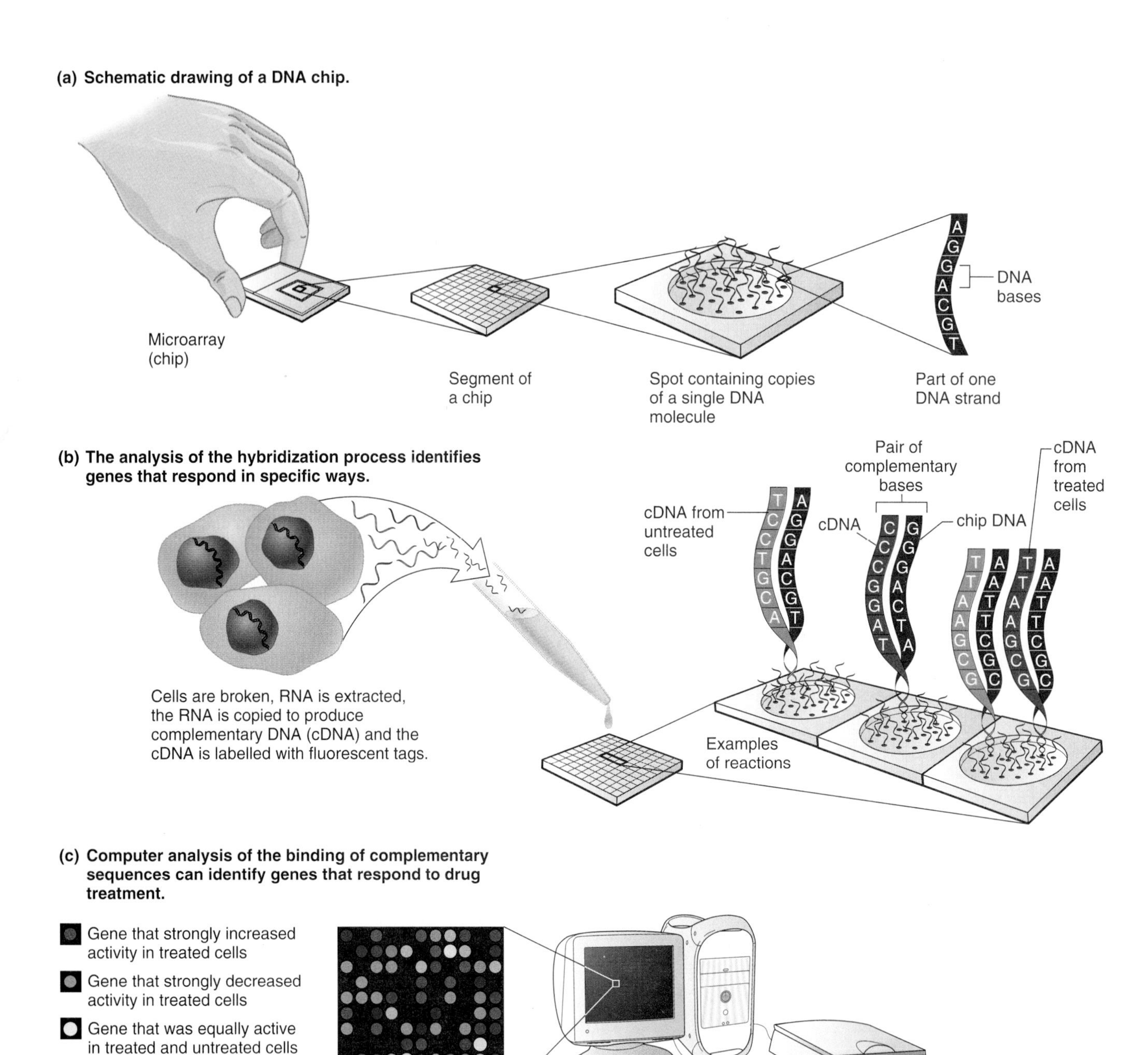

(a) Schematic drawing of a DNA chip.

Microarray
(chip)

Segment of
a chip

Spot containing copies
of a single DNA
molecule

Part of one
DNA strand

DNA
bases

**(b) The analysis of the hybridization process identifies
genes that respond in specific ways.**

Cells are broken, RNA is extracted,
the RNA is copied to produce
complementary DNA (cDNA) and the
cDNA is labelled with fluorescent tags.

cDNA from
untreated
cells

cDNA

Pair of
complementary
bases

chip DNA

cDNA
from
treated
cells

Examples
of reactions

**(c) Computer analysis of the binding of complementary
sequences can identify genes that respond to drug
treatment.**

Gene that strongly increased
activity in treated cells

Gene that strongly decreased
activity in treated cells

Gene that was equally active
in treated and untreated cells

Gene that was inactive
in both groups

A DNA chip
Figure 1.13

Stephen H. Friedn, The Magic of Microarrays, Feb 2002 Scientific American, courtesy Jared Schneidman Designs.

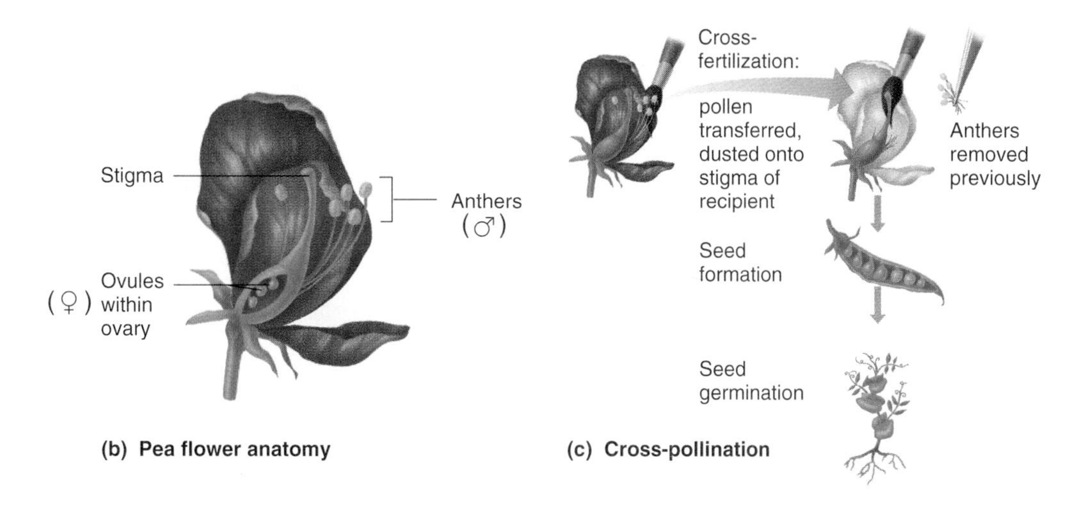

Stigma

Anthers
(♂)

(♀) Ovules
within
ovary

(b) Pea flower anatomy

Cross-
fertilization:

pollen
transferred,
dusted onto
stigma of
recipient

Seed
formation

Seed
germination

Anthers
removed
previously

(c) Cross-pollination

Mendel's experimental organism: The garden pea
Figure 2.7

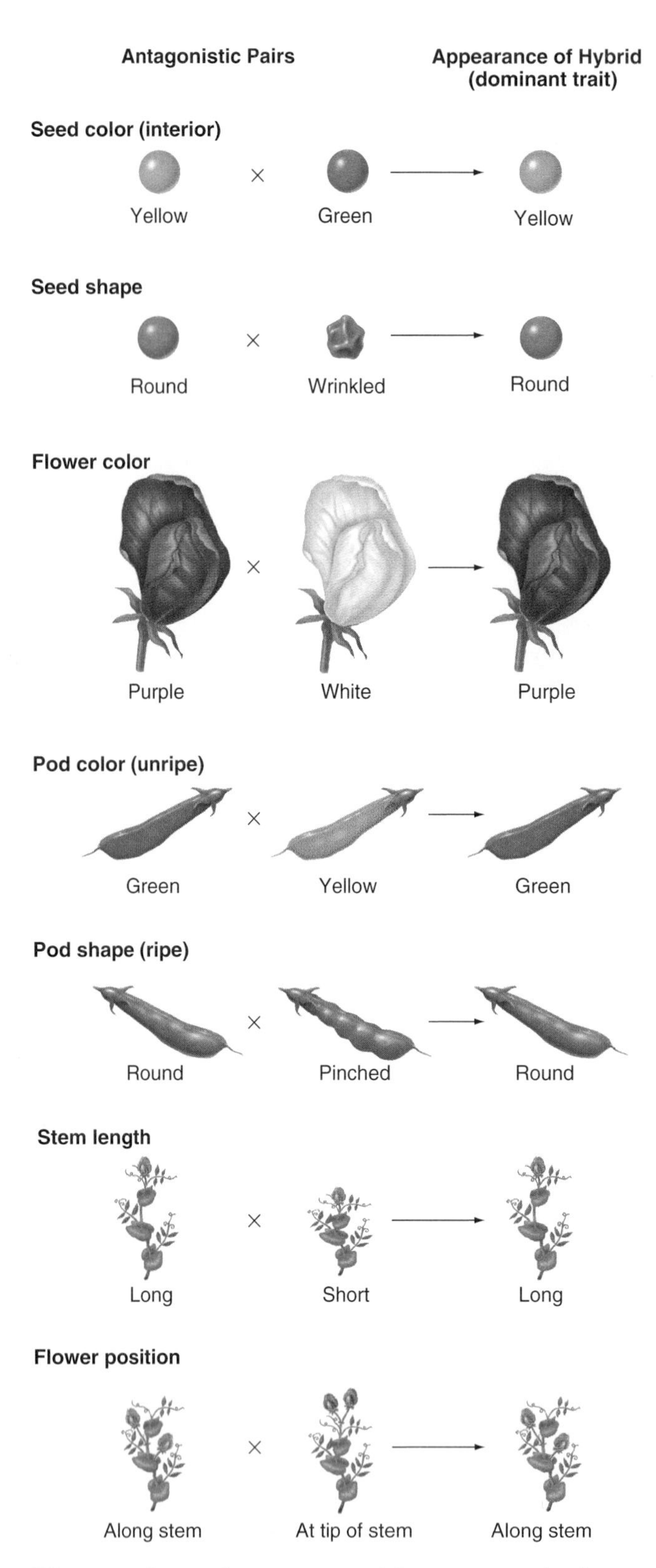

The mating of parents with antagonistic traits produces hybrids
Figure 2.8

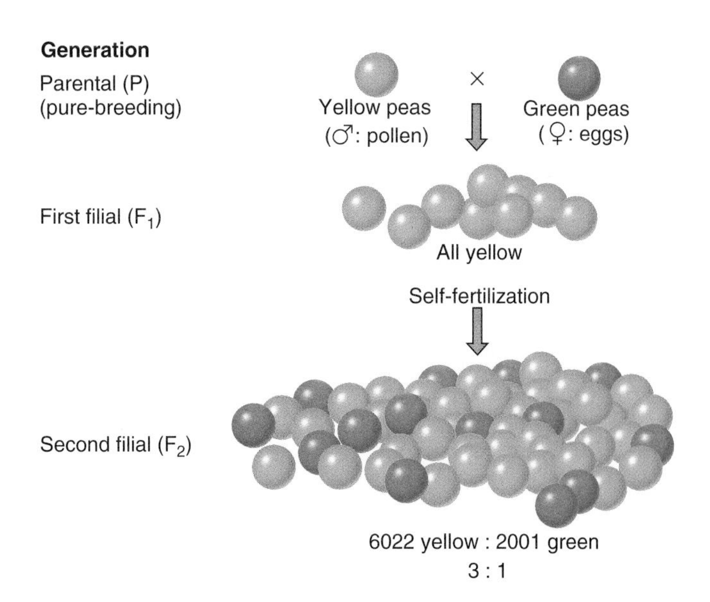

Generation

Parental (P)
(pure-breeding)

Yellow peas
($\male$: pollen)

×

Green peas
($\female$: eggs)

First filial (F_1)

All yellow

Self-fertilization

Second filial (F_2)

6022 yellow : 2001 green
3 : 1

Analyzing a monohybrid cross
Figure 2.9

(a) The two alleles for each trait separate during gamete formation.

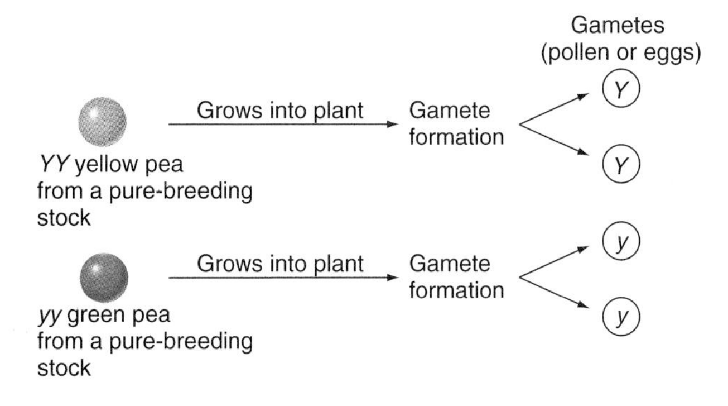

Gametes
(pollen or eggs)

Grows into plant → Gamete formation → Y, Y

YY yellow pea
from a pure-breeding
stock

Grows into plant → Gamete formation → y, y

yy green pea
from a pure-breeding
stock

(b) Two gametes, one from each parent, unite at random at fertilization.

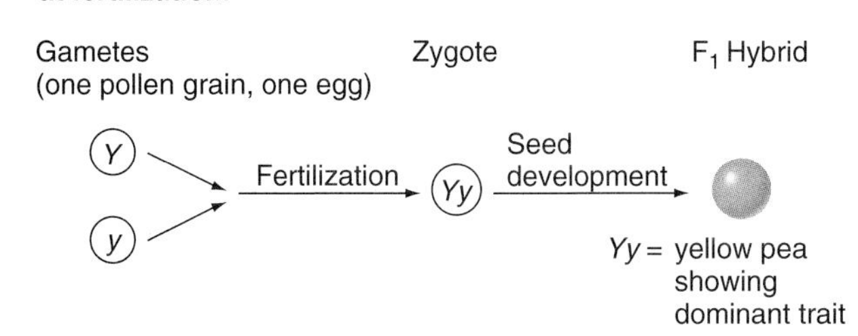

Gametes
(one pollen grain, one egg)

Zygote

F_1 Hybrid

Y, y → Fertilization → Yy → Seed development →

Yy = yellow pea
showing
dominant trait

Y = yellow-determining allele of pea color gene
y = green-determining allele of pea color gene

The law of segregation
Figure 2.10

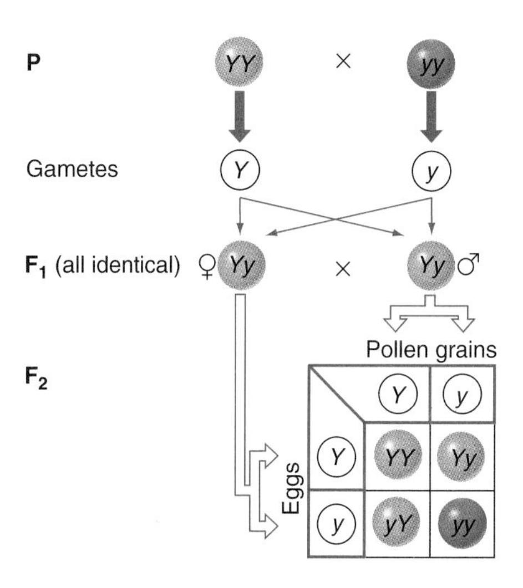

P — YY × yy

Gametes — Y y

F_1 (all identical) — $\female$ Yy × Yy $\male$

F_2

Pollen grains

	Y	y
Y	YY	Yy
y	yY	yy

Eggs

The Punnett square: Visual summary of a cross
Figure 2.11

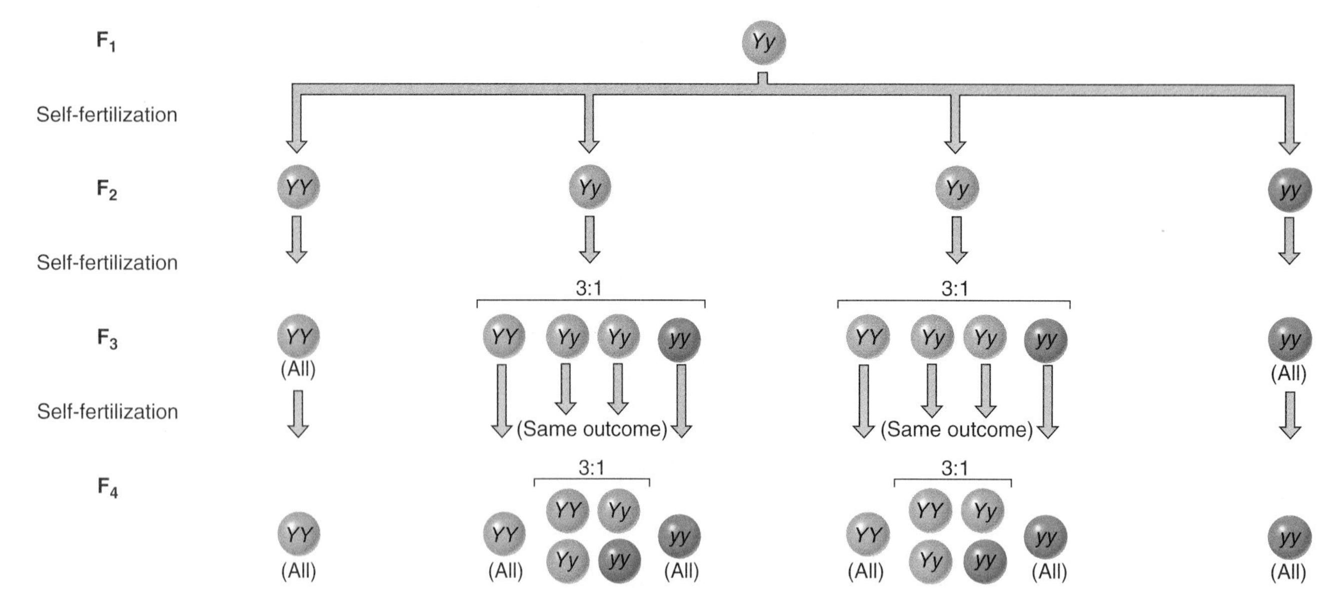

Yellow F₂ peas are of two types: Pure breeding and hybrid
Figure 2.12

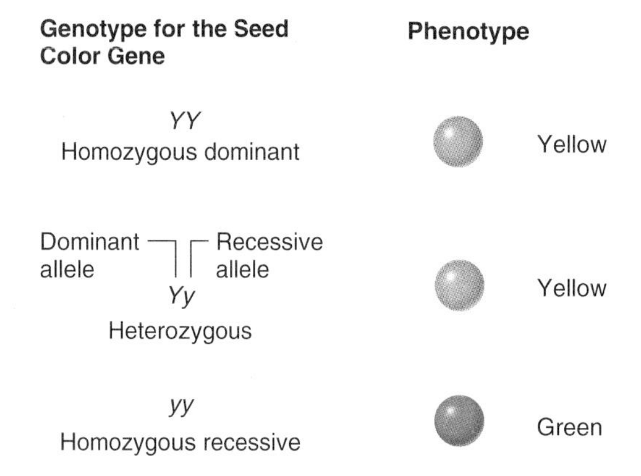

Genotype versus phenotype in homozygotes and heterozygotes
Figure 2.13

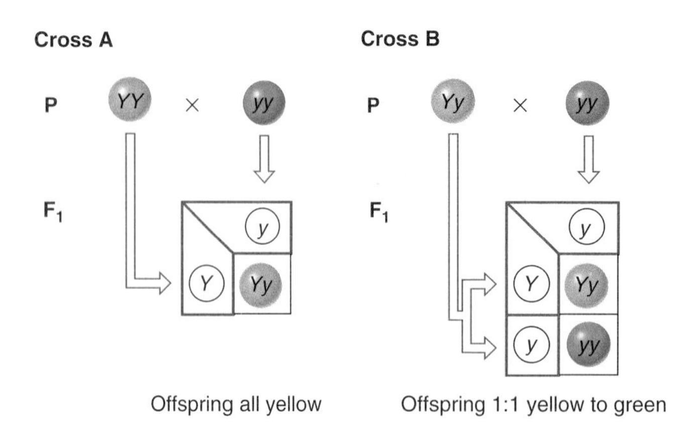

How a testcross reveals genotype
Figure 2.14

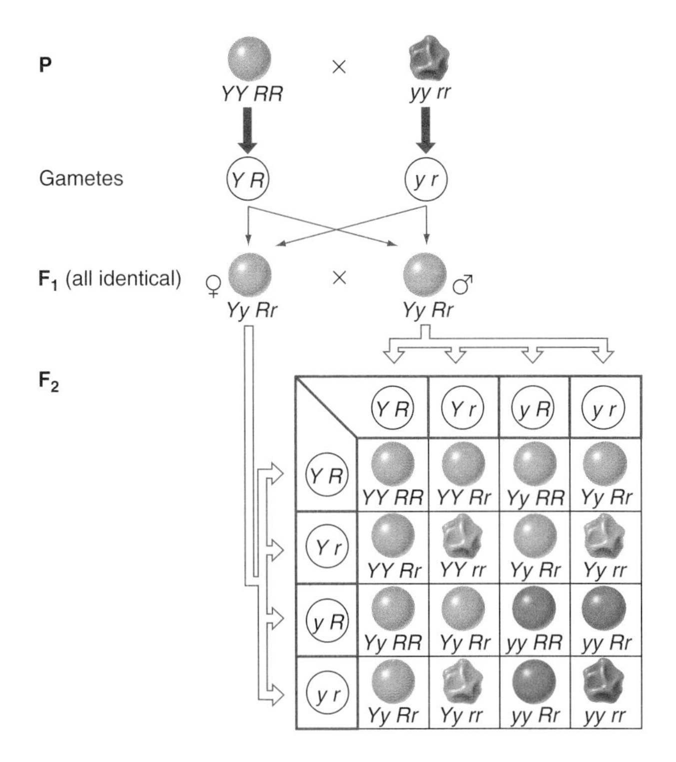

Type	Genotype	Phenotype		Number	Phenotypic ratio
Parental	Y– R–		yellow round	315	9/16
Recombinant	yy R–		green round	108	3/16
Recombinant	Y– rr		yellow wrinkled	101	3/16
Parental	yy rr		green wrinkled	32	1/16

Ratio of yellow (dominant) to green (recessive) = 12:4 or 3:1
Ratio of round (dominant) to wrinkled (recessive) = 12:4 or 3:1

A dihybrid cross produces parental types and recombinant types
Figure 2.15

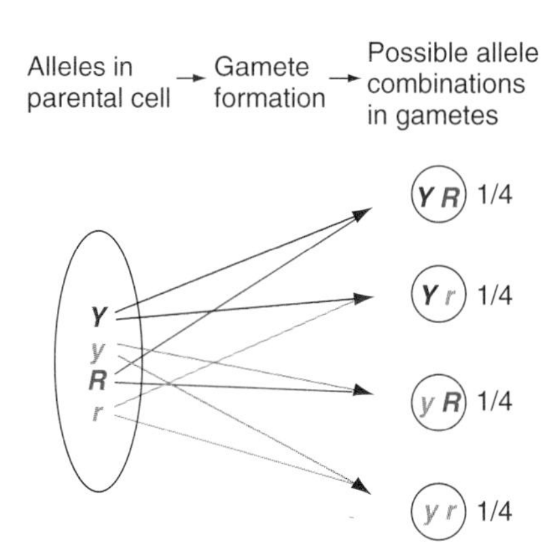

The law of independent assortment
Figure 2.16

Gene 1	Gene 2	Phenotypes
3/4 yellow	3/4 round	9/16 yellow round
	1/4 wrinkled	3/16 yellow wrinkled
1/4 green	3/4 round	3/16 green round
	1/4 wrinkled	1/16 green wrinkled

Following crosses with branched-line diagrams
Figure 2.17

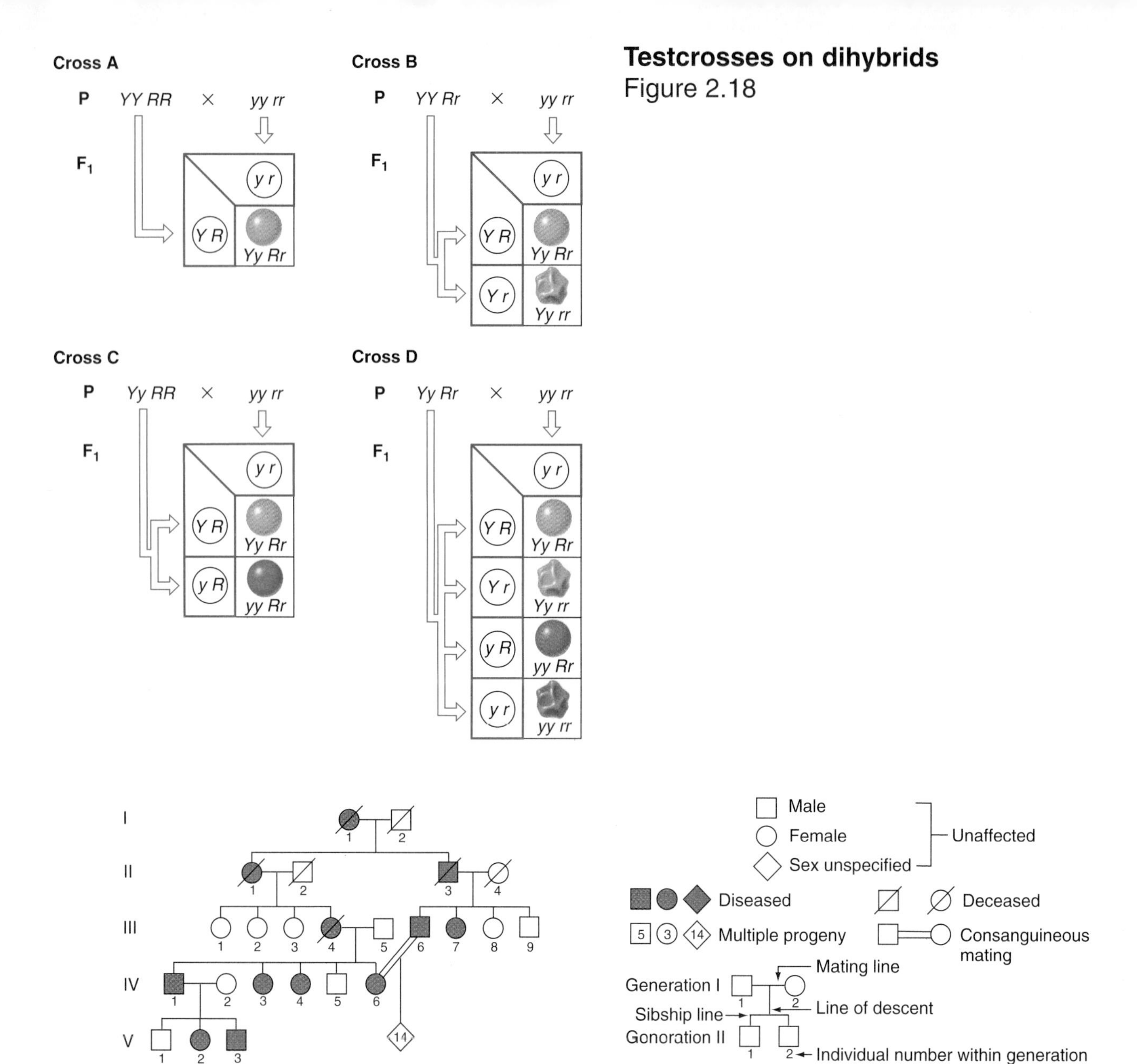

Cross A

P   YY RR   ×   yy rr

F₁

Cross B

P   YY Rr   ×   yy rr

F₁

Cross C

P   Yy RR   ×   yy rr

F₁

Cross D

P   Yy Rr   ×   yy rr

F₁

Testcrosses on dihybrids
Figure 2.18

Huntington disease: A rare dominant trait
Figure 2.20

Cystic fibrosis: A recessive condition
Figure 2.21

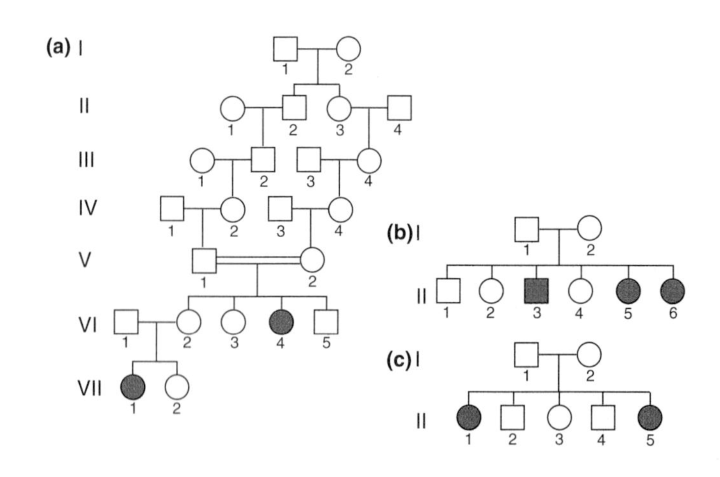

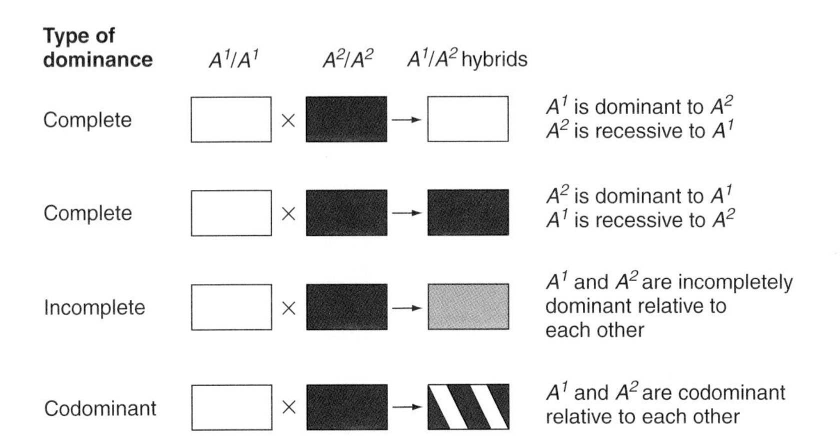

Type of dominance	A^1/A^1	A^2/A^2	A^1/A^2 hybrids	
Complete		$\times$	$\rightarrow$	A^1 is dominant to A^2 A^2 is recessive to A^1
Complete		$\times$	$\rightarrow$	A^2 is dominant to A^1 A^1 is recessive to A^2
Incomplete		$\times$	$\rightarrow$	A^1 and A^2 are incompletely dominant relative to each other
Codominant		$\times$	$\rightarrow$	A^1 and A^2 are codominant relative to each other

Different dominance relationships
Figure 3.2

(b) A Punnett square for incomplete dominance

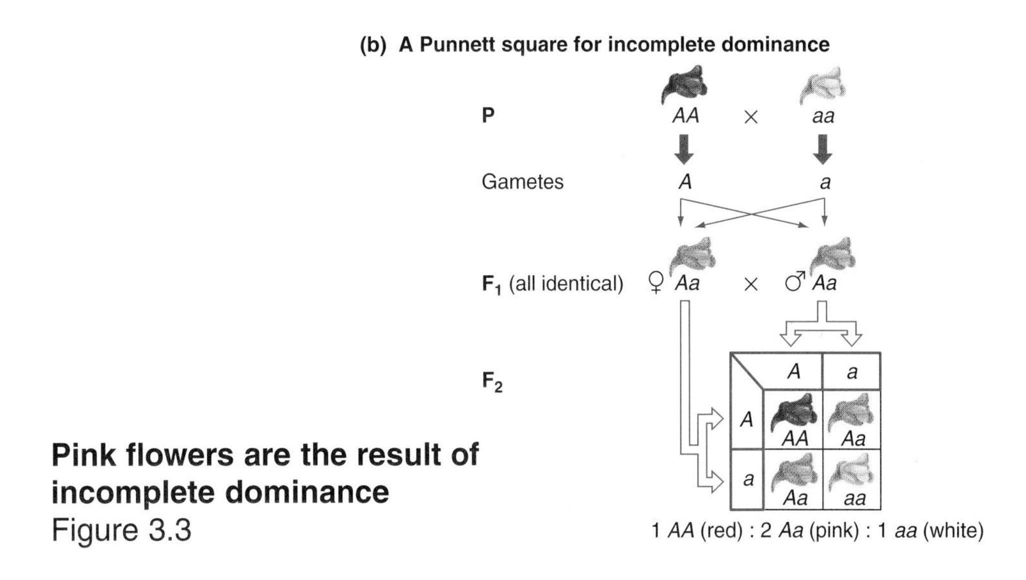

Pink flowers are the result of incomplete dominance
Figure 3.3

1 AA (red) : 2 Aa (pink) : 1 aa (white)

(a) Codominant lentil coat patterns

(b) Codominant blood group alleles

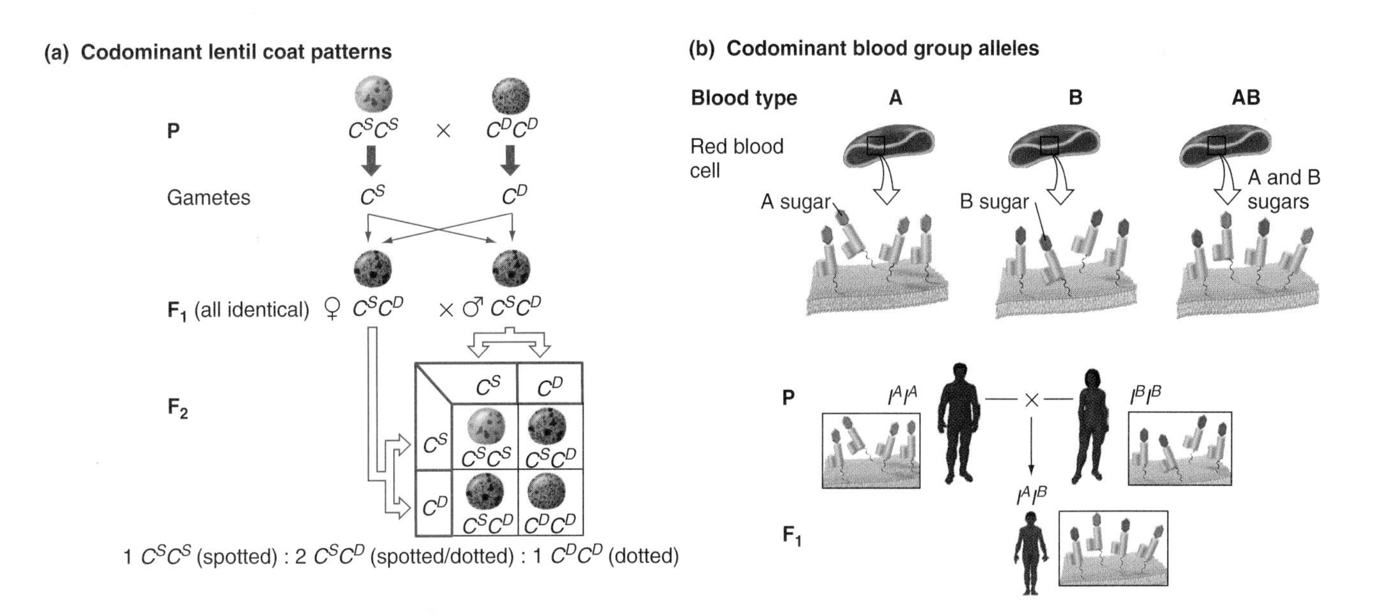

1 C^SC^S (spotted) : 2 C^SC^D (spotted/dotted) : 1 C^DC^D (dotted)

In codominance, F₁ hybrids display the traits of both parents
Figure 3.4

ABO blood types are determined by three alleles of one gene
Figure 3.5

(a)

Genotypes	Corresponding Phenotypes: Type(s) of Molecule on Cell
$I^A I^A$ $I^A i$	A
$I^B I^B$ $I^B i$	B
$I^A I^B$	AB
ii	O

(b)

Blood Type	Antibodies in Serum
A	Antibodies against B
B	Antibodies against A
AB	No antibodies against A or B
O	Antibodies against A and B

(c)

Blood Type of Recipient	Donor Blood Type (Red Cells)			
	A	B	AB	O
A	+	−	−	+
B	−	+	−	+
AB	+	+	+	+
O	−	−	−	+

Parental Generation	F₁ Generation	F₂ Generation	
Parental seed coat pattern in cross Parent 1 × Parent 2	F₁ phenotype	Total F₂ frequencies and phenotypes	Apparent pheno-typic ratio

marbled-1 × clear → marbled-1 → 798 296   3 :1

marbled-2 × clear → marbled-2 → 123 46   3 :1

spotted × clear → spotted → 283 107   3 :1

dotted × clear → dotted → 1,706 522   3 :1

marbled-1 × marbled-2 → marbled-1 → 272 72   3 :1

marbled-1 × spotted → marbled-1 → 499 147   3 :1

marbled-1 × dotted → marbled-1 → 1,597 549   3 :1

marbled-2 × dotted → marbled-2 → 182 70   3 :1

spotted × dotted → spotted/dotted → 168 339 157   1 : 2 : 1

Dominance series: marbled-1 > marbled-2 > spotted = dotted > clear

How to establish the dominance relations between multiple alleles
Figure 3.6

(b) Alleles of the *agouti* gene

Genotype	Phenotype
$A-$	agouti
$a^t a^t$	black/yellow
aa	black

(c) Evidence for a dominance series

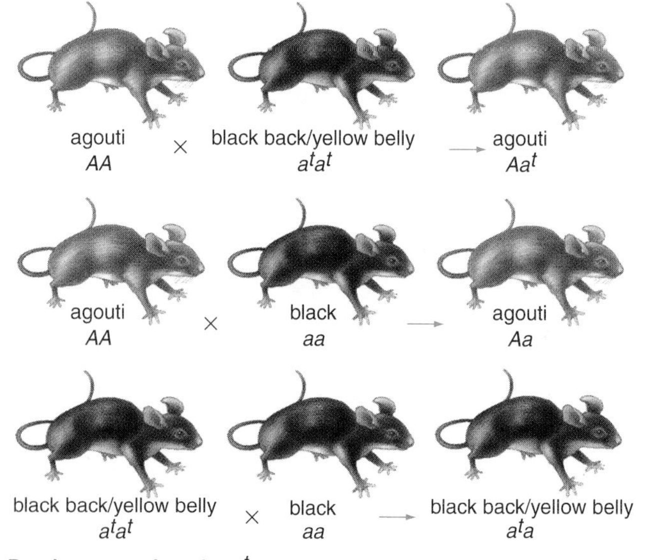

agouti *AA* × black back/yellow belly *a^t a^t* → agouti *Aa^t*

agouti *AA* × black *aa* → agouti *Aa*

black back/yellow belly *a^t a^t* × black *aa* → black back/yellow belly *a^t a*

Dominance series: $A > a^t > a$

The mouse *agouti* gene: One wild-type allele, many mutant alleles
Figure 3.7

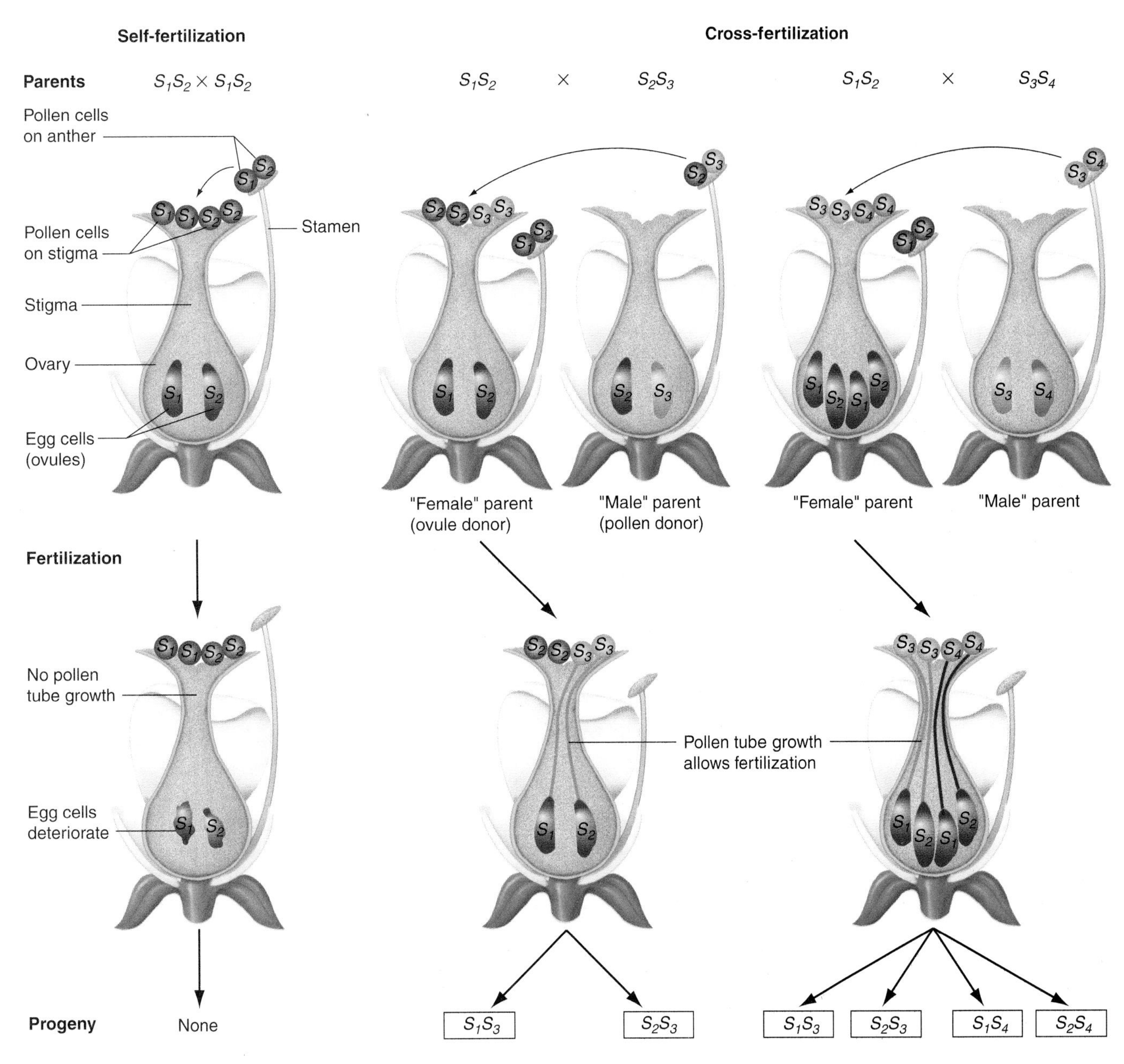

Plant incompatibility systems promote outbreeding and allele proliferation
Figure 3.8

(a) All yellow mice are heterozygotes.

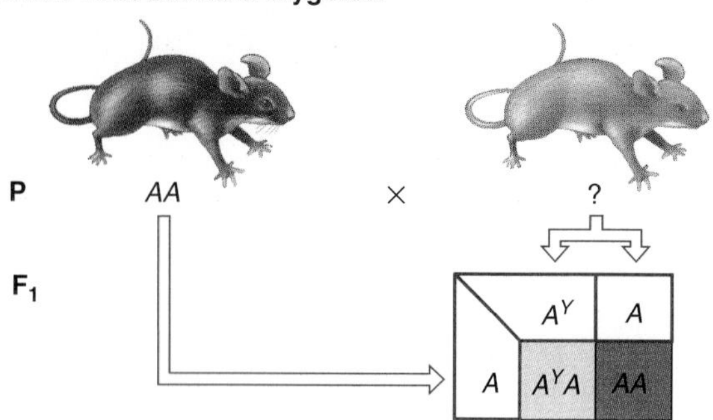

A^Y: A recessive lethal allele that also produces a dominant coat color phenotype
Figure 3.9

(b) Two copies of A^Y cause lethality.

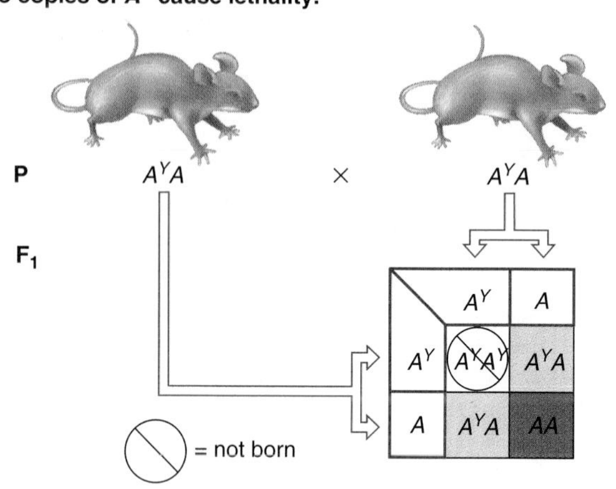

			Dominance Relations at Each Level of Analysis
Phenotypes at Different Levels of Analysis	Normal **AA**	Carrier **AS**	Diseased **SS**	
β-globin polypeptide production				*A* and *S* are codominant
Red blood cell shape at sea level	Normal	Normal	Sickled cells present	*A* is dominant *S* is recessive
Red blood cell concentration at sea level	Normal	Normal	Lower	
Red blood cell shape at high altitudes	Normal	Sickled cells present	Severe sickling	*A* and *S* show incomplete dominance
Red blood cell concentration at high altitudes	Normal	Lower	Very low, anemia	
Susceptibility to malaria	Normal susceptibility	Resistant	Resistant	*S* is dominant *A* is recessive

(b)

Pleiotropy of sickle-cell anemia: Dominance relations vary with the phenotype under consideration
Figure 3.10

(a) A dihybrid cross with lentil coat colors

P AA bb × aa BB

Gametes A b a B

F₁ (all identical) ♀ Aa Bb ♂ Aa Bb

F₂

9	A– B– (brown)
3	A– bb (tan)
3	aa B– (gray)
1	aa bb (green)

	A B	A b	a B	a b
A B	AA BB	AA Bb	Aa BB	Aa Bb
A b	AA Bb	AA bb	Aa Bb	Aa bb
a B	Aa BB	Aa Bb	aa BB	aa Bb
a b	Aa Bb	Aa bb	aa Bb	aa bb

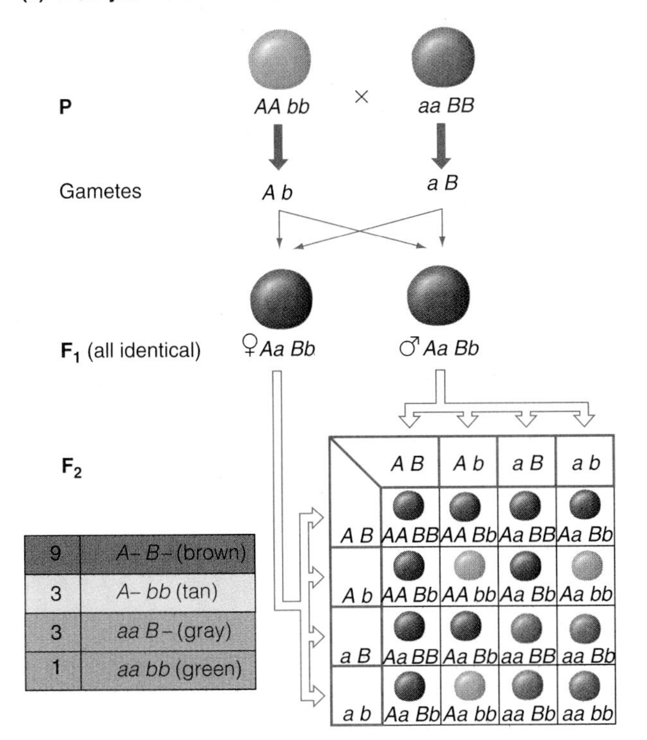

(b) Self-pollination of the F₂ to produce an F₃

Phenotypes of F₂ individual	Observed F₃ phenotypes	Expected proportion of F₂ population*
Green	Green	1/16
Tan	Tan	1/16
Tan	Tan, green	2/16
Gray	Gray, green	2/16
Gray	Gray	1/16
Brown	Brown	1/16
Brown	Brown, tan	2/16
Brown	Brown, gray	2/16
Brown	Brown, gray, tan, green	4/16

*This 1:1:2:2:1:1:2:2:4 F₂ genotypic ratio corresponds to a 9 brown:3 tan:3 gray:1 green F₂ phenotypic ratio.

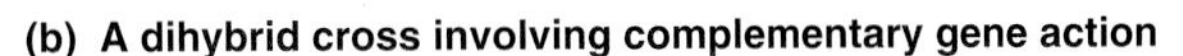

(c) Sorting out the dominance relations by select crosses

Seed coat color of parents	F₂ phenotypes and frequencies	Ratio
Tan × green	231 tan, 85 green	3:1
Gray × green	2586 gray, 867 green	3:1
Brown × gray	964 brown, 312 gray	3:1
Brown × tan	255 brown, 76 tan	3:1
Brown × green	57 brown, 18 gray, 13 tan, 4 green	9:3:3:1

How two genes interact to produce new colors in lentils
Figure 3.11

(b) A dihybrid cross involving complementary gene action

P AA bb × aa BB

Gametes A b a B

F₁ (all identical) ♀ Aa Bb × ♂ Aa Bb

F₂

	A B	A b	a B	a b
A B	AA BB	AA Bb	Aa BB	Aa Bb
A b	AA Bb	AA bb	AaBb	Aa bb
a B	Aa BB	Aa Bb	aa BB	aa Bb
a b	Aa Bb	Aa bb	aa Bb	aa bb

9	A– B– (purple)
7	(3)A– bb ⎤ (3)aa B– ⎬ (white) (1)aa bb ⎦

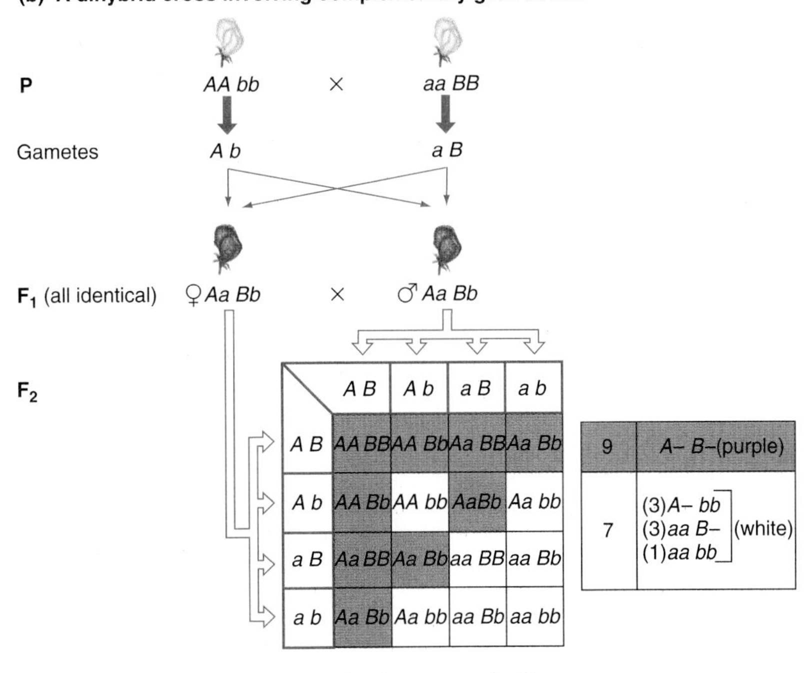

(c) A biochemical model for purple pigment production

Colorless precursor 1 → (Allele A — Pigment change initiated) → Colorless precursor 2 → (Allele B — Pigment change completed) → Purple pigment

Complementary gene action generates color in sweet peas
Figure 3.12

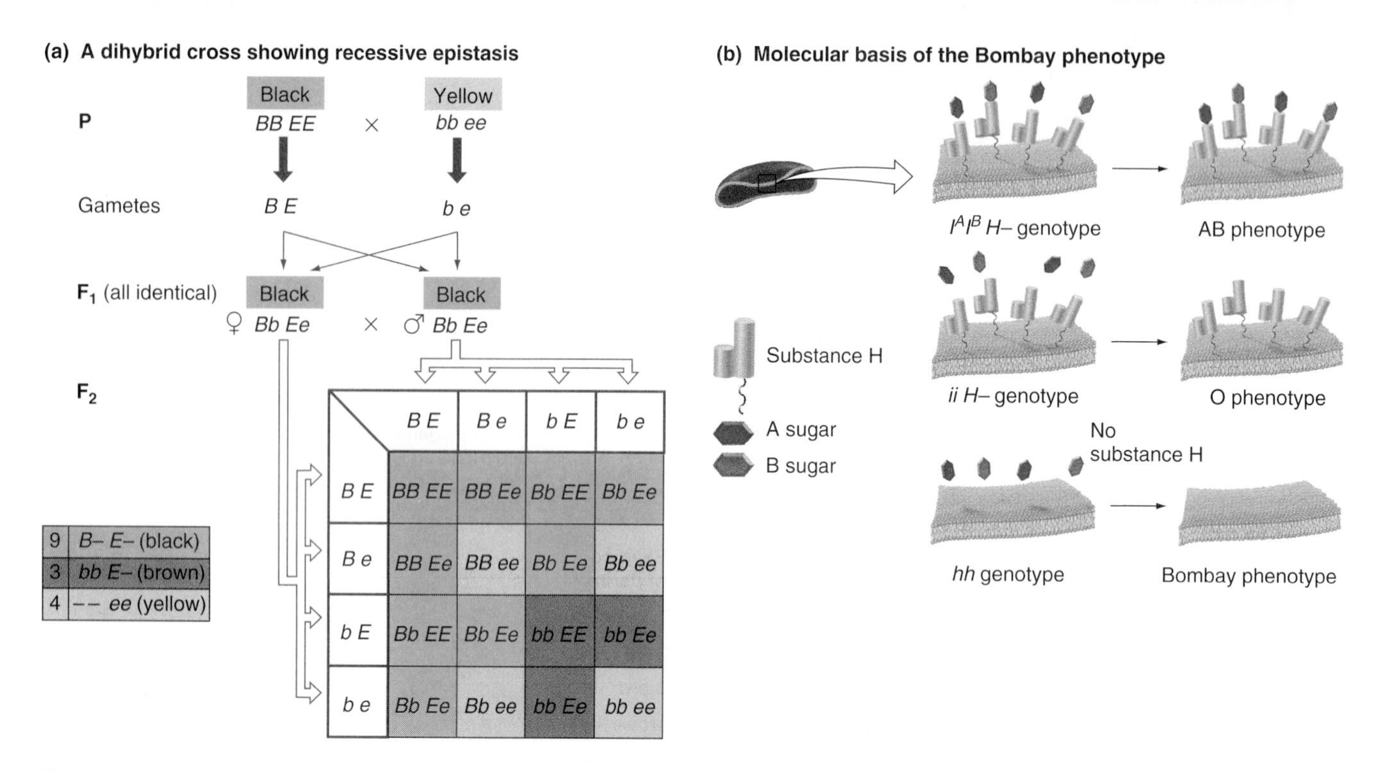

Recessive epistasis: Coat color in Labrador retrievers and a rare human blood type
Figure 3.13

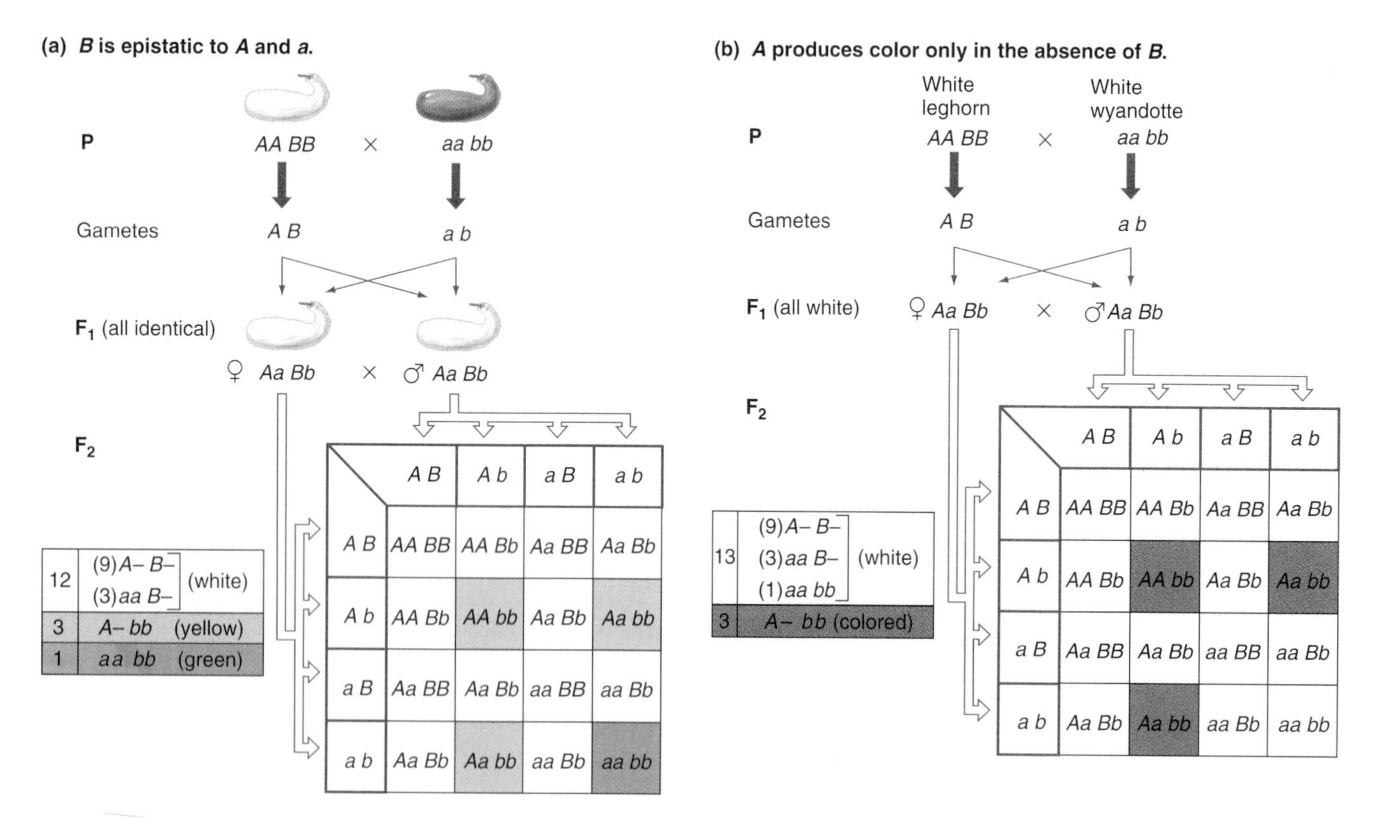

Dominant epistasis produces telltale phenotypic ratios of 12:3:1 or 13:3
Figure 3.14

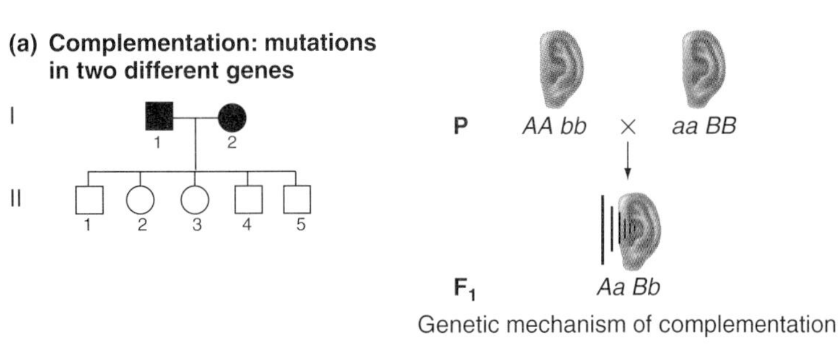

(a) **Complementation: mutations in two different genes**

P $AA\ bb$ × $aa\ BB$

F₁ $Aa\ Bb$

Genetic mechanism of complementation

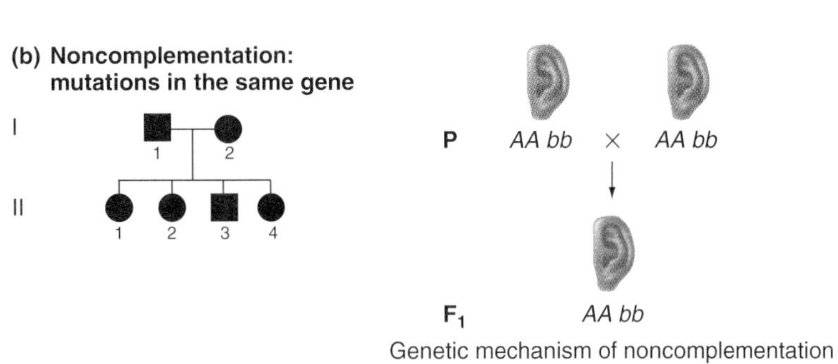

(b) **Noncomplementation: mutations in the same gene**

P $AA\ bb$ × $AA\ bb$

F₁ $AA\ bb$

Genetic mechanism of noncomplementation

Genetic heterogeneity in humans: Mutations in many genes can cause deafness
Figure 3.15

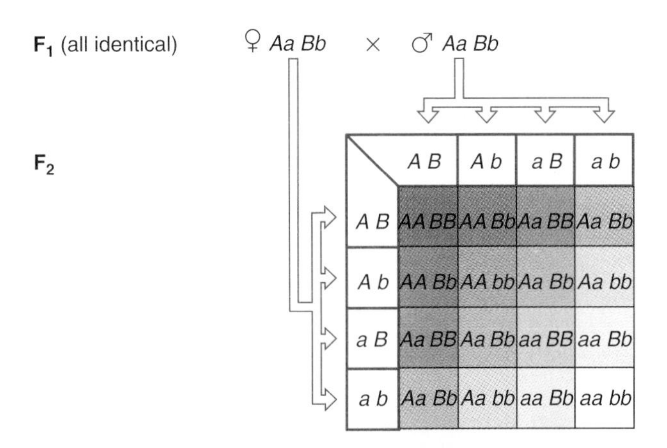

F₁ (all identical)   ♀ $Aa\ Bb$ ×   ♂ $Aa\ Bb$

F₂

	A B	A b	a B	a b
A B	AA BB	AA Bb	Aa BB	Aa Bb
A b	AA Bb	AA bb	Aa Bb	Aa bb
a B	Aa BB	Aa Bb	aa BB	aa Bb
a b	Aa Bb	Aa bb	aa Bb	aa bb

1	AA BB	purple shade 9
2	AA Bb	purple shade 8
2	Aa BB	purple shade 7
1	AA bb	purple shade 6
4	Aa Bb	purple shade 5
1	aa BB	purple shade 4
2	Aa bb	purple shade 3
2	aa Bb	purple shade 2
1	aa bb	purple shade 1 (white)

With incomplete dominance, the interaction of two genes can produce nine different phenotypes for a single trait
Figure 3.16

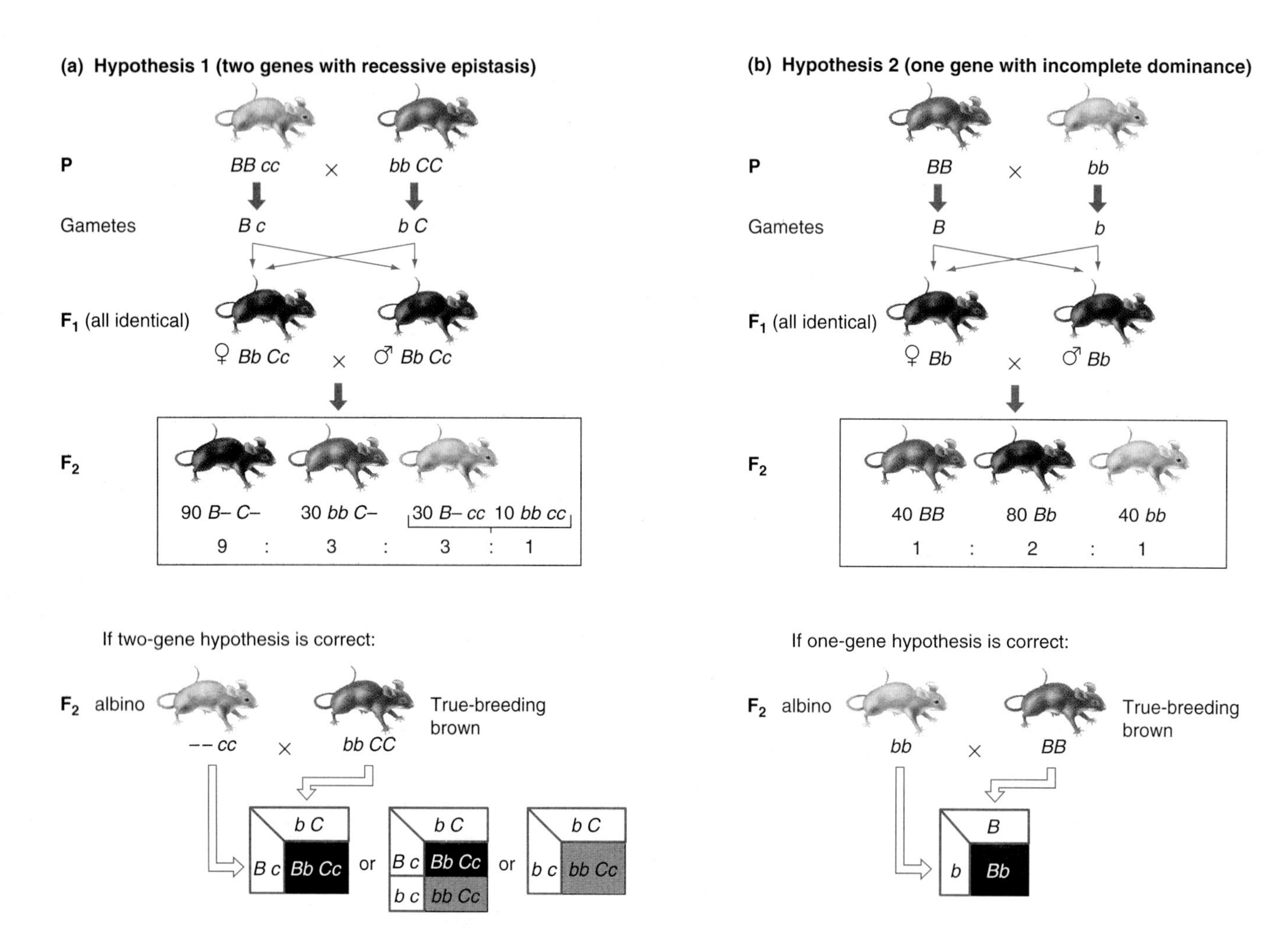

Specific breeding tests can help decide between hypotheses
Figure 3.17

(b) OCA is recessive

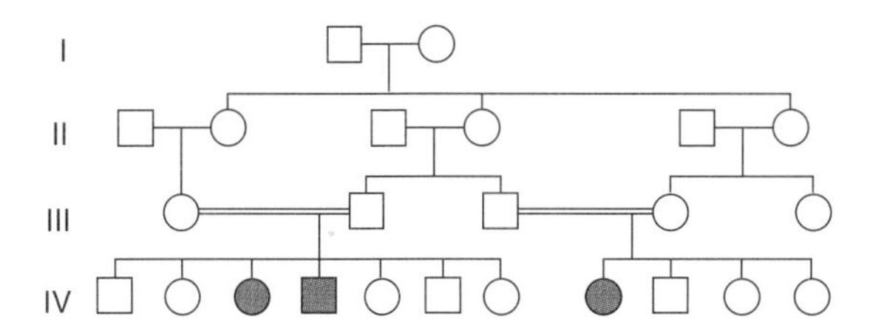

Family pedigrees help unravel the genetic basis of albinism
Figure 3.18

(c) Complementation for albinism

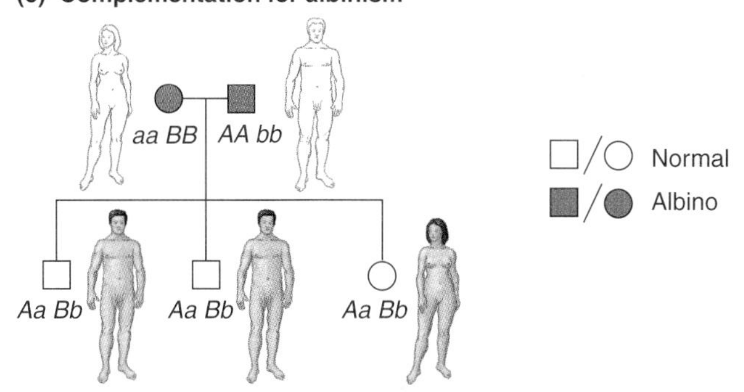

□ / ○   Normal
■ / ●   Albino

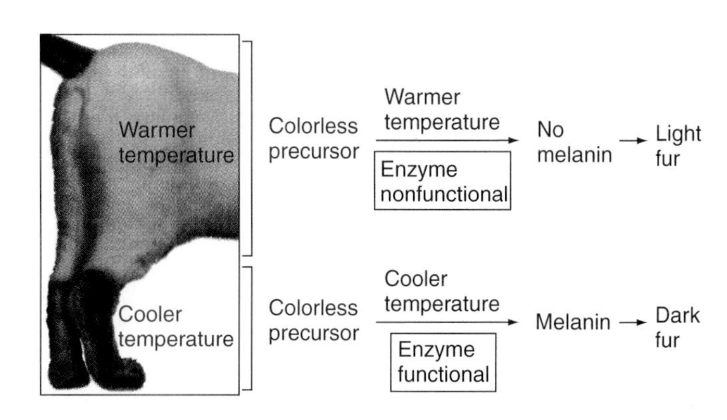

Colorless precursor → Warmer temperature / Enzyme nonfunctional → No melanin → Light fur

Colorless precursor → Cooler temperature / Enzyme functional → Melanin → Dark fur

In Siamese cats, temperature affects coat color
Figure 3.19

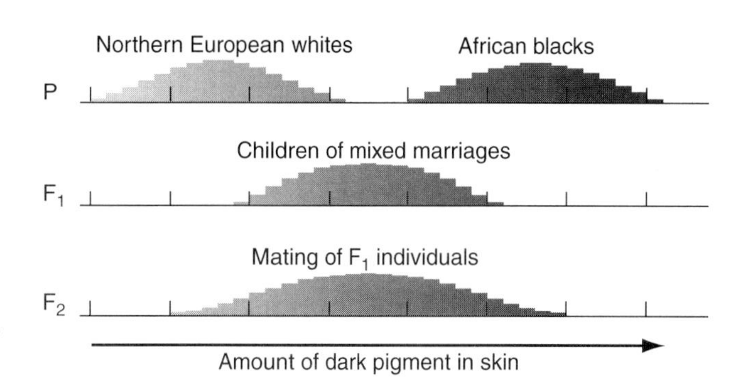

Northern European whites   African blacks

P

Children of mixed marriages

F_1

Mating of F_1 individuals

F_2

Amount of dark pigment in skin

Continuous traits in humans
Figure 3.20

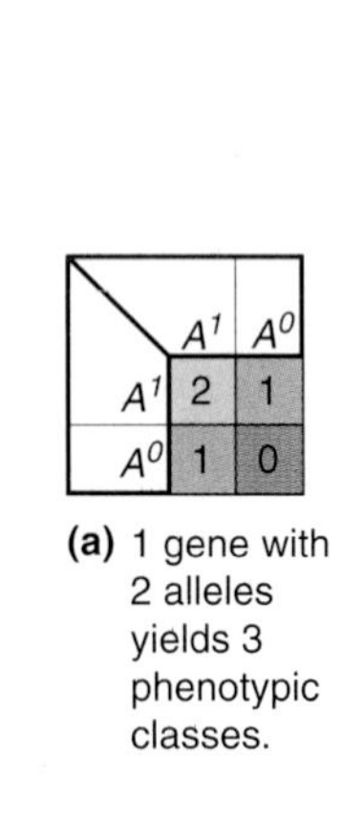

(a) 1 gene with 2 alleles yields 3 phenotypic classes.

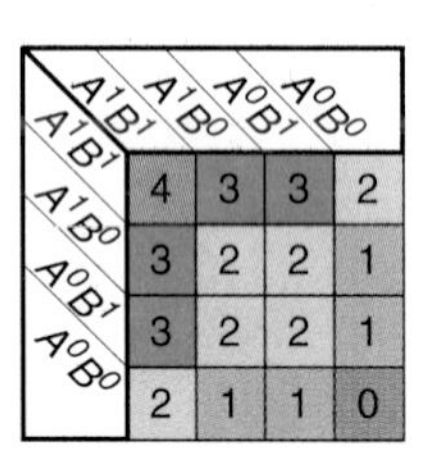

(b) 2 genes with 2 alleles apiece yield 5 phenotypic classes.

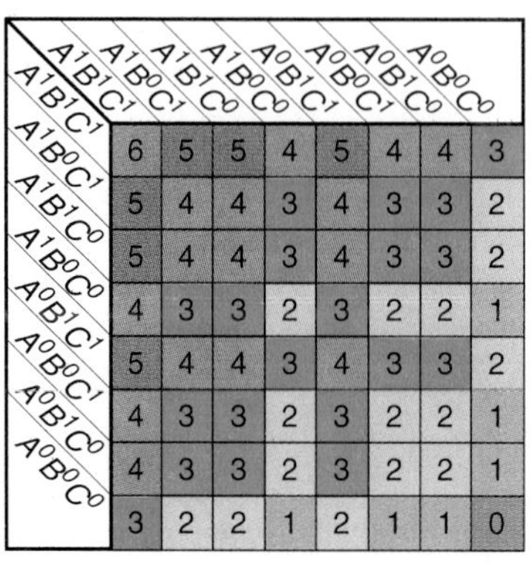

(c) 3 genes with 2 alleles yield 7 phenotypic classes.

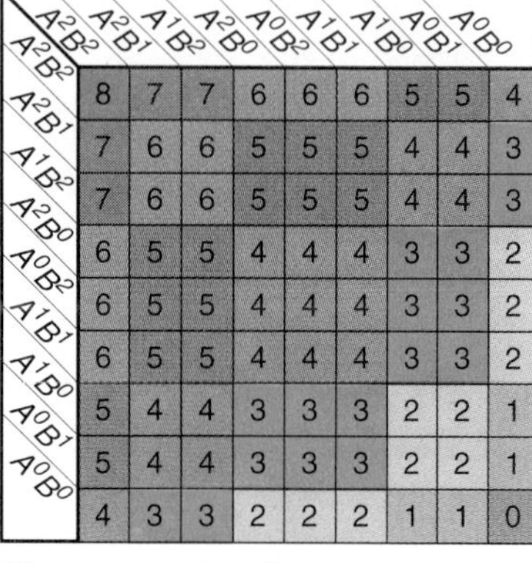

(d) 2 genes with 3 alleles apiece yield 9 phenotypic classes.

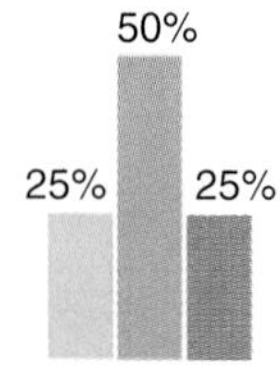

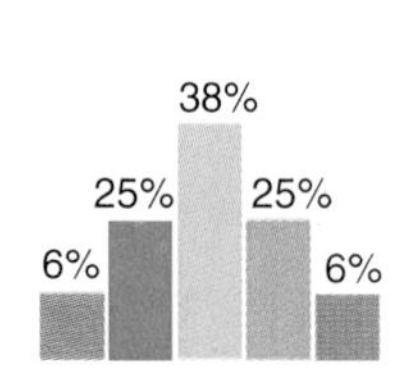

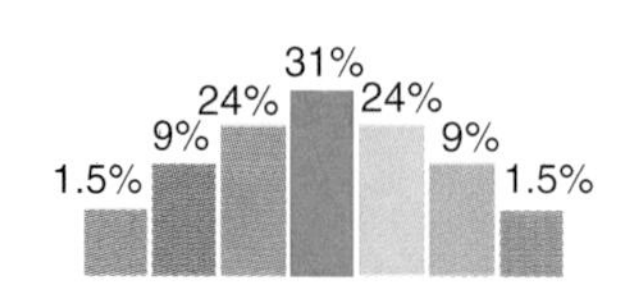

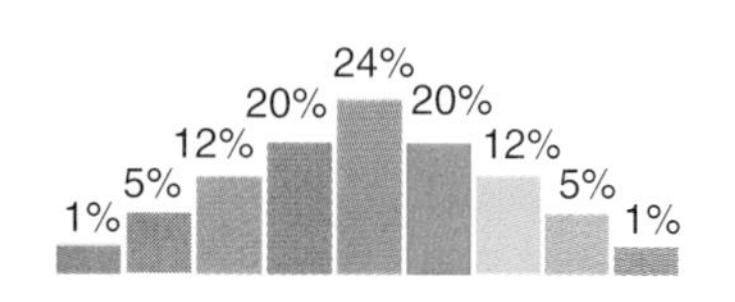

A Mendelian explanation of continuous variation
Figure 3.21

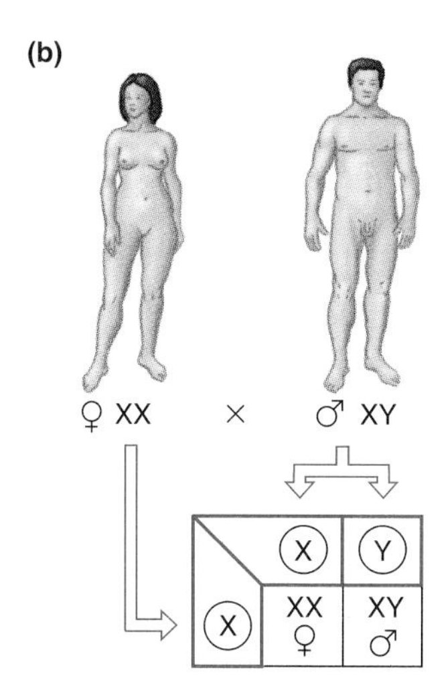

(b)

♀ XX × ♂ XY

	Ⓧ	Ⓨ
Ⓧ	XX ♀	XY ♂
Ⓧ		

How the X and Y chromosomes determine sex in humans
Figure 4.3

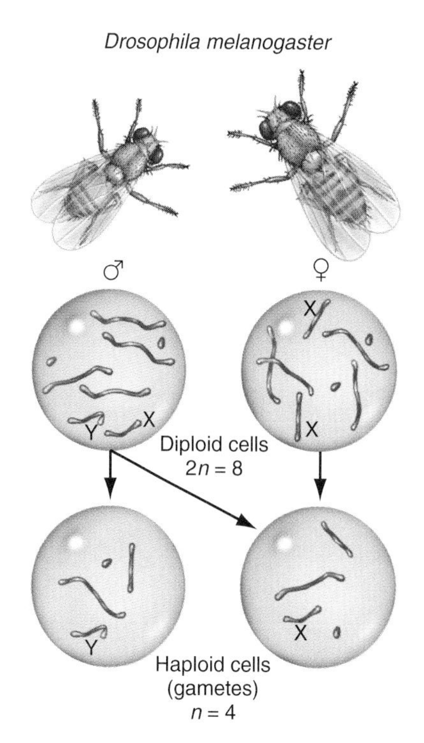

Drosophila melanogaster

♂ ♀

Diploid cells
$2n = 8$

Haploid cells (gametes)
$n = 4$

Diploid versus haploid: $2n$ versus n
Figure 4.4

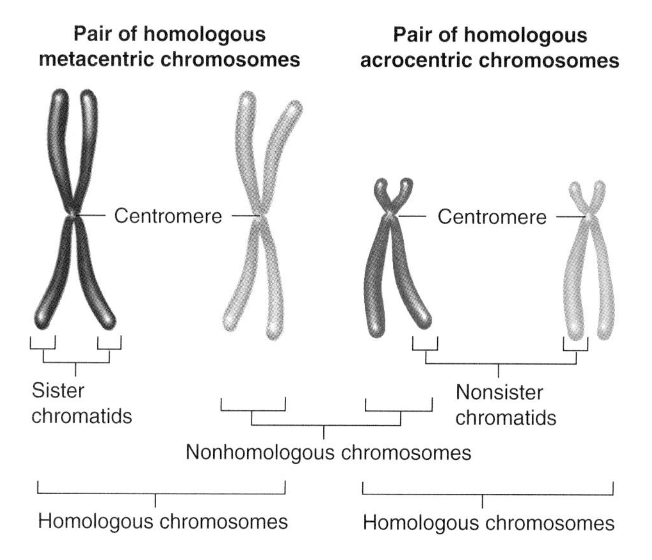

Pair of homologous metacentric chromosomes   Pair of homologous acrocentric chromosomes

Centromere   Centromere

Sister chromatids   Nonsister chromatids

Nonhomologous chromosomes

Homologous chromosomes   Homologous chromosomes

Metaphase chromosomes can be classified by centromere position
Figure 4.5

(a) The cell cycle

M
Mitosis, cytokinesis
G_2
G_1
G_0
Interphase
Chromosome duplication
S

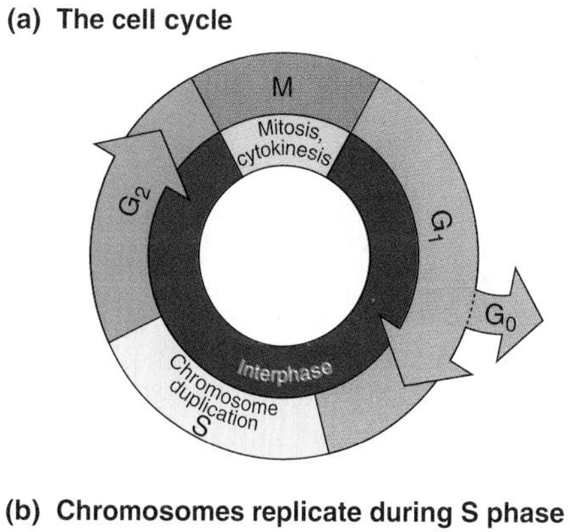

(b) Chromosomes replicate during S phase

G_1: interphase, gap before duplication

S: DNA synthesis and chromosome duplication

G_2: interphase, gap before mitosis

A *a* *B* *b*

A A *a a* *B B* *b b*

The cell cycle: An alternation between interphase and mitosis
Figure 4.7

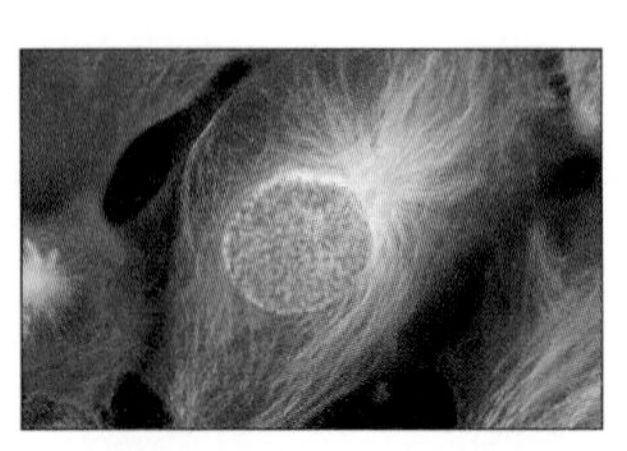
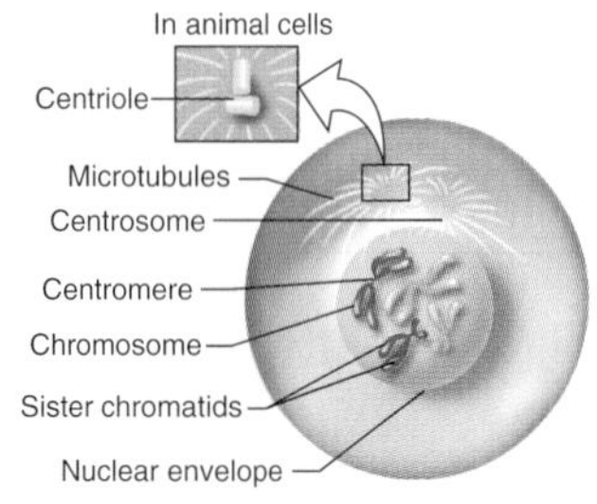

In animal cells
Centriole
Microtubules
Centrosome
Centromere
Chromosome
Sister chromatids
Nuclear envelope

(a) Prophase: (1) Chromosomes condense and become visible; (2) centrosomes move apart toward opposite poles and generate new microtubules; (3) nucleoli begin to disappear.

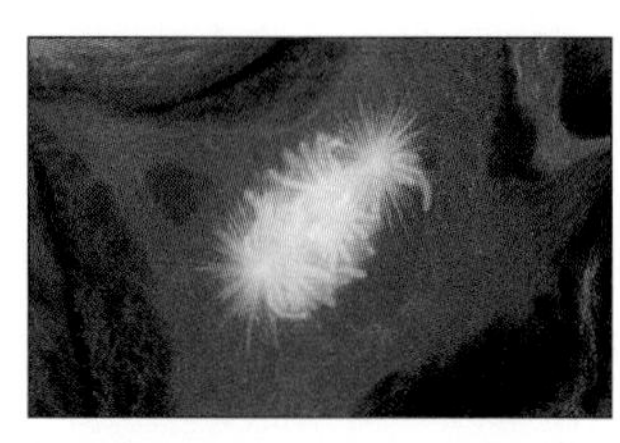
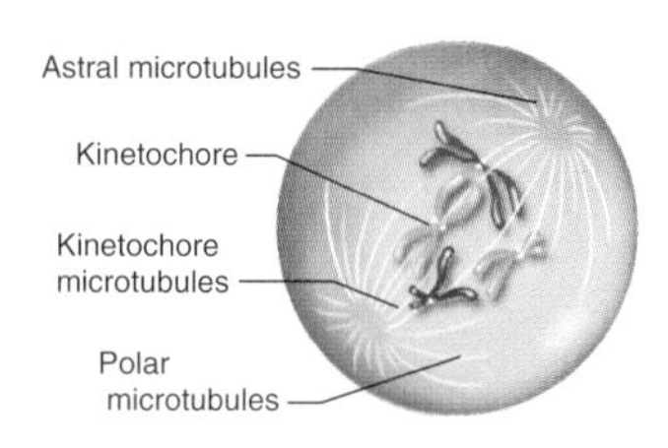

Astral microtubules
Kinetochore
Kinetochore microtubules
Polar microtubules

(b) Prometaphase: (1) Nuclear envelope breaks down; (2) microtubules from the centrosomes invade the nucleus; (3) sister chromatids attach to microtubules from opposite centrosomes.

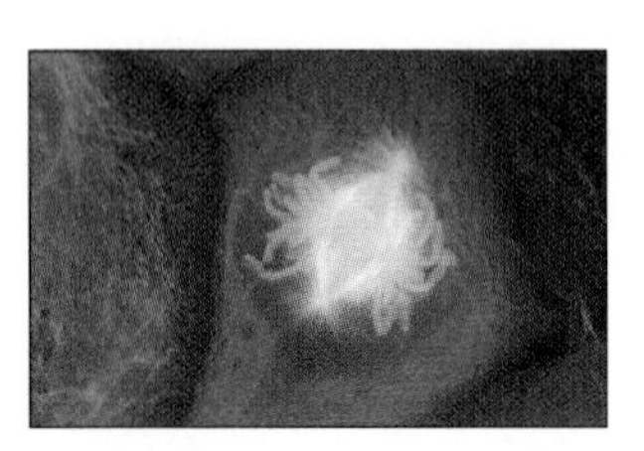
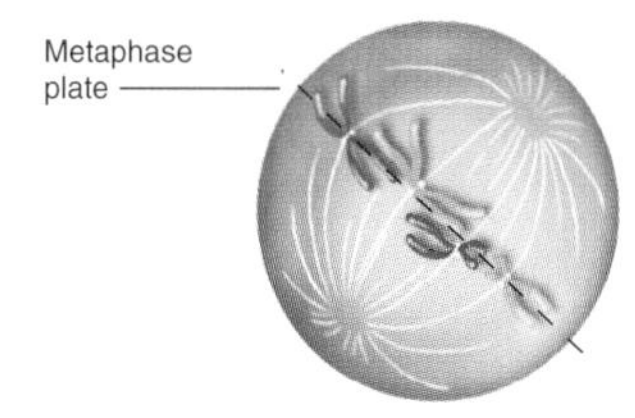

Metaphase plate

(c) Metaphase: Chromosomes align on the metaphase plate with sister chromatids facing opposite poles.

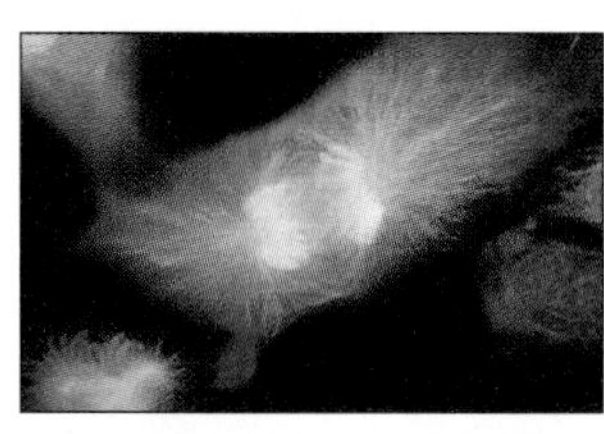
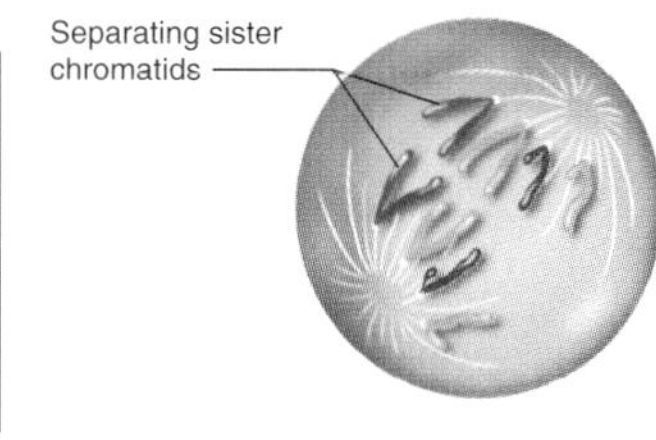
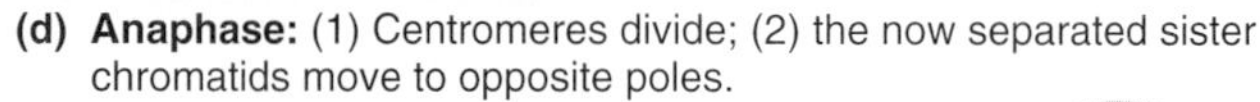

Separating sister chromatids

(d) Anaphase: (1) Centromeres divide; (2) the now separated sister chromatids move to opposite poles.

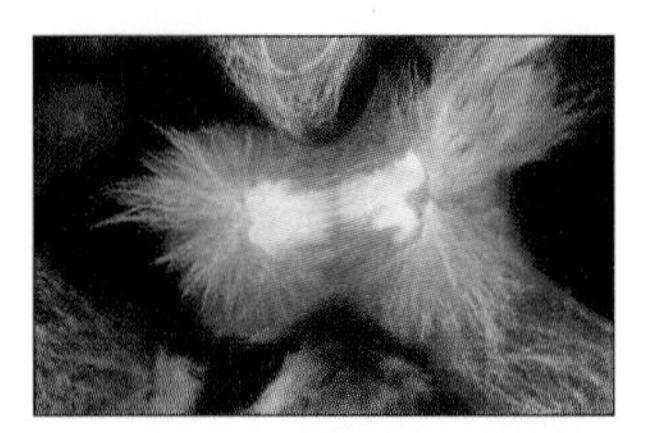
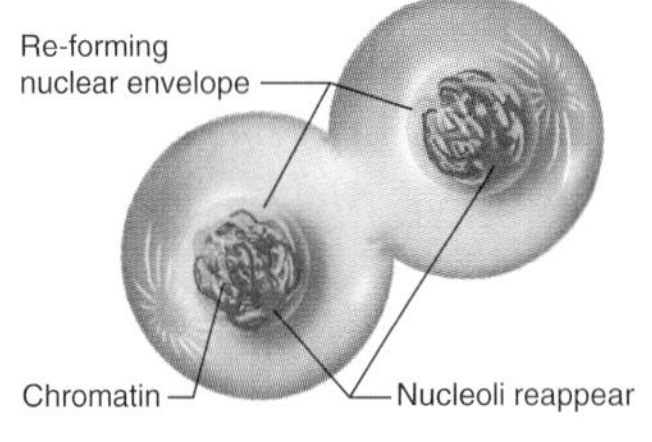

Re-forming nuclear envelope
Chromatin
Nucleoli reappear

(e) Telophase: (1) Nuclear membranes and nucleoli re-form; (2) spindle fibers disappear; (3) chromosomes uncoil and become a tangle of chromatin.

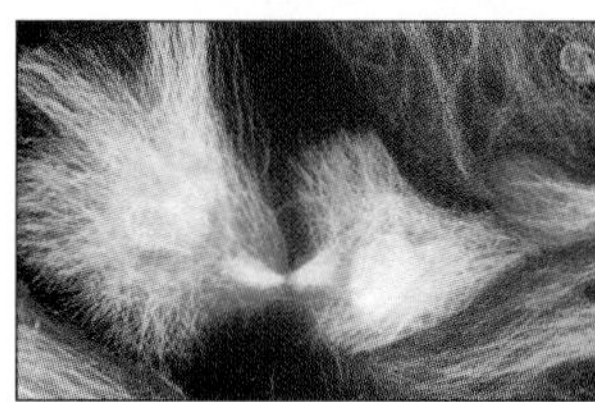

(f) Cytokinesis: The cytoplasm divides, splitting the elongated parent cell into two daughter cells with identical nuclei.

Mitosis maintains the chromosome number of the parent cell nucleus in the two daughter nuclei

Figure 4.8

a-f: Photographs by Dr. Conly L. Reider, Division of Molecular Medicine, Wadsworth Center, NYS Dept. of Health, Albany, NY

(a) Cytokinesis in an animal cell

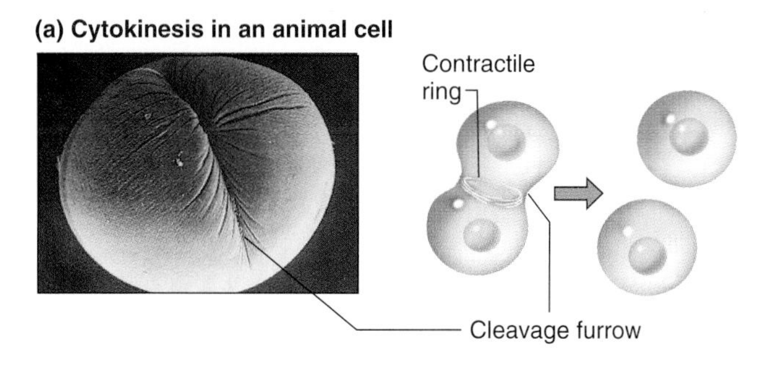

Contractile ring

Cleavage furrow

(b) Cytokinesis in a plant cell

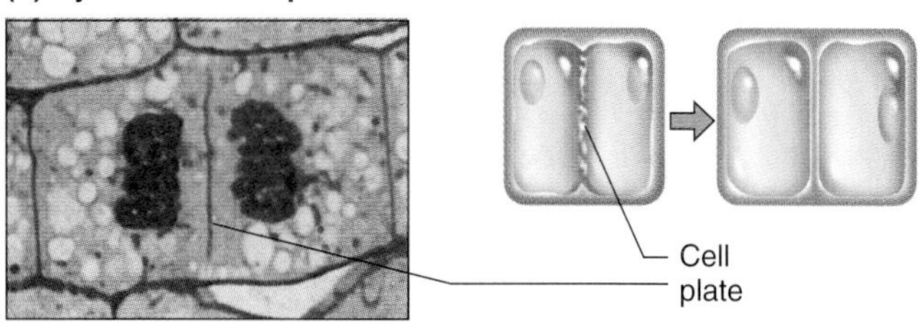

Cell plate

Cytokinesis: The cytoplasm divides, producing two daughter cells
Figure 4.9

a: © David M. Phillip/Visuals Unlimited; **b:** © R. Calentine/ Visuals Unlimited

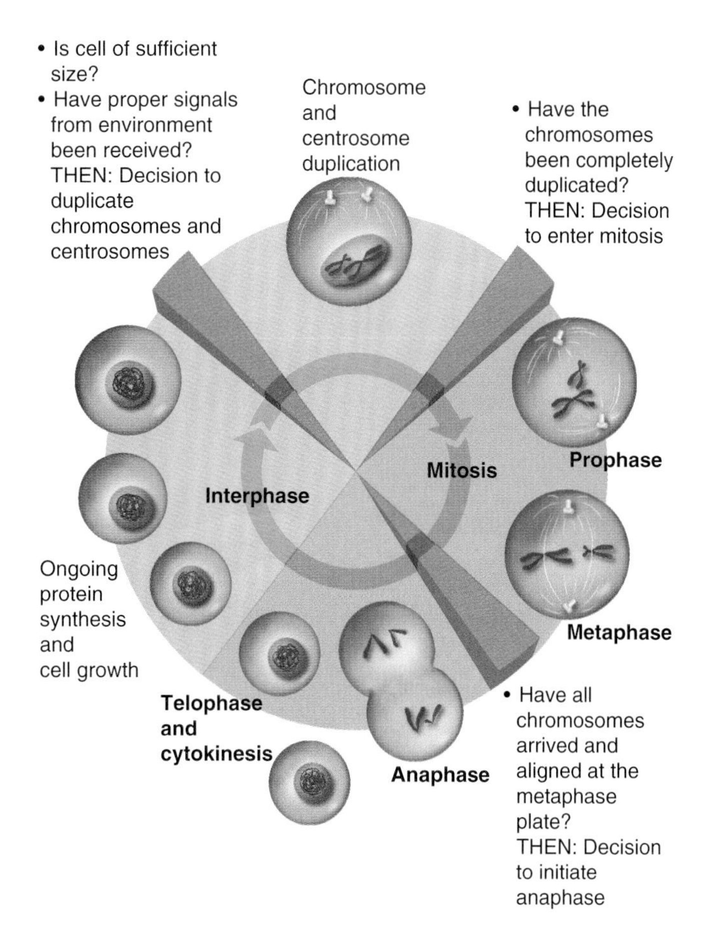

- Is cell of sufficient size?
- Have proper signals from environment been received? THEN: Decision to duplicate chromosomes and centrosomes

Chromosome and centrosome duplication

- Have the chromosomes been completely duplicated? THEN: Decision to enter mitosis

Mitosis

Prophase

Interphase

Metaphase

Ongoing protein synthesis and cell growth

Telophase and cytokinesis

Anaphase

- Have all chromosomes arrived and aligned at the metaphase plate? THEN: Decision to initiate anaphase

Checkpoints help regulate the cell cycle
Figure 4.11

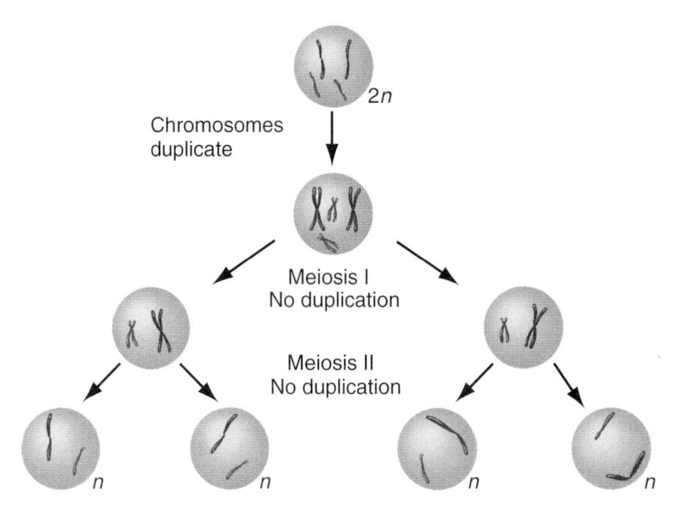

An overview of meiosis: The chromosomes replicate once, while the nuclei divide twice
Figure 4.12

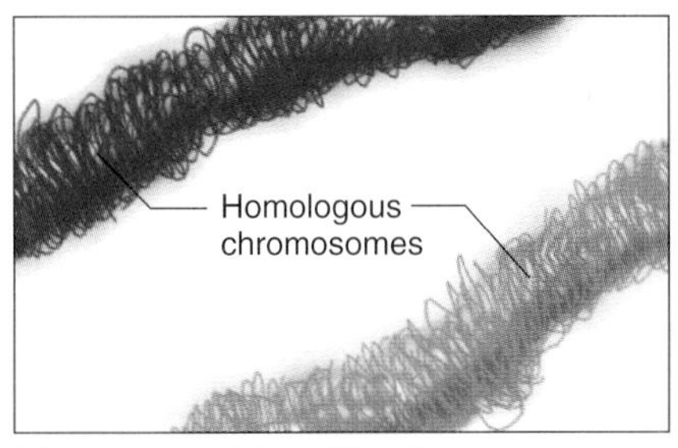

(a) Leptotene: Threadlike chromosomes begin to condense and thicken, becoming visible as discrete structures. Although the chromosomes have duplicated, the sister chromatids of each chromosome are not yet visible in the microscope.

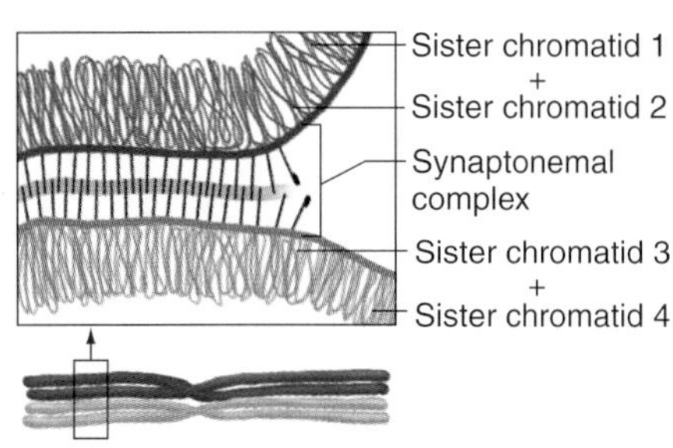

(b) Zygotene: Chromosomes are clearly visible and begin active pairing with homologous chromosomes along the synaptonemal complex to form a bivalent, or tetrad.

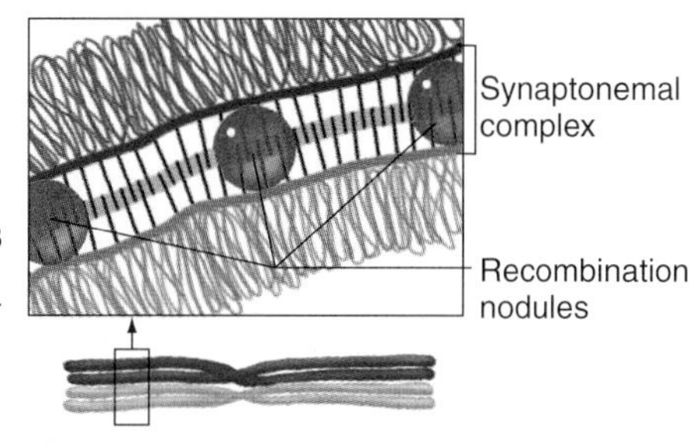

(c) Pachytene: Full synapsis of homologues. Recombination nodules appear along the synaptonemal complex.

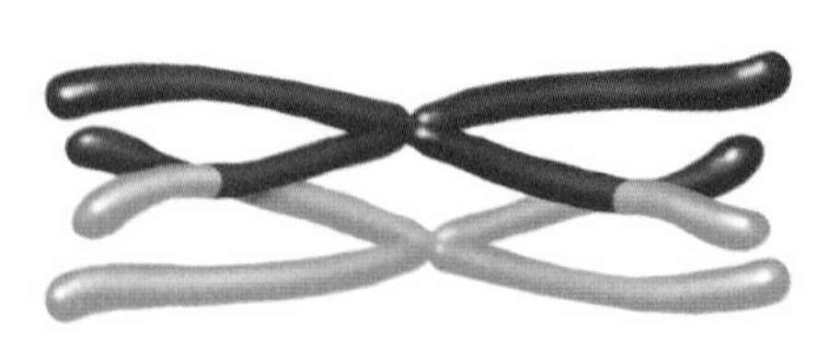

(d) Diplotene: Bivalent appears to pull apart slightly but remains connected at crossover sites, called chiasmata.

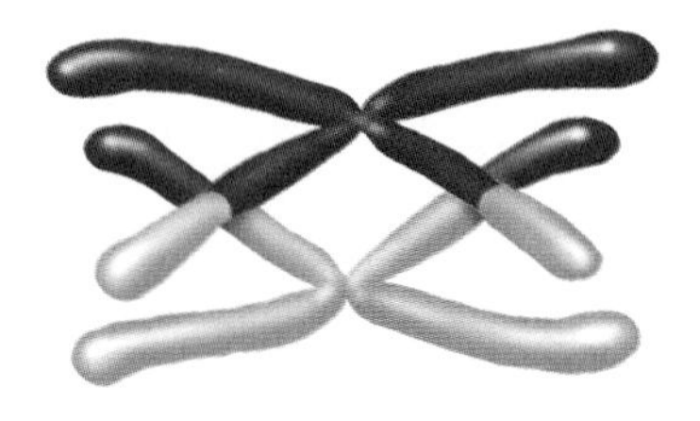

(e) Diakinesis: Further condensation of chromatids. Nonsister chromatids that have exchanged parts by crossing-over remain closely associated at chiasmata.

Prophase I of meiosis at very high magnification
Figure 4.14

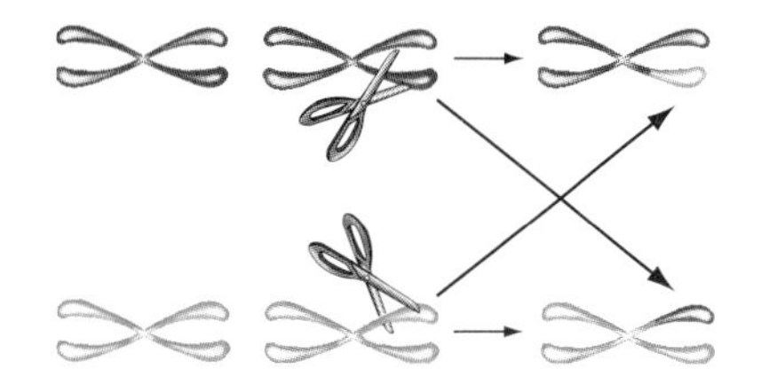

How crossing-over produces recombined chromosomes
Figure 4.15

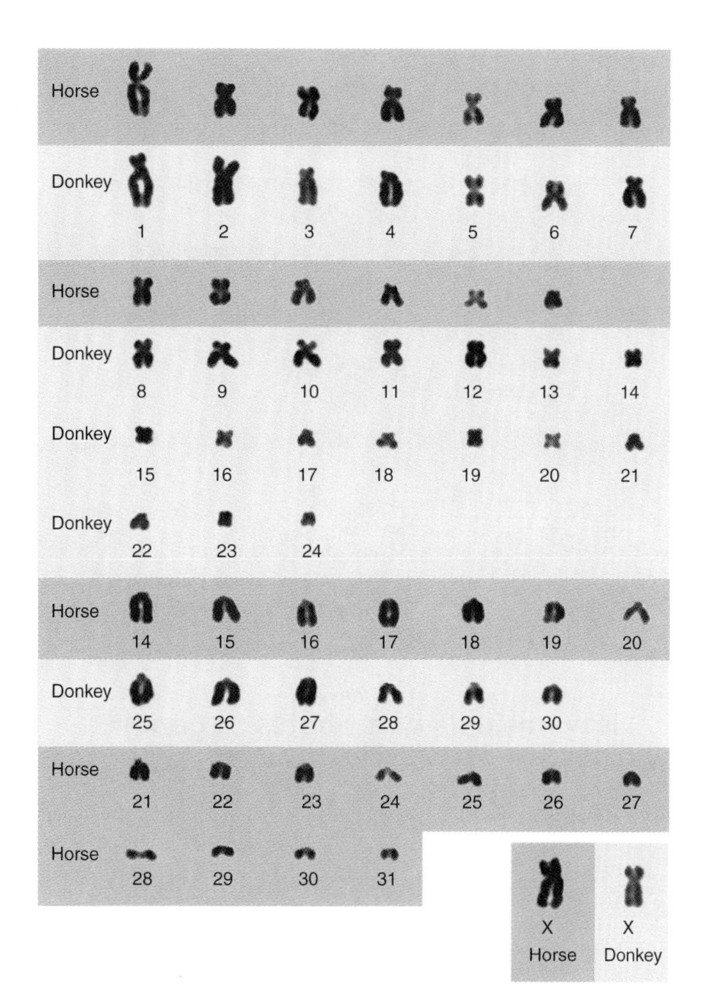

Hybrid sterility: When chromosomes cannot pair during meiosis I, they will segregate improperly
Figure 4.16

© Dr. Leona Chemnick, Dr. Oliver Ryder/San Diego Zoo, Center for Reproduction of Endangered Species

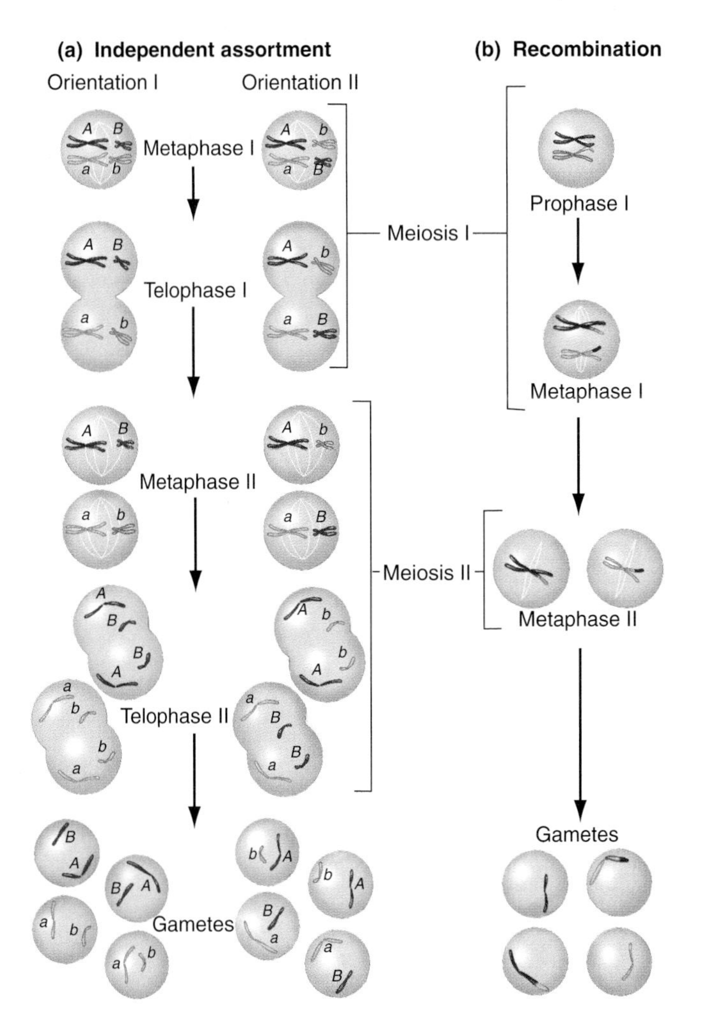

(a) Independent assortment

Orientation I   Orientation II

Metaphase I

Telophase I

Meiosis I

Metaphase II

Meiosis II

Telophase II

Gametes

(b) Recombination

Prophase I

Meiosis I

Metaphase I

Meiosis II

Metaphase II

Gametes

How meiosis contributes to genetic diversity
Figure 4.17

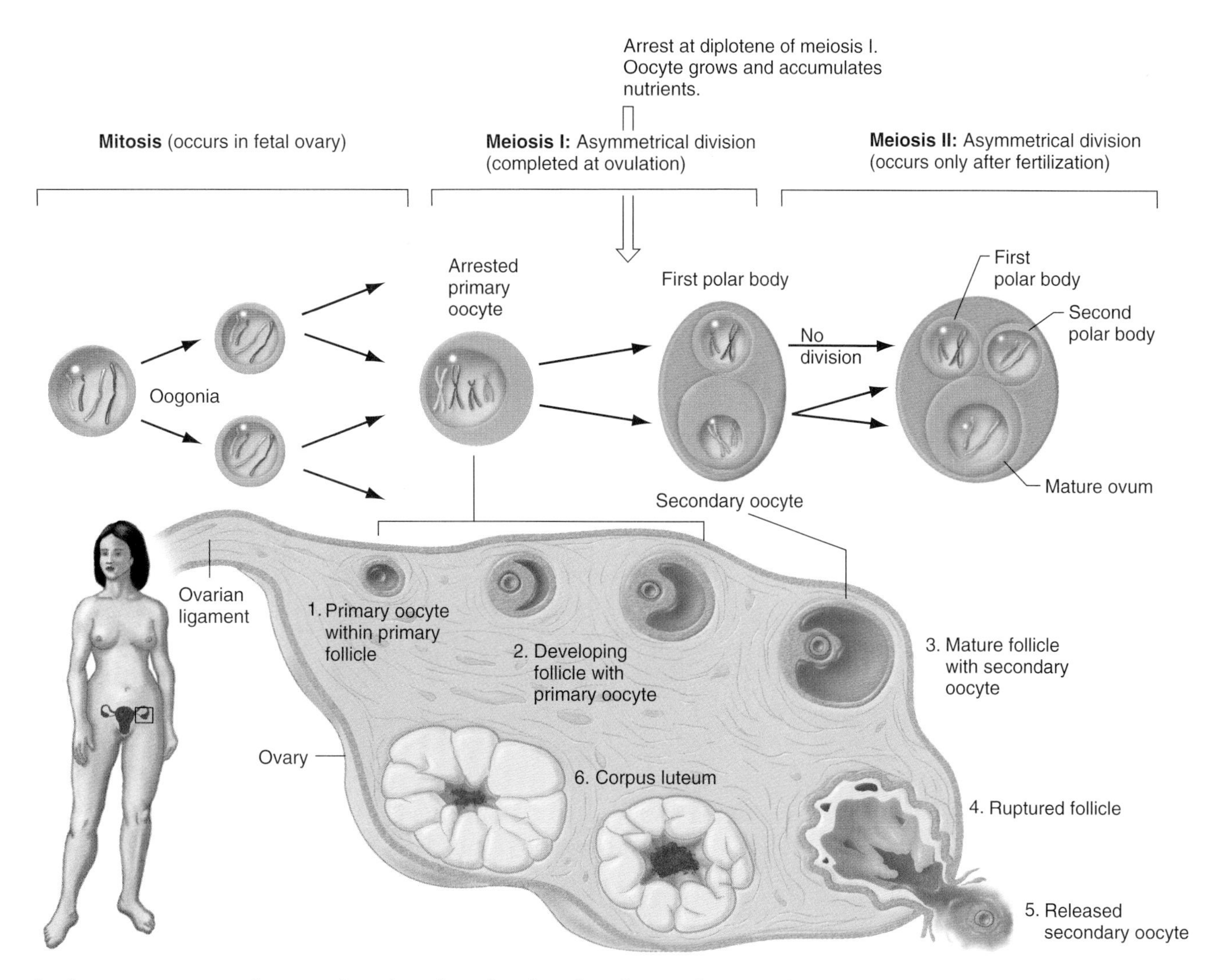

In humans, egg formation begins in the fetal ovaries and arrests during the prophase of meiosis I

Figure 4.18

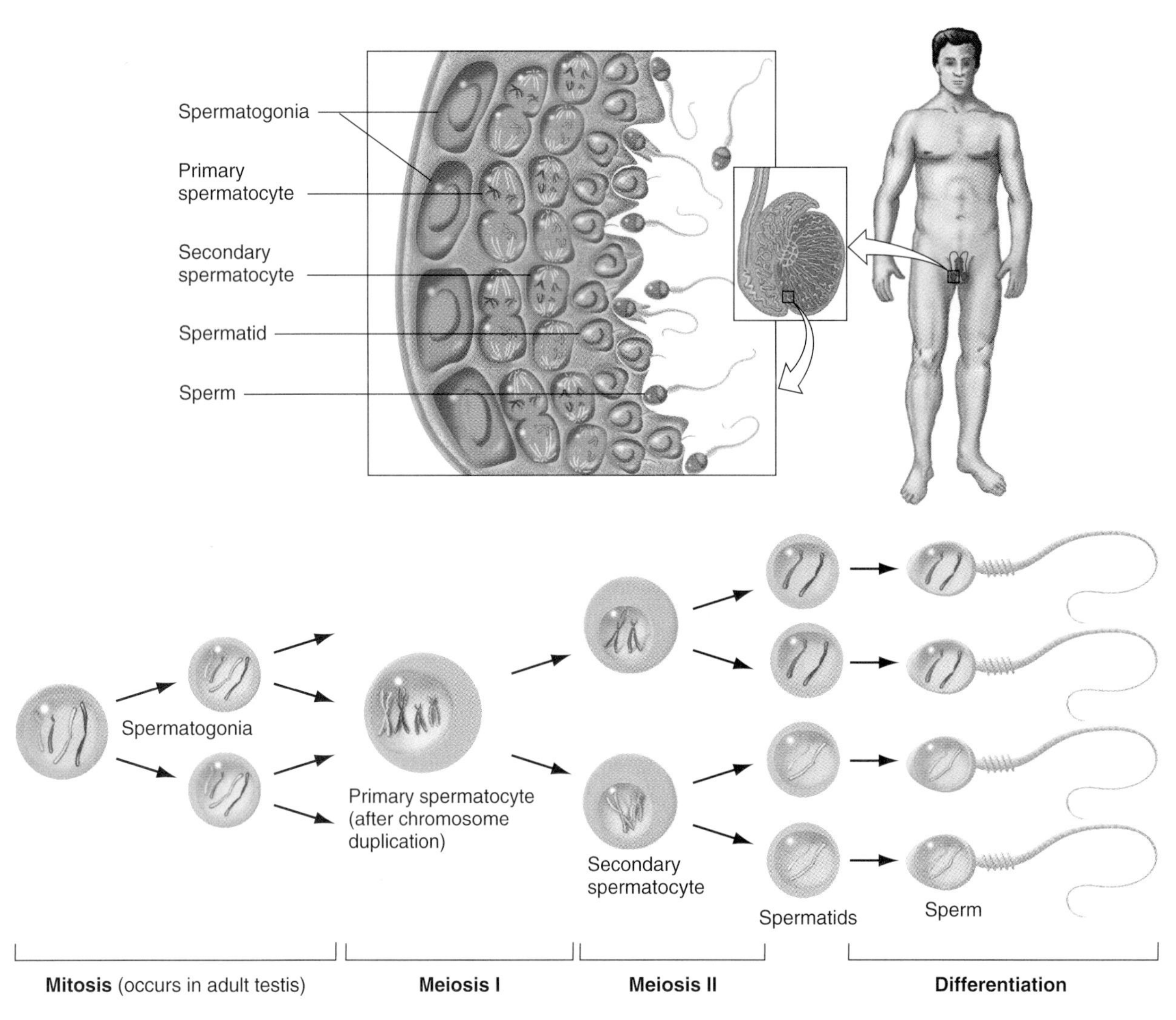

Spermatogonia

Primary spermatocyte

Secondary spermatocyte

Spermatid

Sperm

Spermatogonia

Primary spermatocyte (after chromosome duplication)

Secondary spermatocyte

Spermatids

Sperm

Mitosis (occurs in adult testis)   **Meiosis I**   **Meiosis II**   **Differentiation**

Human sperm form continuously in the testes after puberty
Figure 4.19

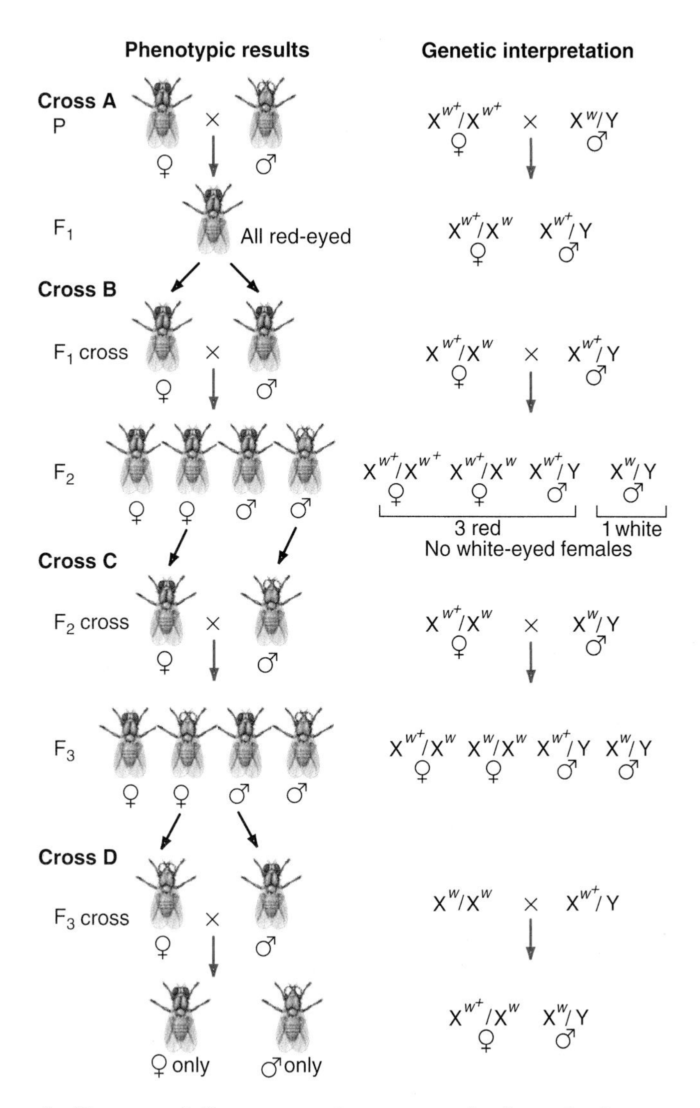

**A *Drosophila* eye color gene is located on
the X chromosome**
Figure 4.20

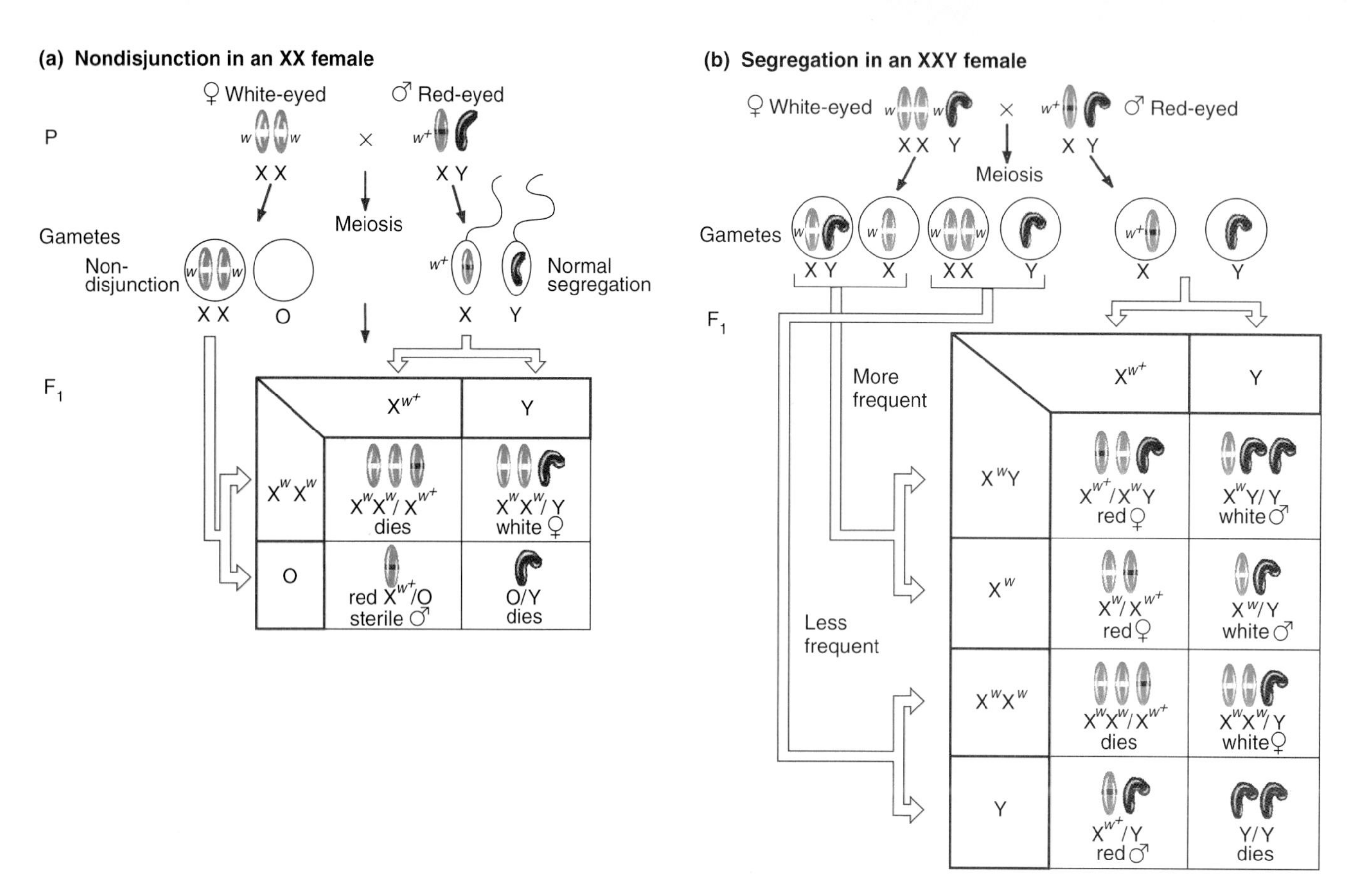

Nondisjunction: Rare mistakes in meiosis help confirm the chromosome theory
Figure 4.21

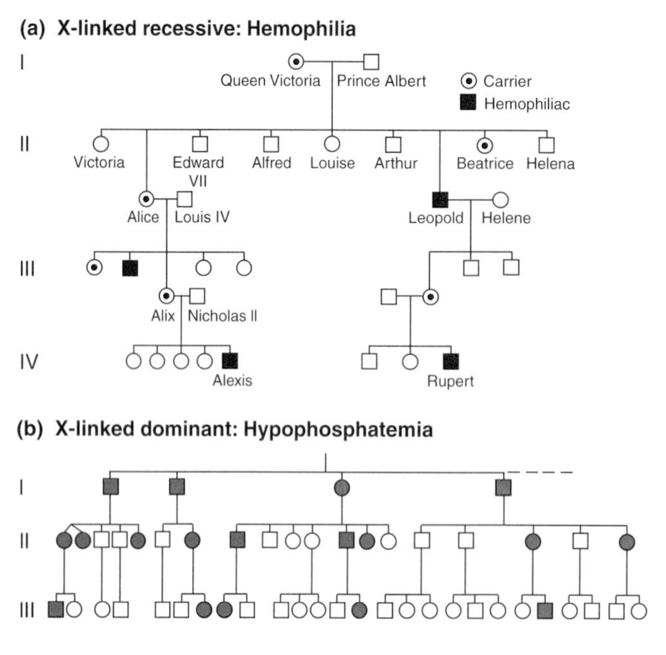

X-linked traits may be recessive or dominant
Figure 4.23

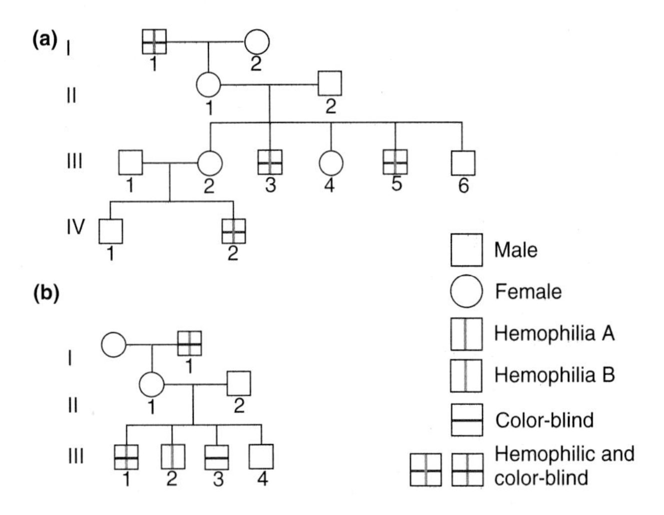

(a)

(b)

Pedigrees indicate that colorblindness and the two forms of hemophilia are X-linked traits

Figure 5.1

Legend:
- Male (□)
- Female (○)
- Hemophilia A
- Hemophilia B
- Color-blind
- Hemophilic and color-blind

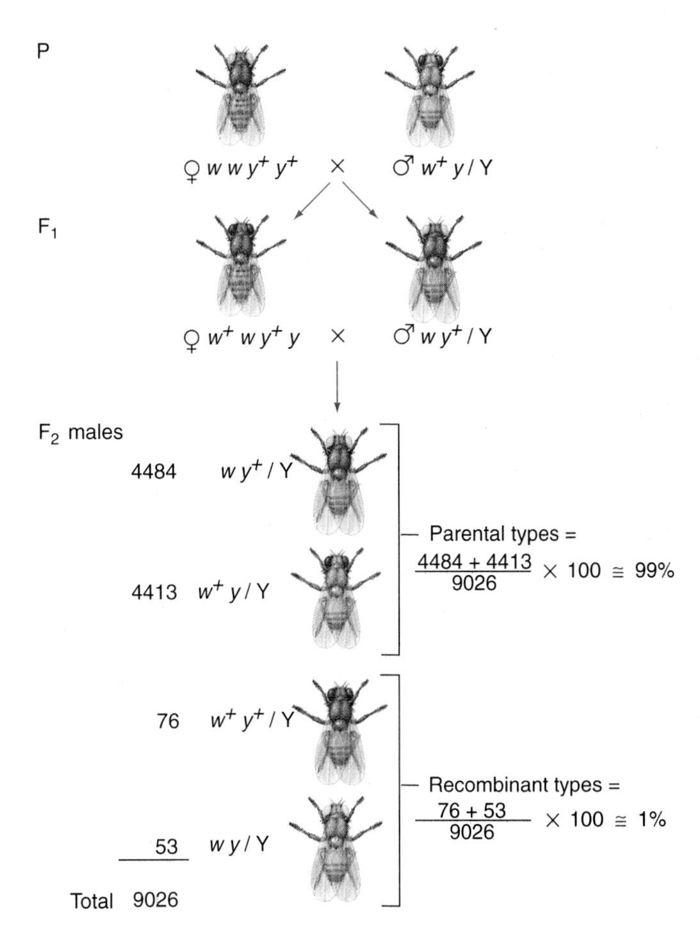

P

♀ $w\ w\ y^+\ y^+$ × ♂ $w^+\ y\ /\ Y$

F₁

♀ $w^+\ w\ y^+\ y$ × ♂ $w\ y^+\ /\ Y$

F₂ males

4484 $w\ y^+\ /\ Y$

4413 $w^+\ y\ /\ Y$

Parental types = $\dfrac{4484 + 4413}{9026} \times 100 \cong 99\%$

76   $w^+\ y^+\ /\ Y$

53   $w\ y\ /\ Y$

Total   9026

Recombinant types = $\dfrac{76 + 53}{9026} \times 100 \cong 1\%$

When genes are linked, parental combinations outnumber recombinant types

Figure 5.2

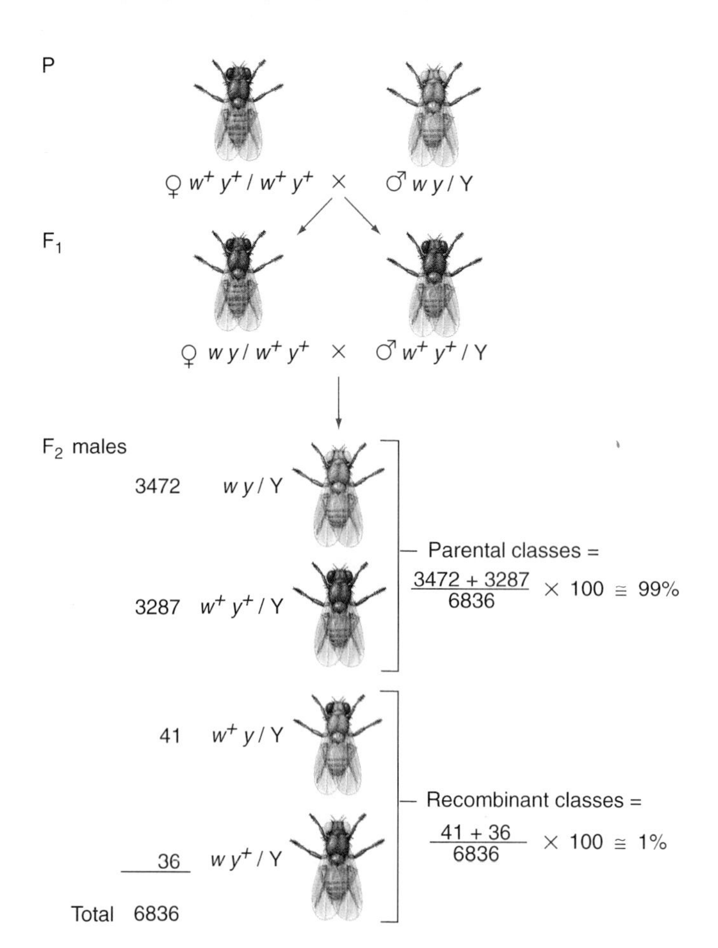

Designations of "parental" and "recombinant" relate to past history
Figure 5.3

The recombination frequency depends on the gene pair
Figure 5.4

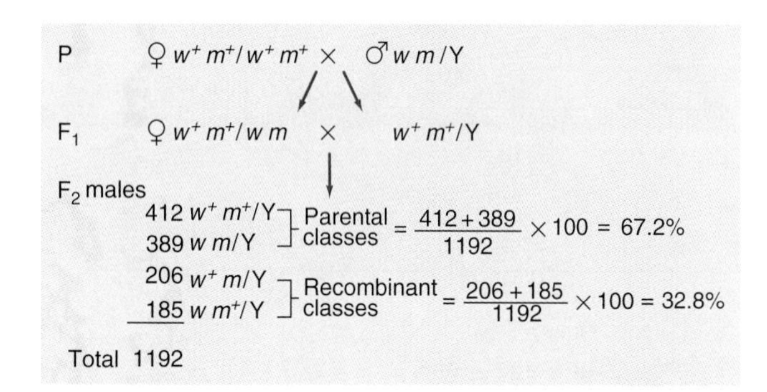

P   ♀ $w^+ m^+/w^+ m^+$ × ♂ $w m$/Y

F₁   ♀ $w^+ m^+/w m$ × $w^+ m^+$/Y

F₂ males

412 $w^+ m^+$/Y ⎤ Parental = $\frac{412+389}{1192}$ × 100 = 67.2%
389 $w m$/Y   ⎦ classes

206 $w^+ m$/Y ⎤ Recombinant = $\frac{206+185}{1192}$ × 100 = 32.8%
185 $w m^+$/Y ⎦ classes

Total 1192

P   ♀ $b c^+/b c^+$ × ♂ $b^+ c/b^+ c$

F₁ (all identical) $b c^+/b^+ c$

Testcross   ♀ $b c^+/b^+ c$ × ♂ $b c/b c$

Testcross 2934 b $c^+/b c$ ⎤ Parental = $\frac{2934 + 2768}{7419}$ × 100 = 77%
progeny   2768 $b^+ c/b c$ ⎦ classes

              871 $b c/b c$ ⎤ Recombinant = $\frac{871+846}{7419}$ × 100 = 23%
              846 $b^+ c^+/b c$ ⎦ classes

Total 7419

Autosomal genes can also exhibit linkage
Figure 5.5

Cross: *A B / a b* × *a b / a b*

Progeny	Experiment 1	Experiment 2
A B	17	34
a b	14	28
A b	8	16
A B	11	22
Total	50	100

Class	Observed / Expected		Observed / Expected	
Parentals	31	25	62	50
Recombination	19	25	38	50

Applying the chi square test to see if genes *A* and *B* are linked

Figure 5.6

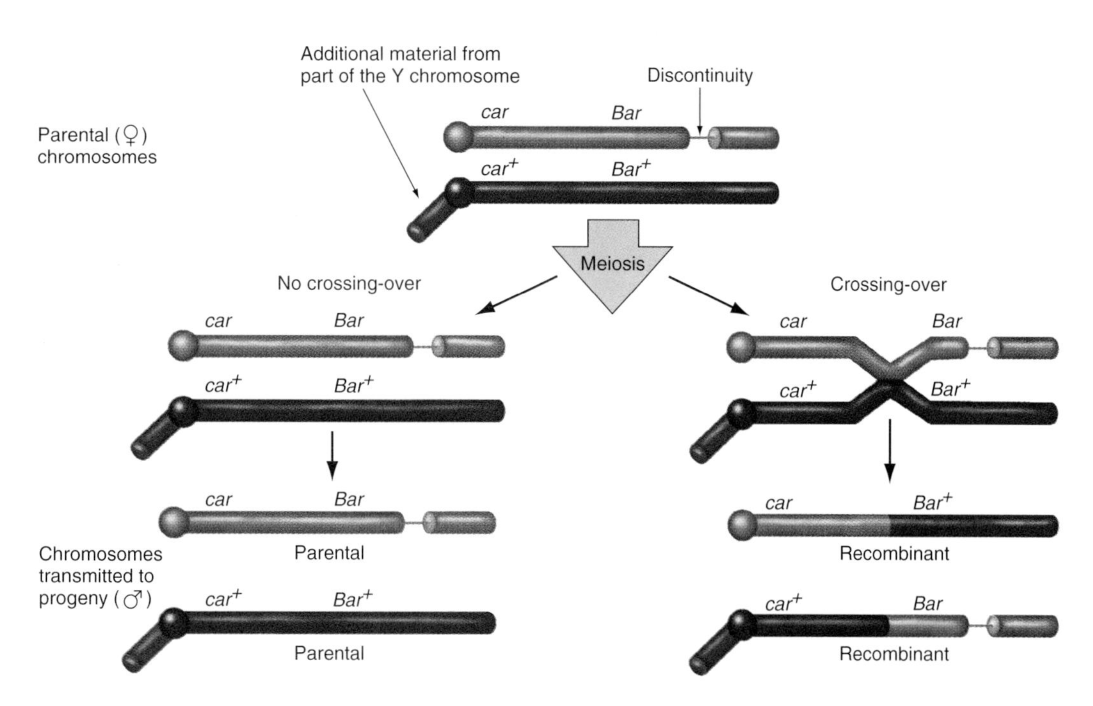

Evidence that recombination results from reciprocal exchanges between homologous chromosomes

Figure 5.7

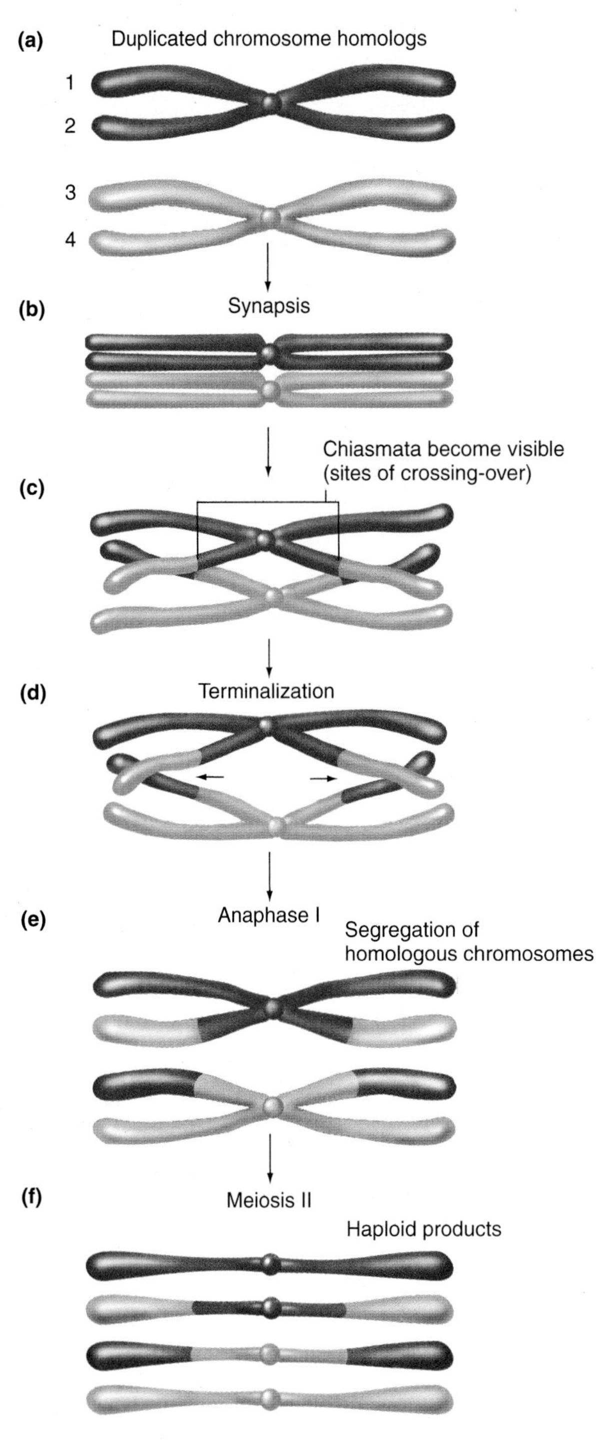

(a) Duplicated chromosome homologs

1
2

3
4

(b) Synapsis

Chiasmata become visible
(sites of crossing-over)
(c)

(d) Terminalization

(e) Anaphase I
Segregation of
homologous chromosomes

(f) Meiosis II
Haploid products

**Recombination through the light
microscope**
Figure 5.8

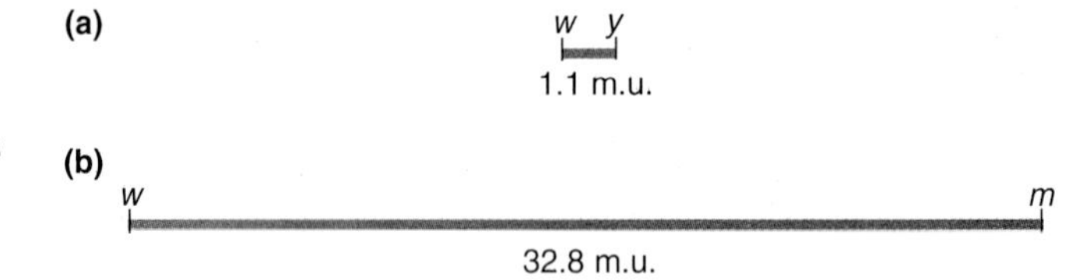

**Recombination frequencies are the basis
of genetic maps**
Figure 5.9

(a)
w y
1.1 m.u.

(b)
w   m
32.8 m.u.

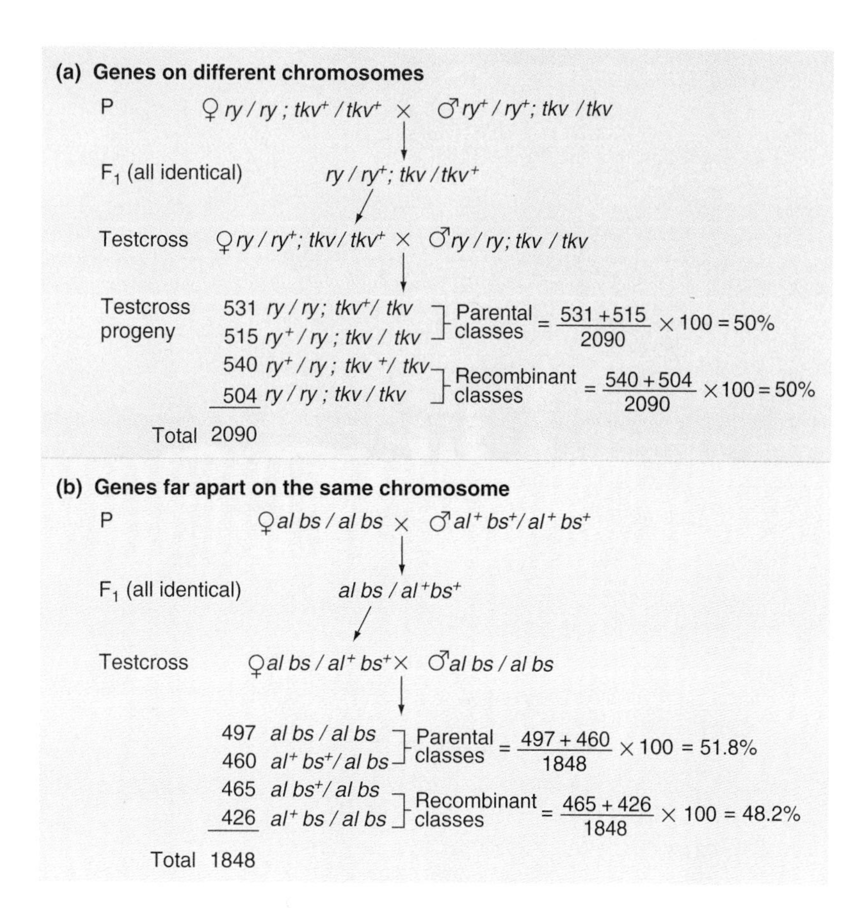

Unlinked genes show a recombination frequency of 50%

Figure 5.10

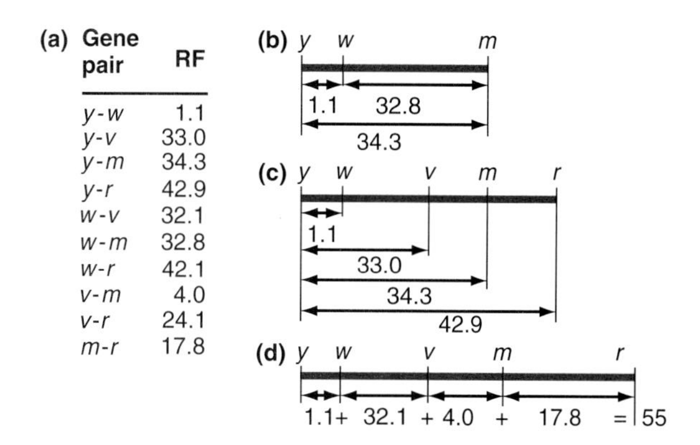

Mapping genes by comparisons of two-point crosses

Figure 5.11

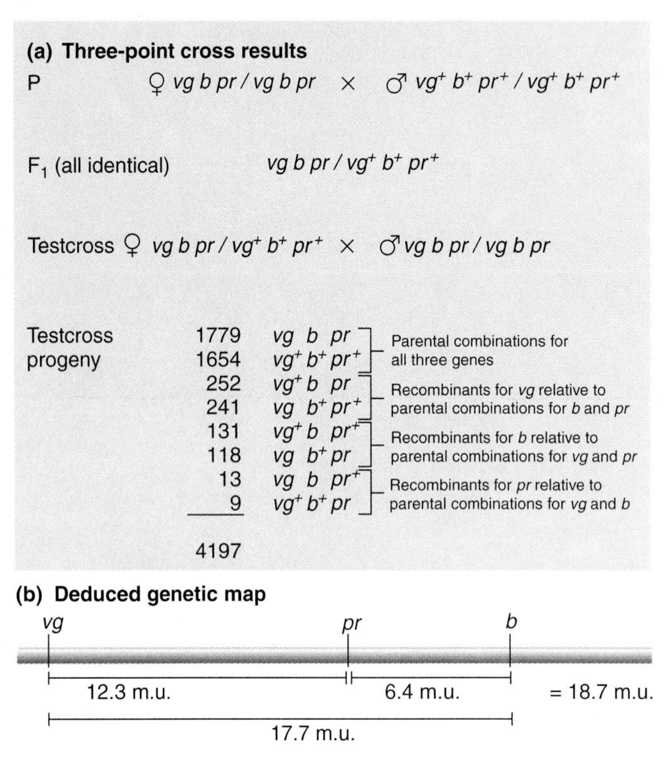

(a) **Three-point cross results**

P   ♀ *vg b pr / vg b pr*   ×   ♂ *vg⁺ b⁺ pr⁺ / vg⁺ b⁺ pr⁺*

F₁ (all identical)   *vg b pr / vg⁺ b⁺ pr⁺*

Testcross ♀ *vg b pr / vg⁺ b⁺ pr⁺*   ×   ♂ *vg b pr / vg b pr*

Testcross progeny		
1779	*vg b pr*	Parental combinations for all three genes
1654	*vg⁺ b⁺ pr⁺*	
252	*vg⁺ b pr*	Recombinants for *vg* relative to parental combinations for *b* and *pr*
241	*vg b⁺ pr⁺*	
131	*vg⁺ b pr⁺*	Recombinants for *b* relative to parental combinations for *vg* and *pr*
118	*vg b⁺ pr*	
13	*vg b pr⁺*	Recombinants for *pr* relative to parental combinations for *vg* and *b*
9	*vg⁺ b⁺ pr*	
4197		

(b) **Deduced genetic map**

vg ———— 12.3 m.u. ———— pr —— 6.4 m.u. —— b   = 18.7 m.u.

|—————— 17.7 m.u. ——————|

Analyzing the results of a three-point cross
Figure 5.12

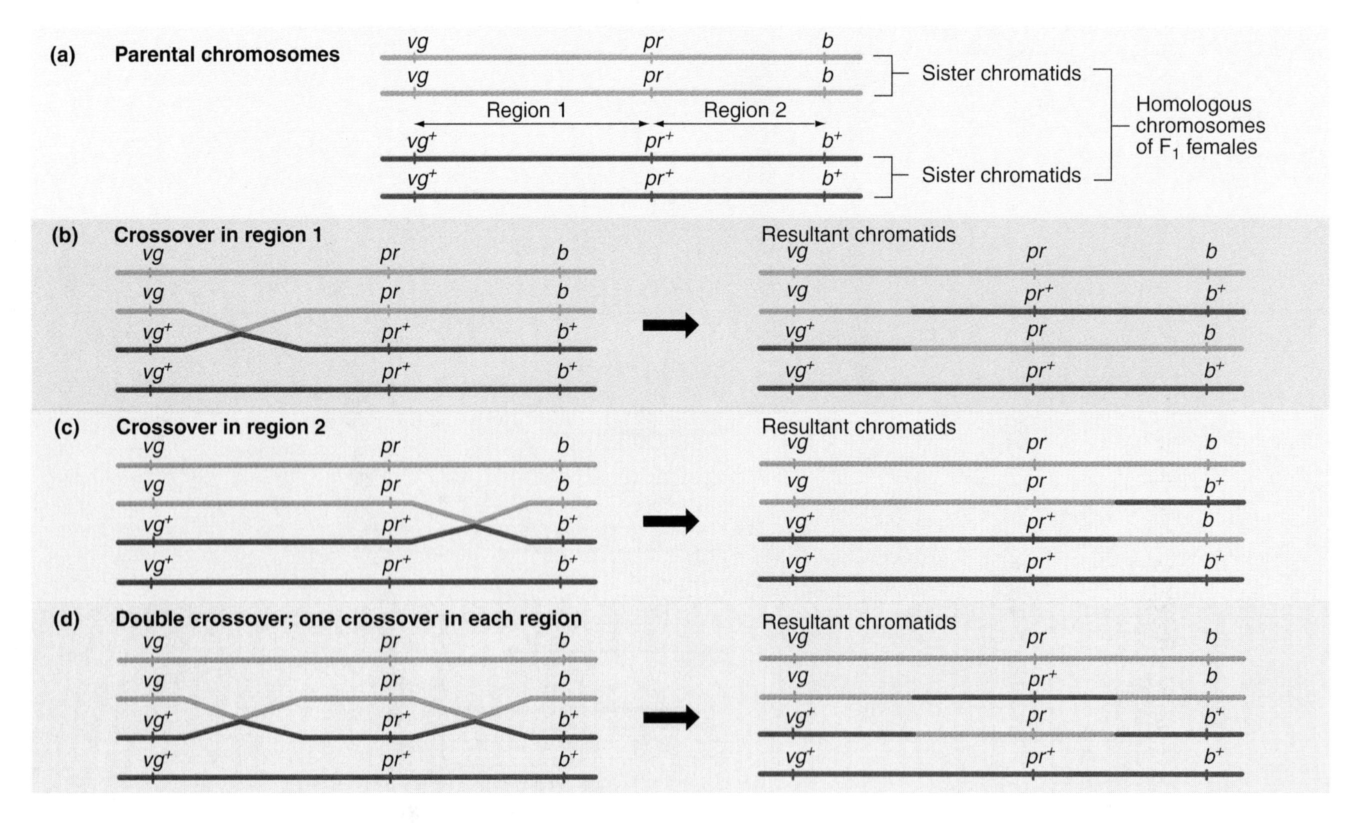

(a) **Parental chromosomes**

vg ———— pr ———— b   Sister chromatids

vg ———— pr ———— b   Homologous chromosomes of F₁ females

Region 1 → Region 2

vg⁺ ———— pr⁺ ———— b⁺

vg⁺ ———— pr⁺ ———— b⁺   Sister chromatids

(b) **Crossover in region 1**   Resultant chromatids

(c) **Crossover in region 2**   Resultant chromatids

(d) **Double crossover; one crossover in each region**   Resultant chromatids

Inferring the location of a crossover event
Figure 5.13

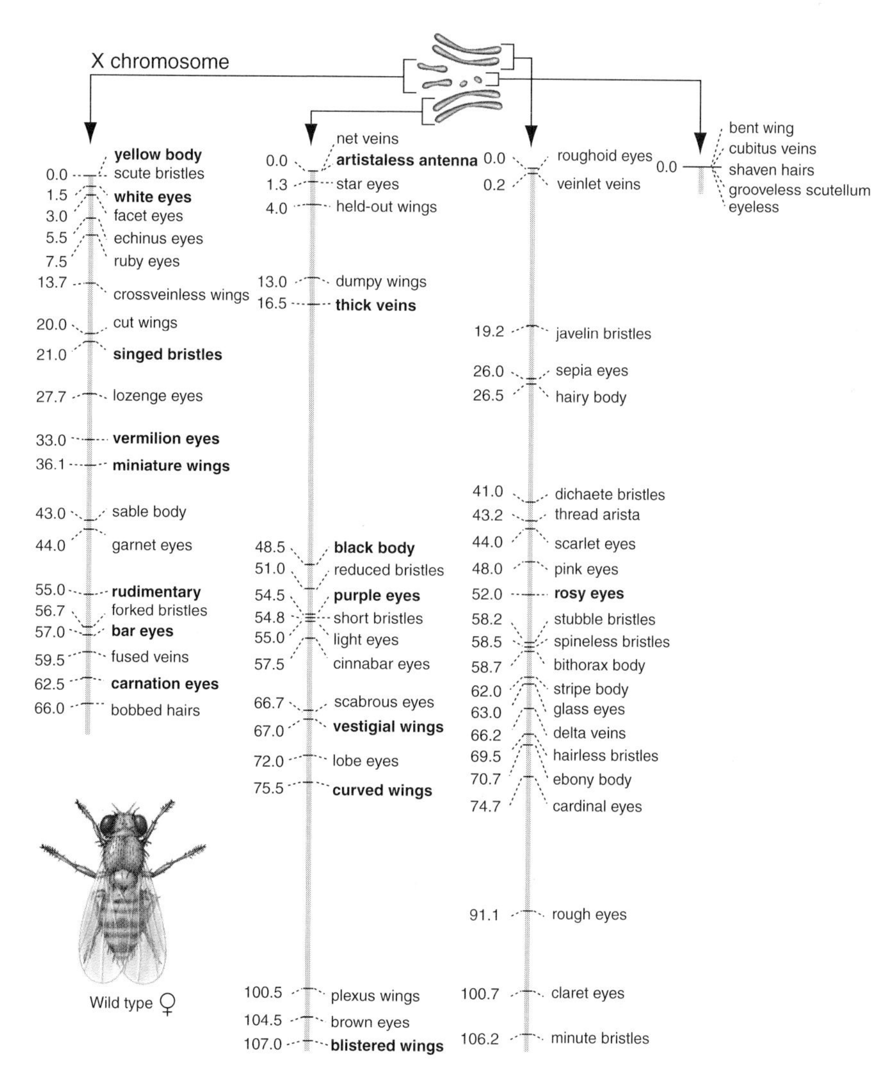

$♀ w^+ w\ y^+ y\ m^+ m \quad \times \quad ♂ X / Y$

Before data analysis, you do not
know the gene order or allele
combination on each chromosome.

Male progeny

2278	$w^+ y^+ m$ /Y	Parental class (noncrossover)
2157	$w\ y\ m^+$ /Y	
1203	$w\ y\ m$ /Y	Crossover in region 2 (between w and m)
1092	$w^+ y^+ m^+$ /Y	
49	$w^+ y\ m$ /Y	Crossover in region 1 (between y and m)
41	$w\ y^+ m^+$ /Y	
2	$w^+ y\ m^+$ /Y	Double crossovers
1	$w\ y^+ m$ /Y	
6823		

After data analysis, you can conclude that
the gene order and allele combinations on the
X chromosomes of the F_1 females were $y\ w\ m^+ / y^+\ w^+\ m$.

How three-point crosses verify Sturtevant's map
Figure 5.14

Drosophila melanogaster has four linkage groups
Figure 5.15

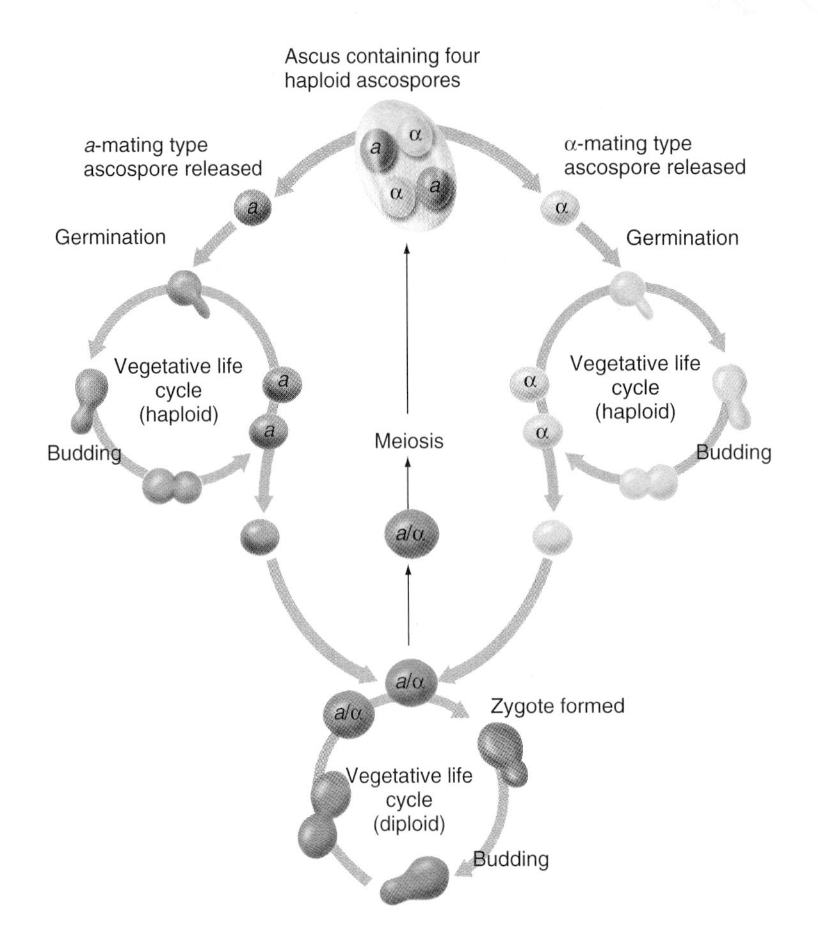

Ascus containing four
haploid ascospores

a-mating type
ascospore released

α-mating type
ascospore released

Germination

Germination

Vegetative life
cycle
(haploid)

Vegetative life
cycle
(haploid)

Budding

Budding

Meiosis

a/α

a/α

Zygote formed

a/α

Vegetative life
cycle
(diploid)

Budding

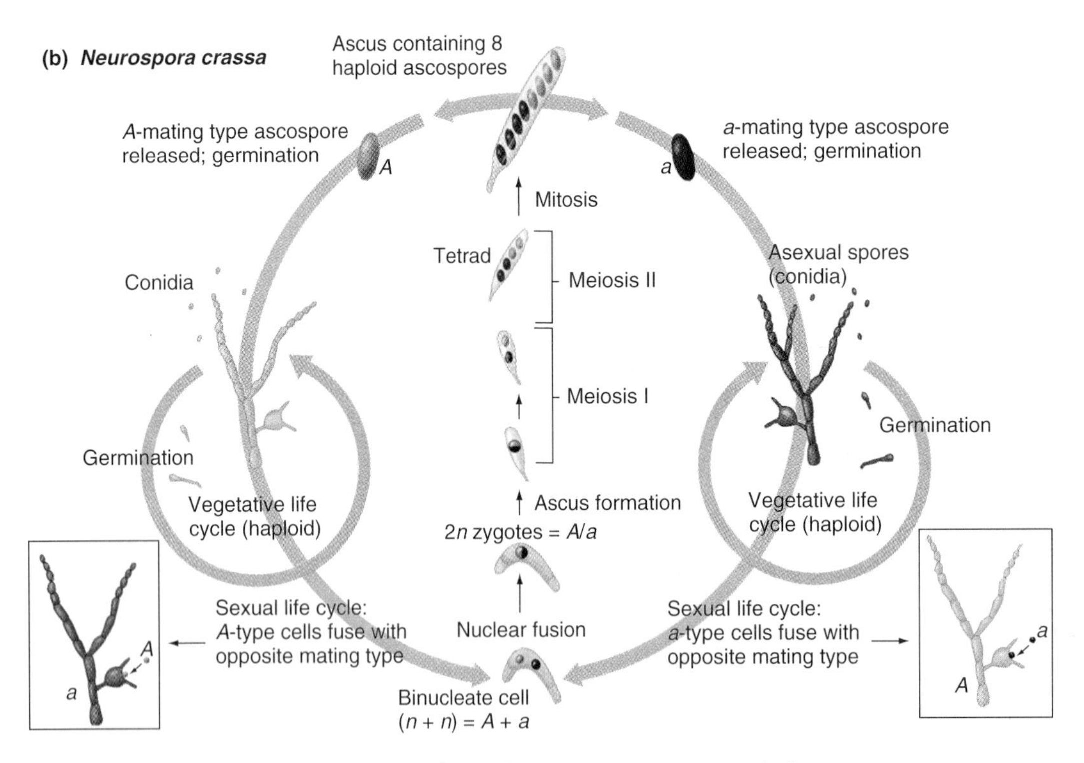

(b) *Neurospora crassa*

Ascus containing 8
haploid ascospores

A-mating type ascospore
released; germination

a-mating type ascospore
released; germination

A

a

Mitosis

Conidia

Tetrad

Meiosis II

Asexual spores
(conidia)

Meiosis I

Germination

Germination

Vegetative life
cycle (haploid)

Ascus formation

Vegetative life
cycle (haploid)

2*n* zygotes = *A*/*a*

A

a

a

A

Sexual life cycle:
A-type cells fuse with
opposite mating type

Nuclear fusion

Sexual life cycle:
a-type cells fuse with
opposite mating type

Binucleate cell
(*n* + *n*) = *A* + *a*

**The life cycles of the yeast *Saccharomyces cerevisiae*
and the bread mold *Neurospora crassa***

Figure 5.16

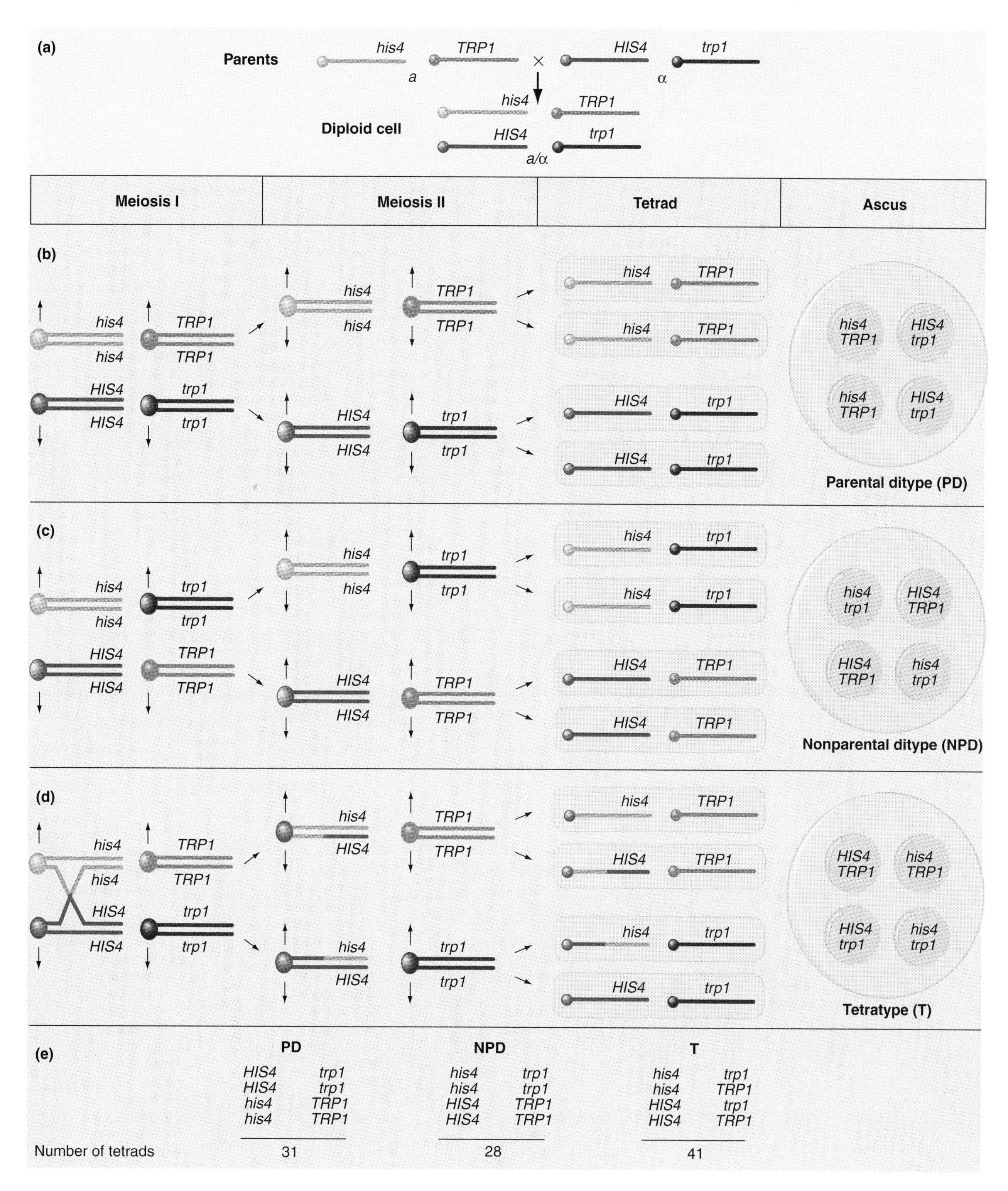

How meiosis can generate three kinds of tetrads when two genes are on different chromosomes

Figure 5.17

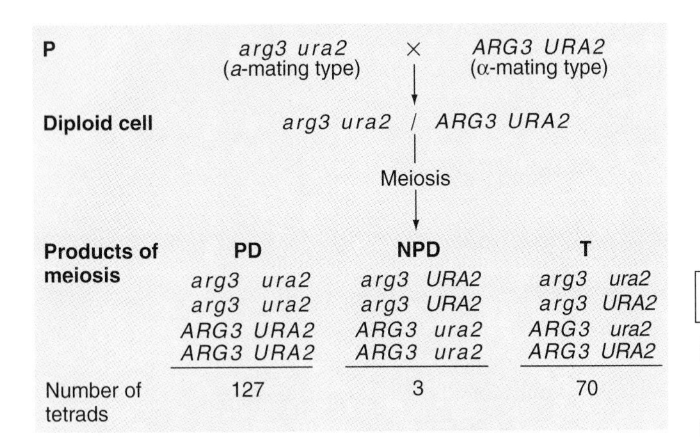

Products of meiosis	PD	NPD	T
	arg3 ura2	arg3 URA2	arg3 ura2
	arg3 ura2	arg3 URA2	arg3 URA2
	ARG3 URA2	ARG3 ura2	ARG3 ura2
	ARG3 URA2	ARG3 ura2	ARG3 URA2
Number of tetrads	127	3	70

When Genes are Linked, PDs Exceed NPDs
Figure 5.18

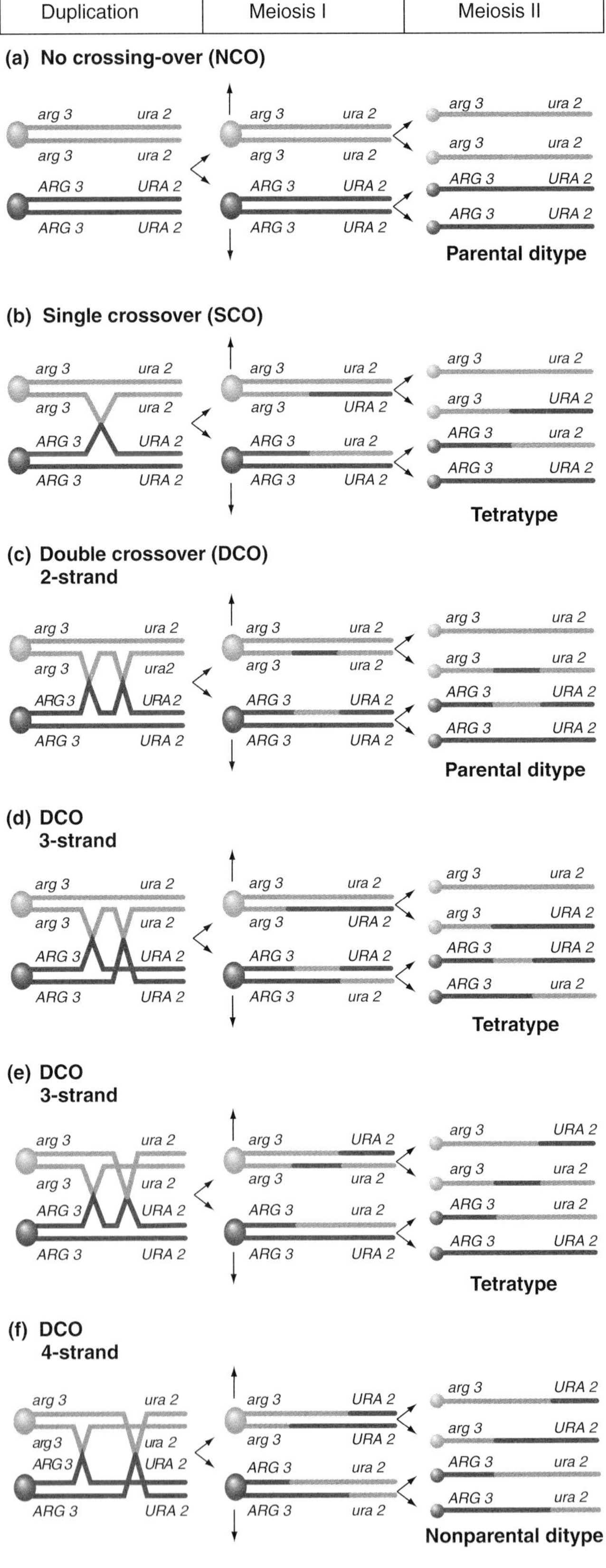

How crossovers between linked genes generate different tetrads
Figure 5.19

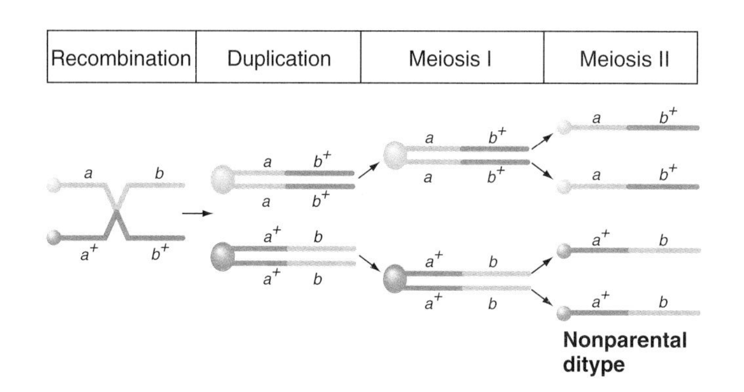

Recombination	Duplication	Meiosis I	Meiosis II

A mistaken model: Recombination before chromosome replication
Figure 5.20

Nonparental ditype

How ordered tetrads form
Figure 5.22

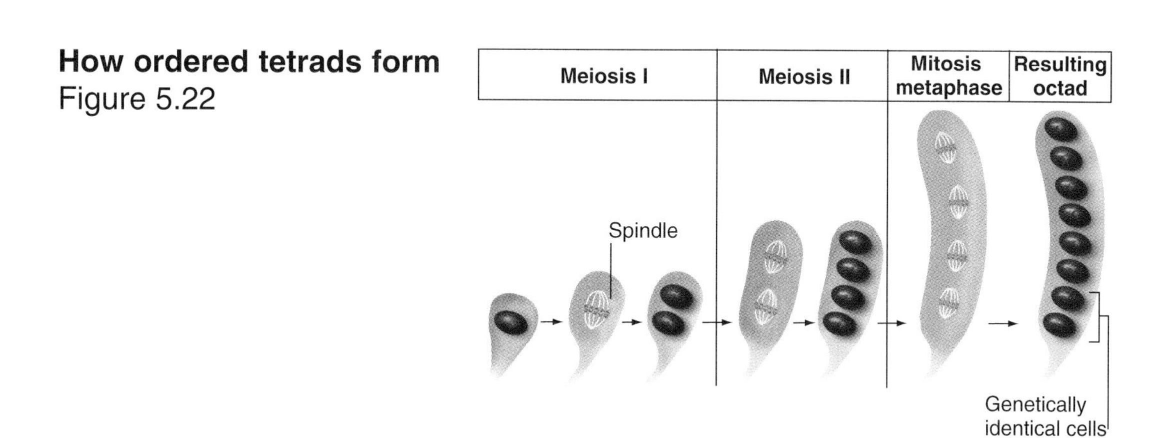

Meiosis I	Meiosis II	Mitosis metaphase	Resulting octad

Spindle

Genetically identical cells

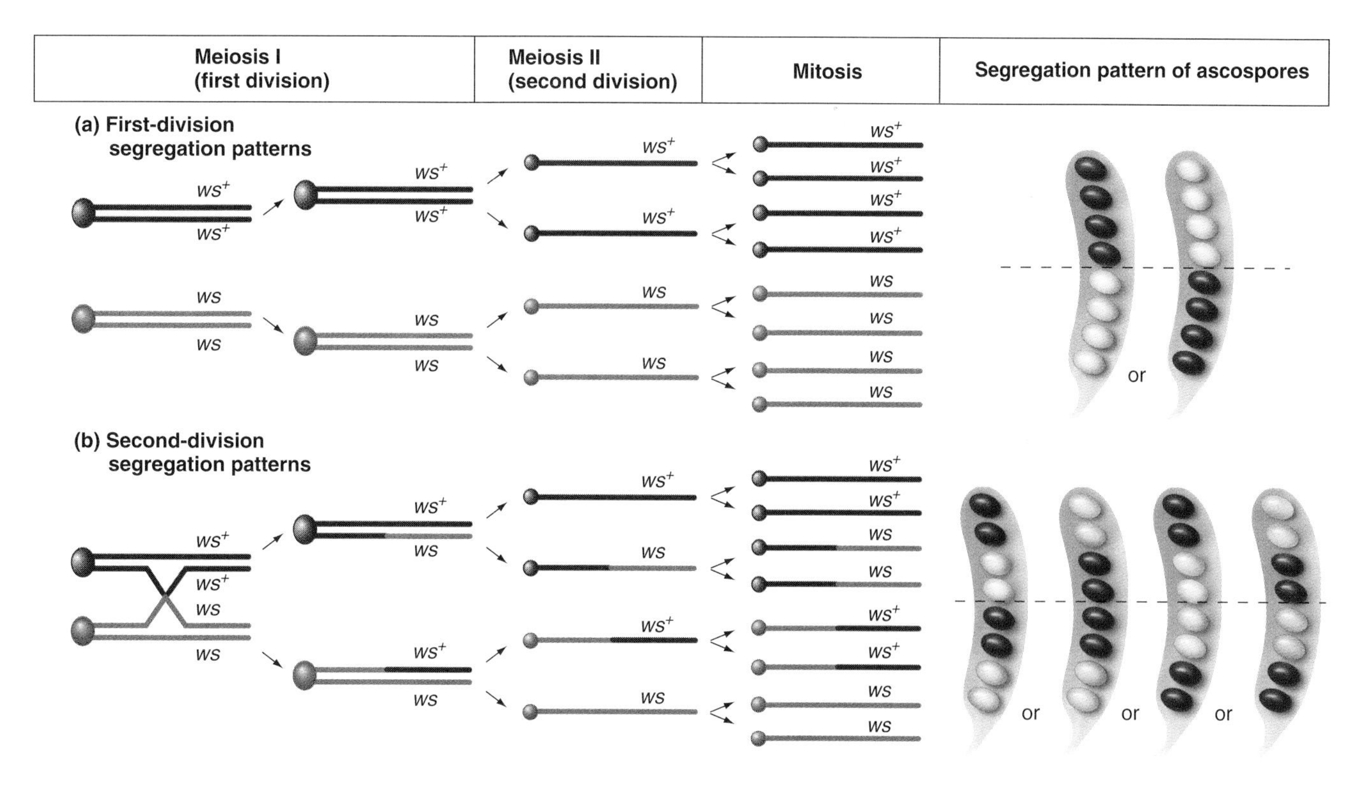

Meiosis I (first division)	Meiosis II (second division)	Mitosis	Segregation pattern of ascospores

(a) First-division segregation patterns

(b) Second-division segregation patterns

Two segregation patterns in ordered asci
Figure 5.23

(a) A *Neurospora* cross

Tetrad group	A	B	C	D	E	F	G
Segregation pattern	*thr arg*	*thr arg*	*thr arg*	*thr arg⁺*	*thr arg⁺*	*thr arg⁺*	*thr arg*
	thr arg	*thr⁺ arg*	*thr arg⁺*	*thr⁺ arg*	*thr⁺ arg*	*thr arg⁺*	*thr⁺ arg⁺*
	thr⁺ arg⁺	*thr⁺ arg⁺*	*thr⁺ arg*	*thr⁺ arg⁺*	*thr⁺ arg*	*thr⁺ arg*	*thr⁺ arg⁺*
	thr⁺ arg⁺	*thr arg⁺*	*thr⁺ arg⁺*	*thr arg*	*thr arg⁺*	*thr⁺ arg*	*thr arg*
Total in group	72	16	11	2	2	1	1

(b) Corresponding genetic map

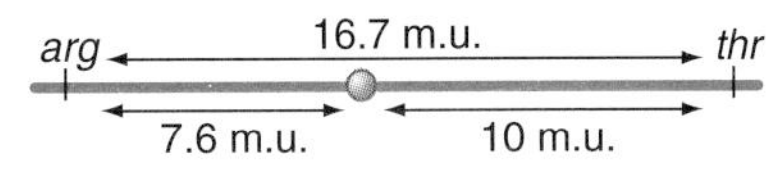

arg ←———— 16.7 m.u. ————→ thr
←— 7.6 m.u. —→ ←—— 10 m.u. ——→

Genetic mapping by ordered-tetrad analysis: An example
Figure 5.24

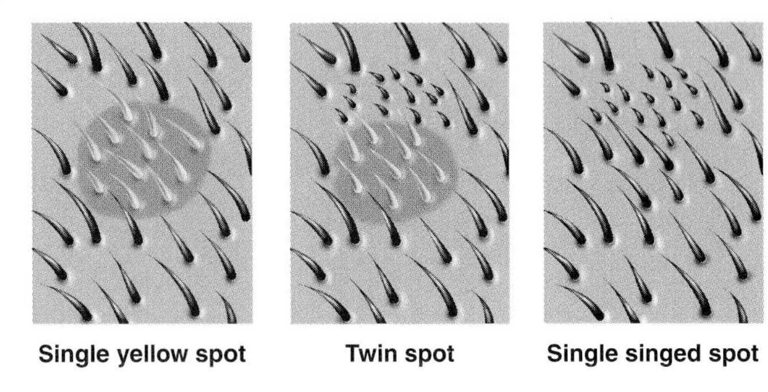

Single yellow spot   **Twin spot**   **Single singed spot**

Twin spots: A form of genetic mosaicism
Figure 5.25

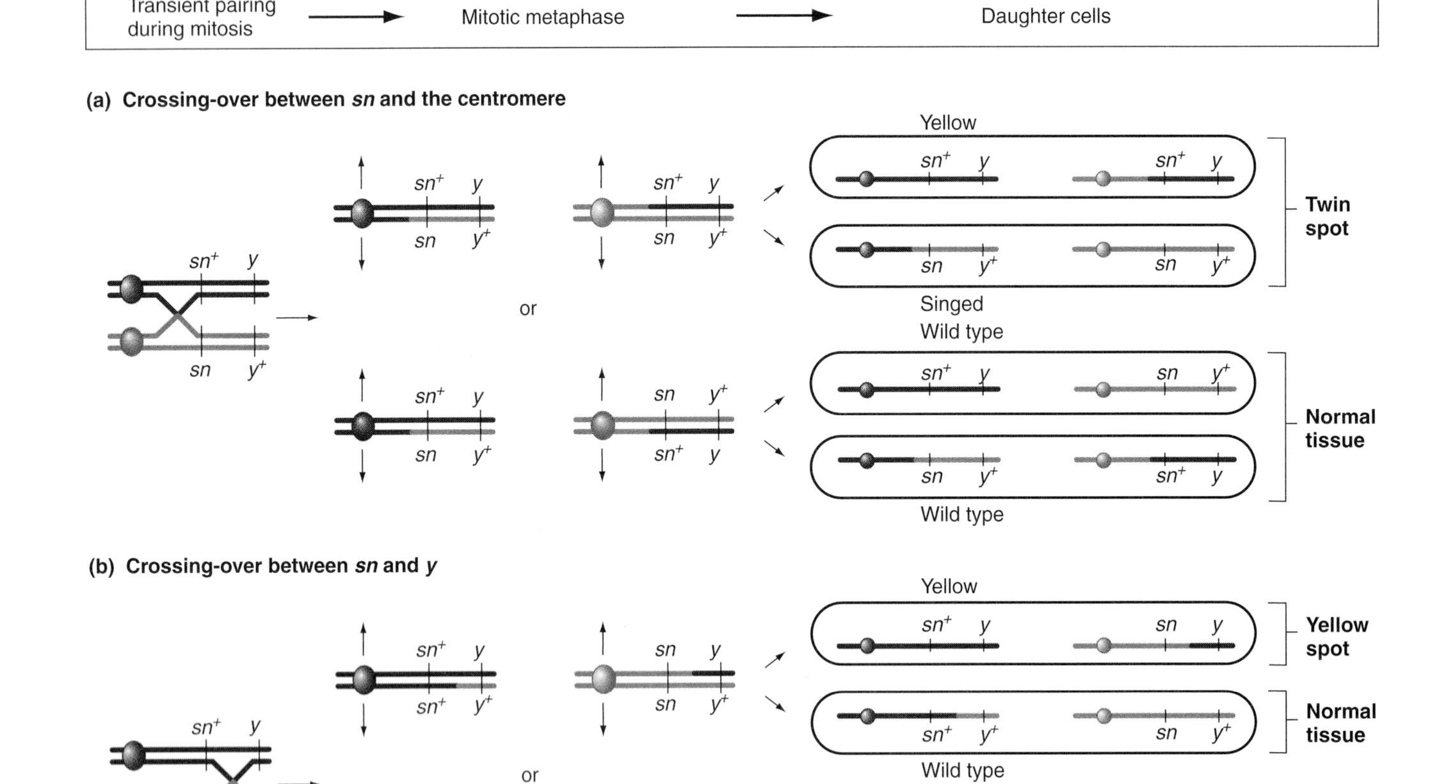

Mitotic crossing-over
Figure 5.26

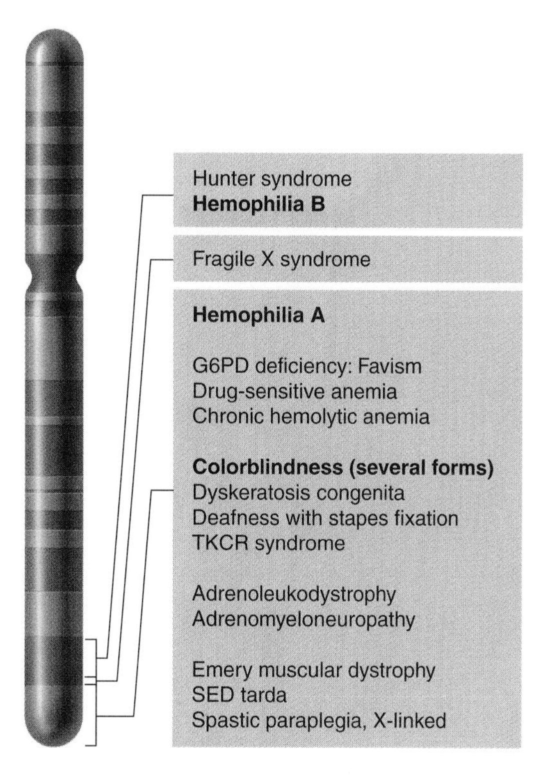

Hunter syndrome
Hemophilia B

Fragile X syndrome

Hemophilia A

G6PD deficiency: Favism
Drug-sensitive anemia
Chronic hemolytic anemia

Colorblindness (several forms)
Dyskeratosis congenita
Deafness with stapes fixation
TKCR syndrome

Adrenoleukodystrophy
Adrenomyeloneuropathy

Emery muscular dystrophy
SED tarda
Spastic paraplegia, X-linked

**A genetic map of part of the
human X chromosome**
Figure 5.28

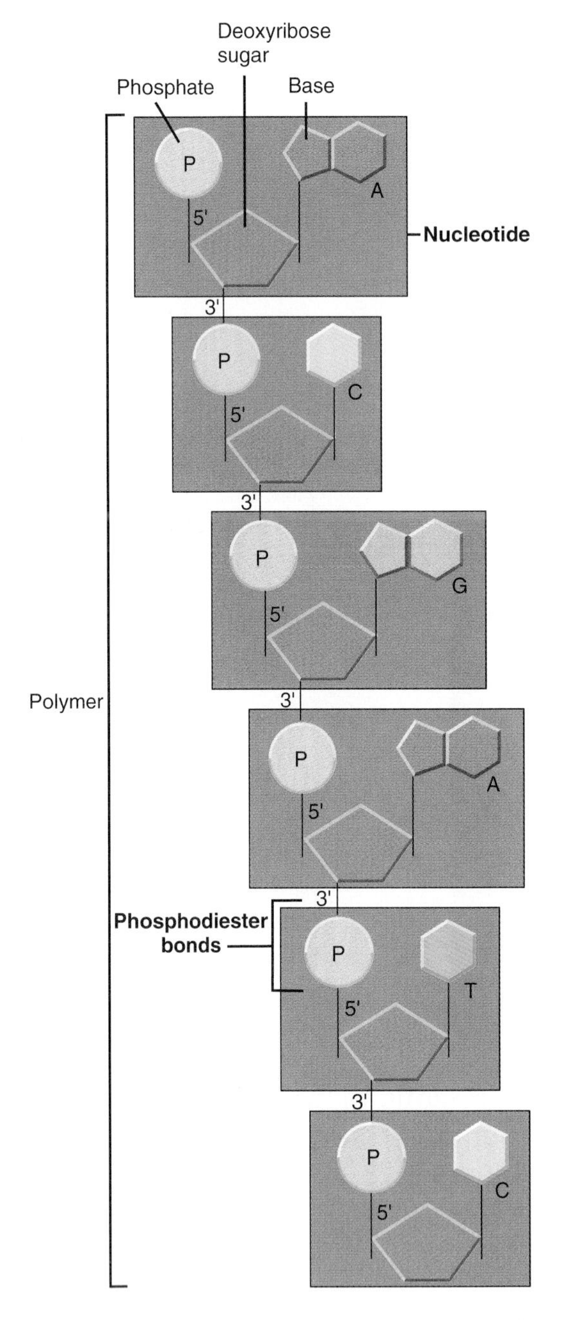

The chemical composition of DNA
Figure 6.2

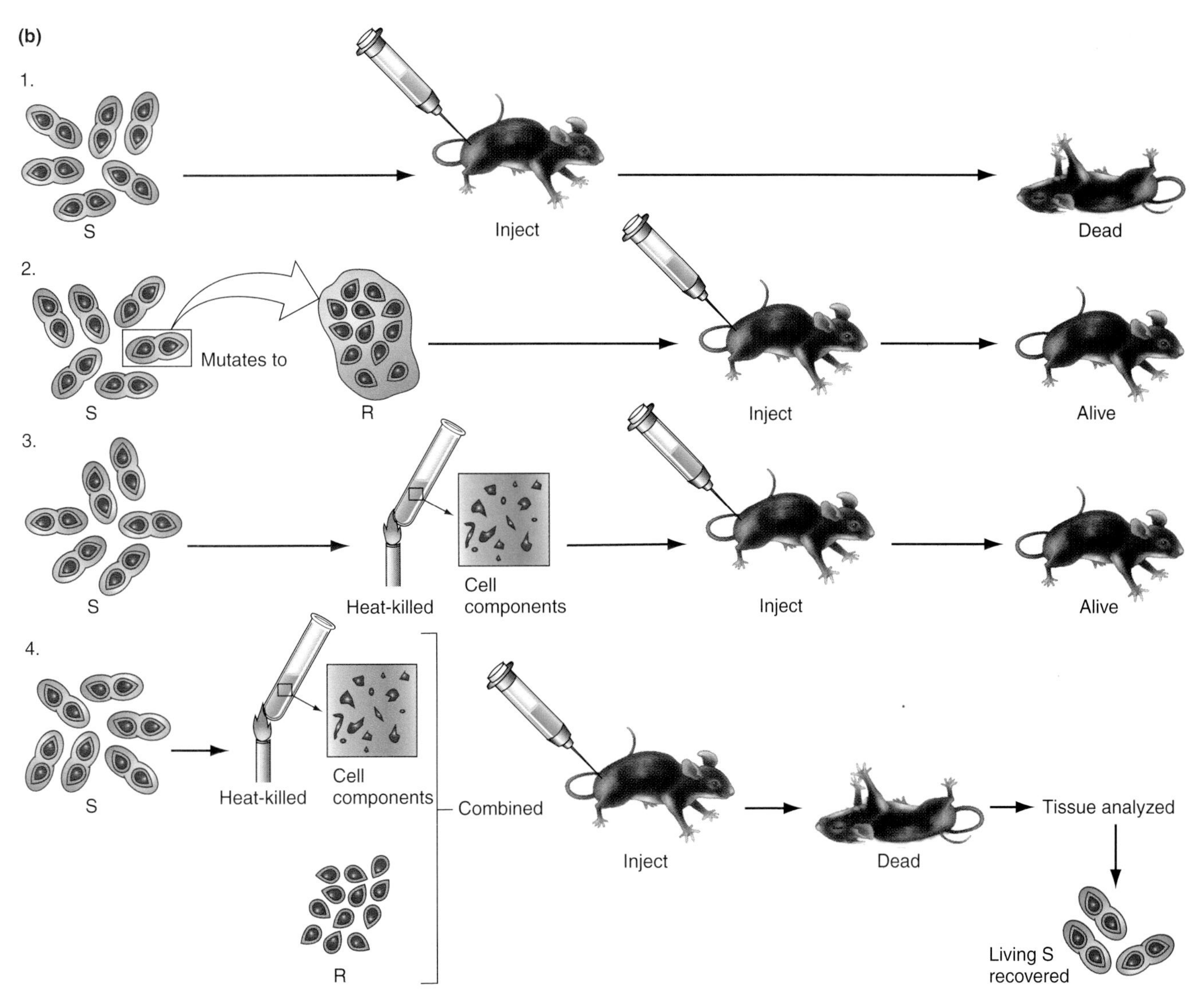

(b)

1.

S Inject Dead

2.

S Mutates to R Inject Alive

3.

S Heat-killed Cell components Inject Alive

4.

S Heat-killed Cell components

R
Combined Inject Dead Tissue analyzed

Living S recovered

Griffith's demonstration of bacterial transformation
Figure 6.3

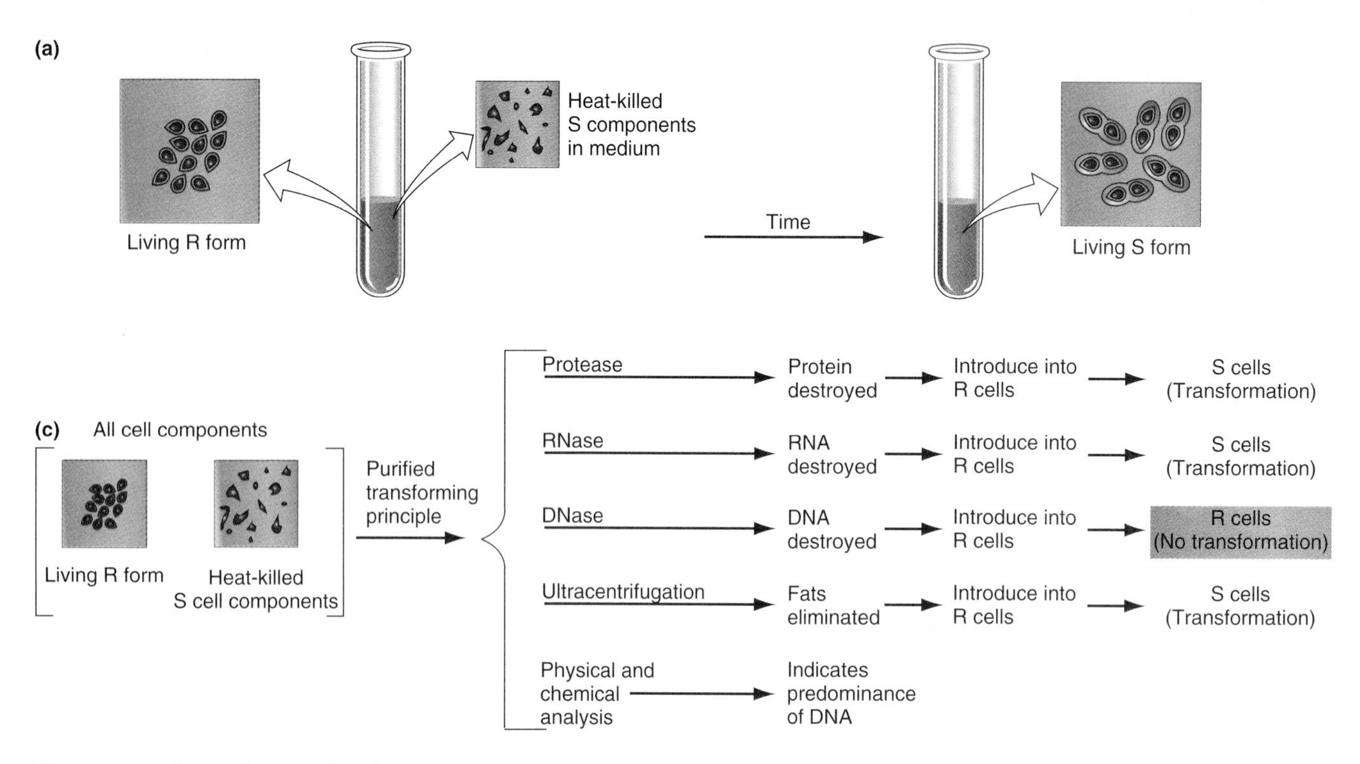

The transforming principle is DNA: Experimental confirmation
Figure 6.4

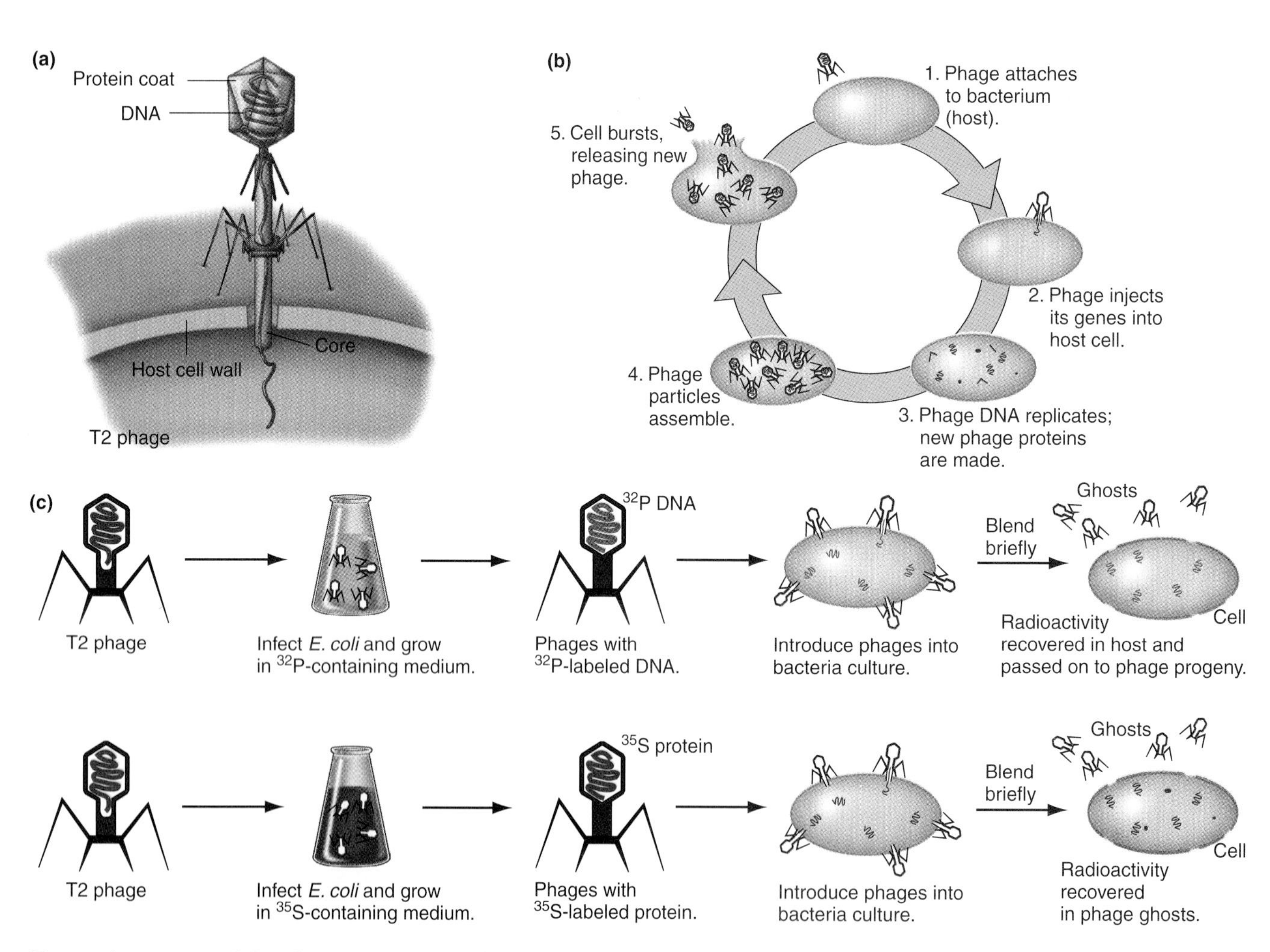

Experiments with viruses provide convincing evidence that genes are made of DNA
Figure 6.5

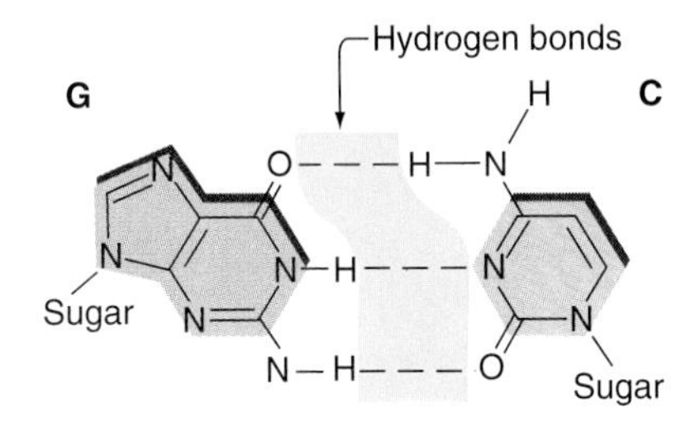

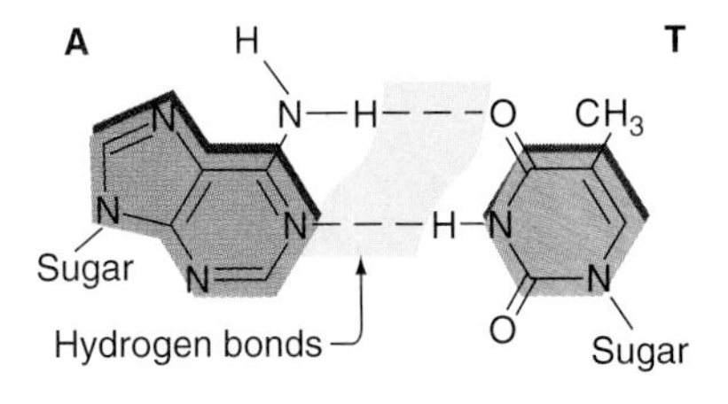

Complementary base pairing
Figure 6.8

(a)

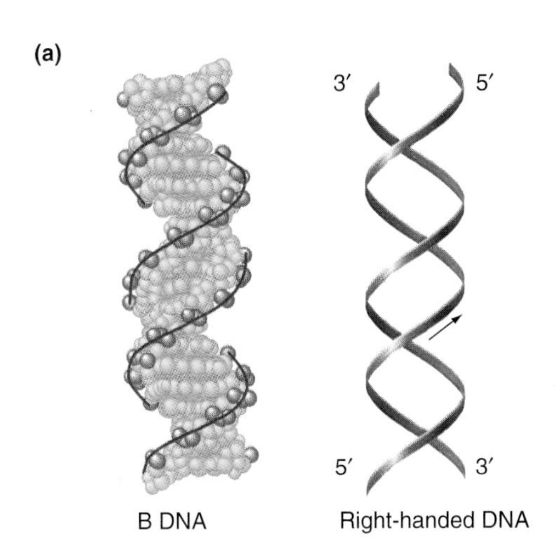

B DNA   Right-handed DNA

(b)

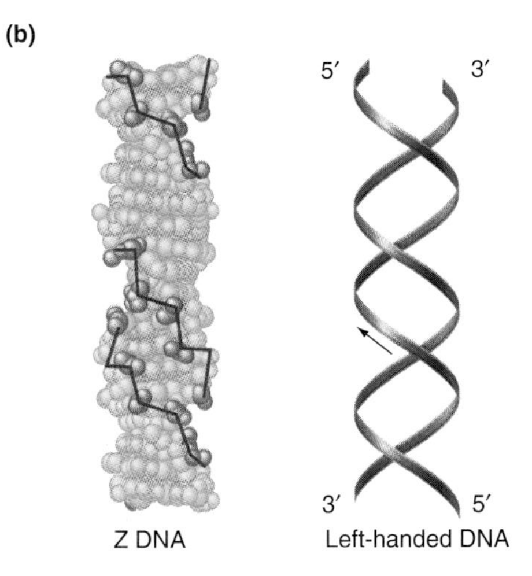

Z DNA   Left-handed DNA

Z DNA is one variant of the double helix
Figure 6.10

(a)

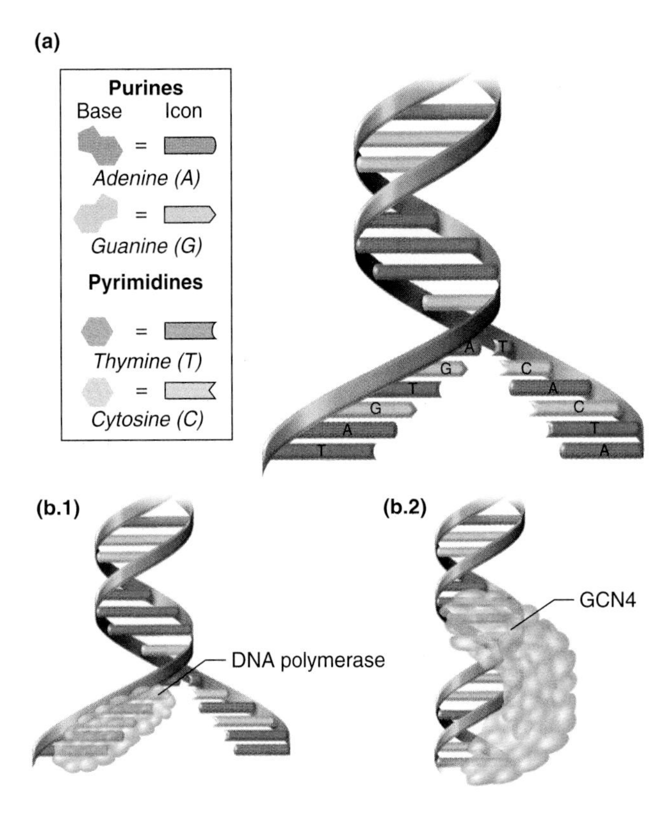

Purines

Base	Icon

Adenine (A)

Guanine (G)

Pyrimidines

Thymine (T)

Cytosine (C)

(b.1)   **(b.2)**

DNA polymerase

GCN4

DNA stores information in the sequence of its bases
Figure 6.12

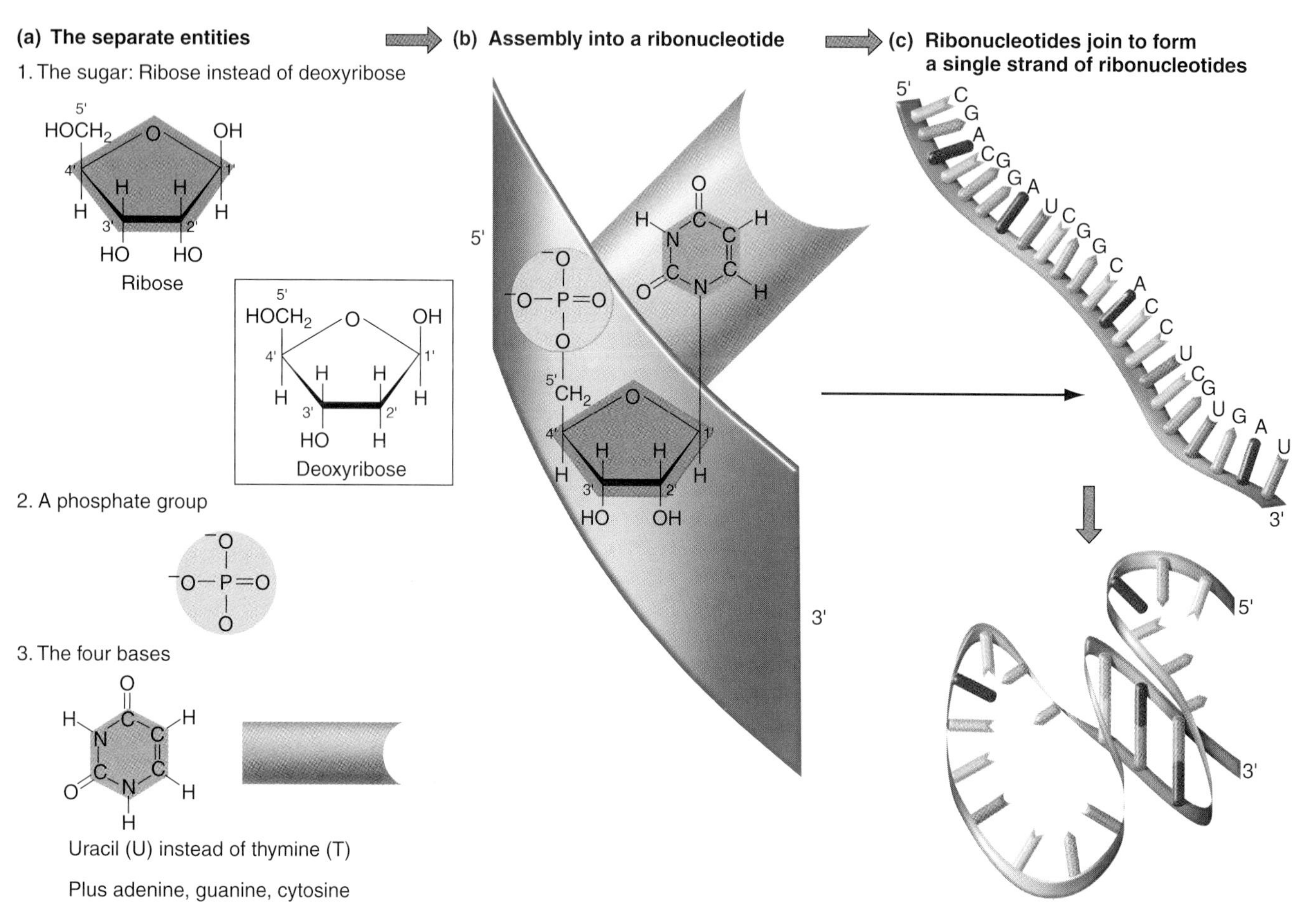

(a) The separate entities

1. The sugar: Ribose instead of deoxyribose

Ribose

Deoxyribose

2. A phosphate group

3. The four bases

Uracil (U) instead of thymine (T)

Plus adenine, guanine, cytosine

(b) Assembly into a ribonucleotide

(c) Ribonucleotides join to form a single strand of ribonucleotides

RNA: Chemical constituents and complex folding pattern
Figure 6.13

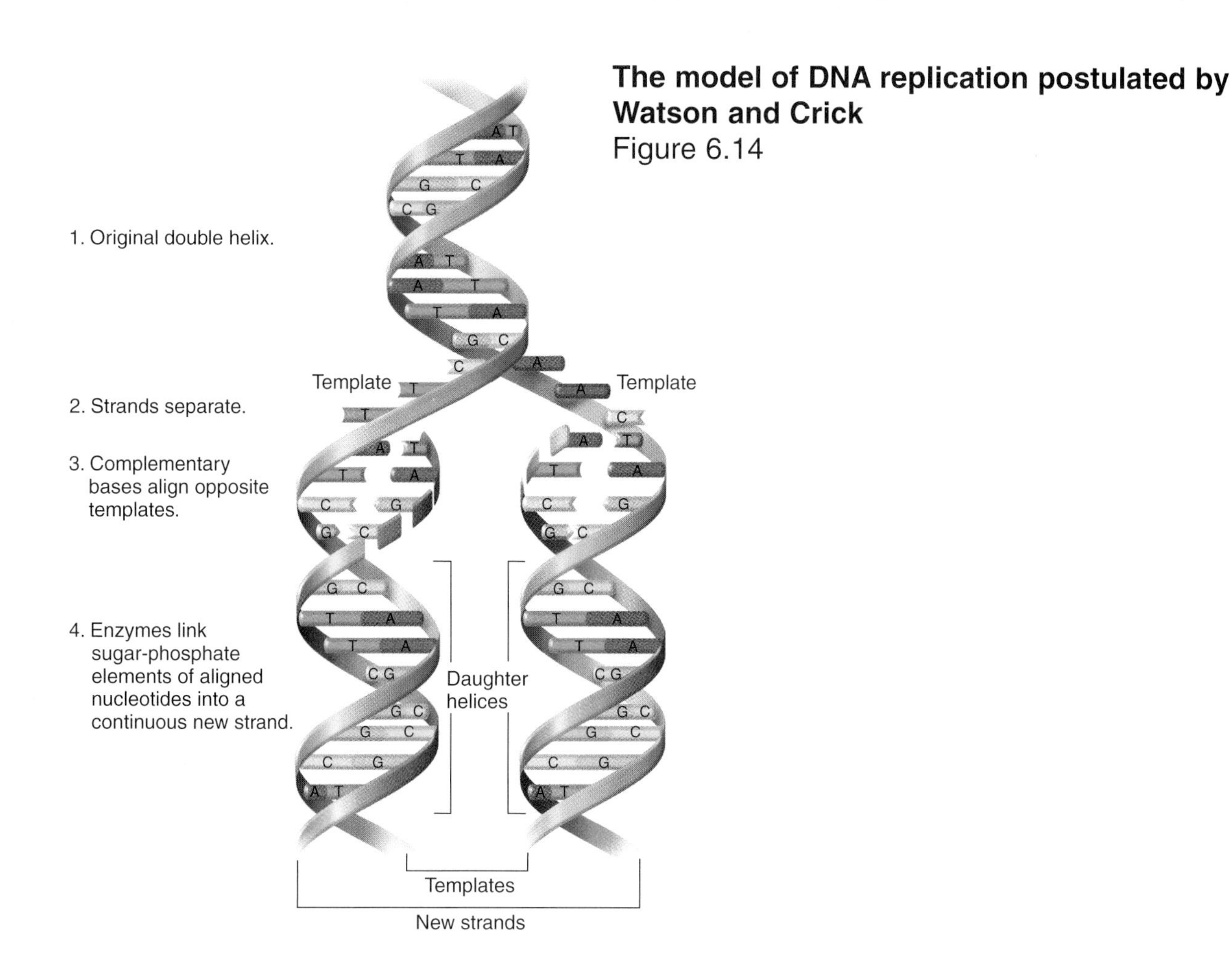

The model of DNA replication postulated by Watson and Crick
Figure 6.14

1. Original double helix.

2. Strands separate.

3. Complementary bases align opposite templates.

4. Enzymes link sugar-phosphate elements of aligned nucleotides into a continuous new strand.

Template

Template

Daughter helices

Templates

New strands

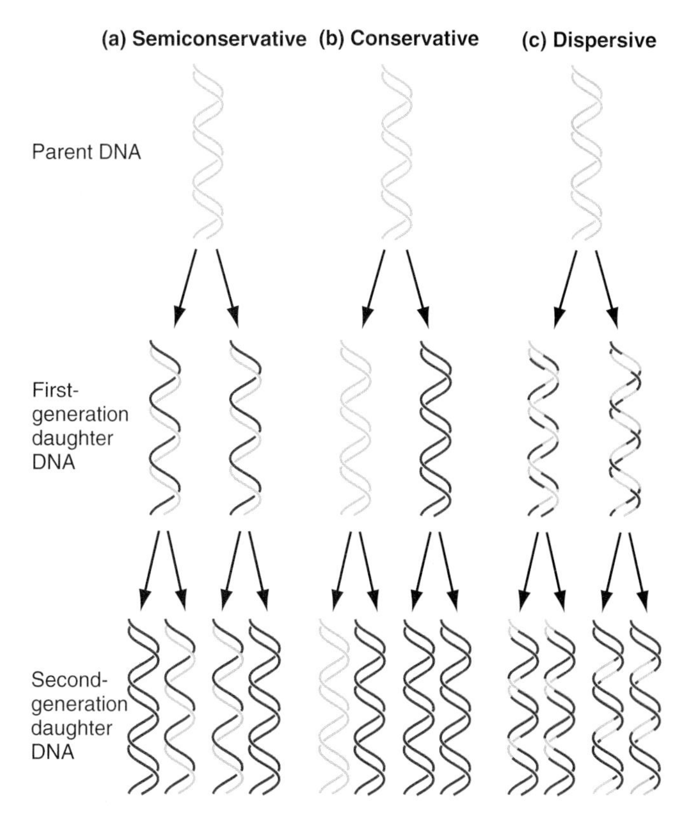

(a) Semiconservative **(b) Conservative** **(c) Dispersive**

Parent DNA

First-generation daughter DNA

Second-generation daughter DNA

Three possible models of DNA replication
Figure 6.15

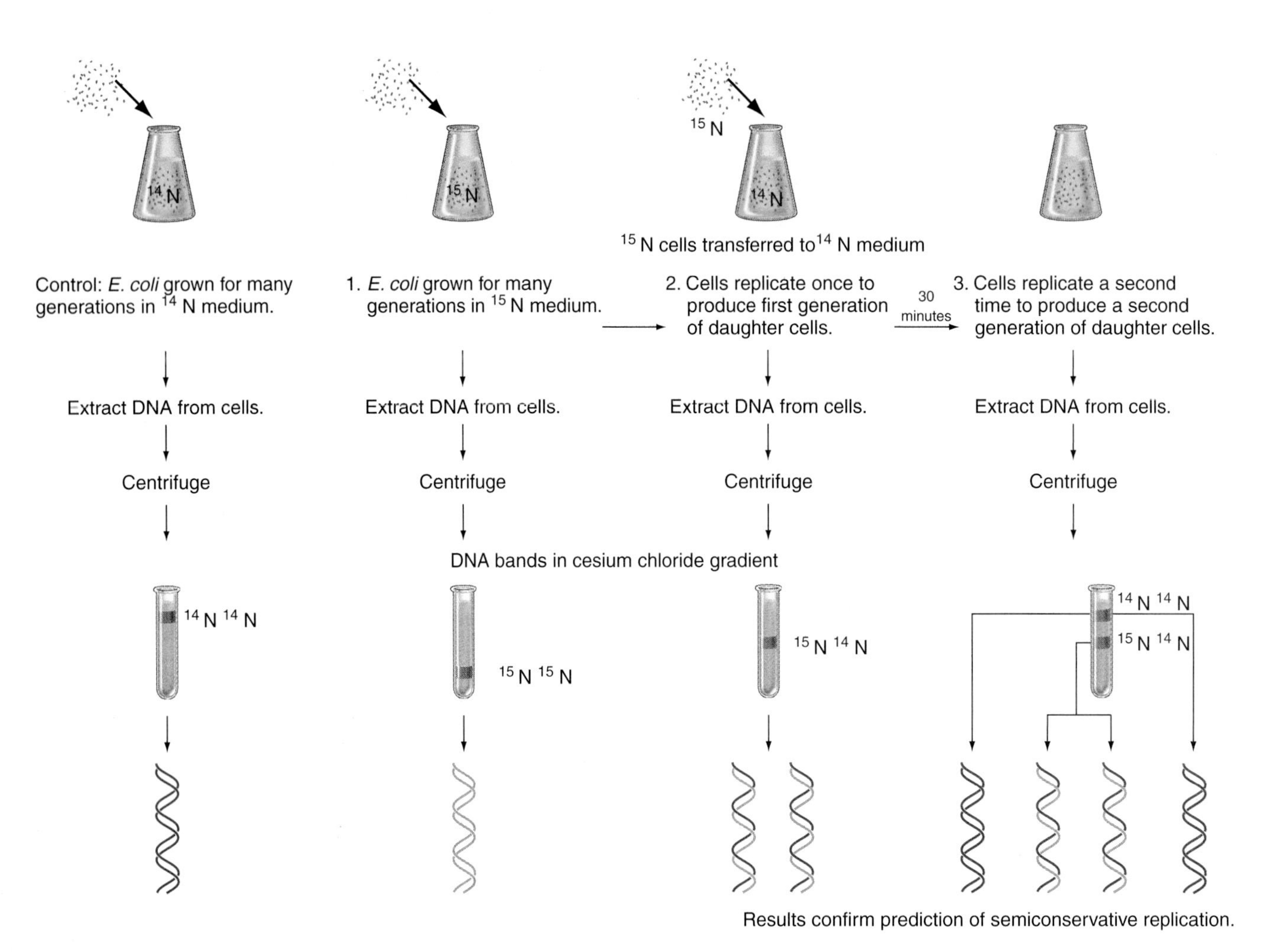

Control: *E. coli* grown for many generations in ^{14}N medium.

1. *E. coli* grown for many generations in ^{15}N medium.

2. Cells replicate once to produce first generation of daughter cells.

3. Cells replicate a second time to produce a second generation of daughter cells.

30 minutes

^{15}N cells transferred to ^{14}N medium

Extract DNA from cells.

Extract DNA from cells.

Extract DNA from cells.

Extract DNA from cells.

Centrifuge

Centrifuge

Centrifuge

Centrifuge

DNA bands in cesium chloride gradient

^{14}N ^{14}N

^{15}N ^{15}N

^{15}N ^{14}N

^{14}N ^{14}N
^{15}N ^{14}N

Results confirm prediction of semiconservative replication.

How the Meselson-Stahl experiment confirmed semiconservative replication
Figure 6.16

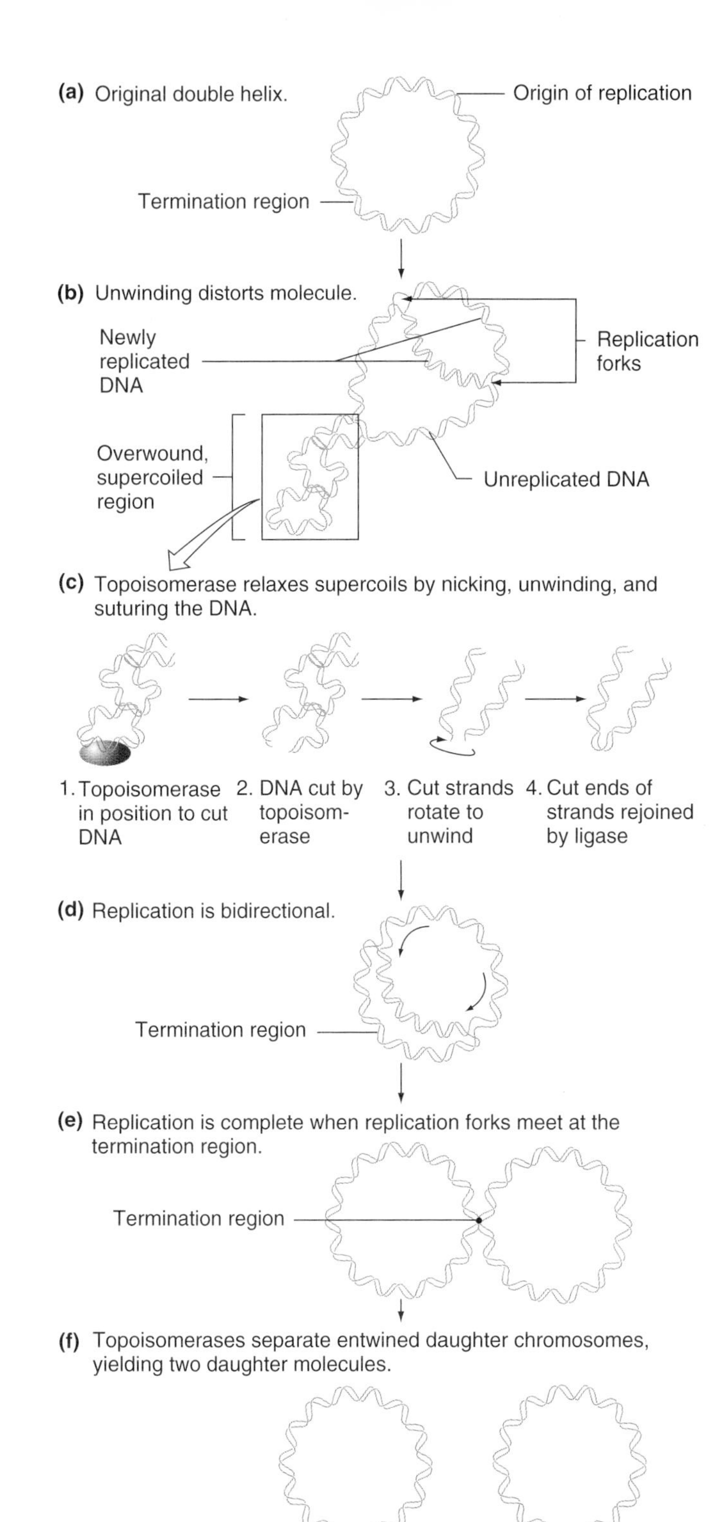

(a) Original double helix.

Origin of replication

Termination region

The bidirectional replication of a circular bacterial chromosome: An overview
Figure 6.18

(b) Unwinding distorts molecule.

Newly replicated DNA

Replication forks

Overwound, supercoiled region

Unreplicated DNA

(c) Topoisomerase relaxes supercoils by nicking, unwinding, and suturing the DNA.

1. Topoisomerase in position to cut DNA
2. DNA cut by topoisomerase
3. Cut strands rotate to unwind
4. Cut ends of strands rejoined by ligase

(d) Replication is bidirectional.

Termination region

(e) Replication is complete when replication forks meet at the termination region.

Termination region

(f) Topoisomerases separate entwined daughter chromosomes, yielding two daughter molecules.

DNA molecules break and rejoin during recombination: The experimental evidence
Figure 6.19

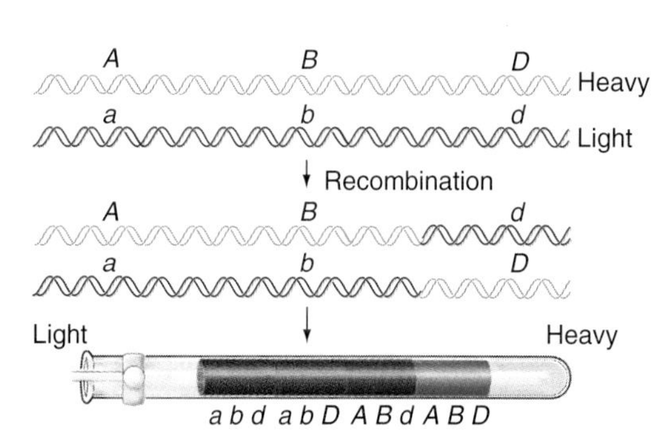

A   B   D   Heavy
a   b   d   Light

Recombination

A   B   d
a   b   D

Light   Heavy

a b d a b D A B d A B D

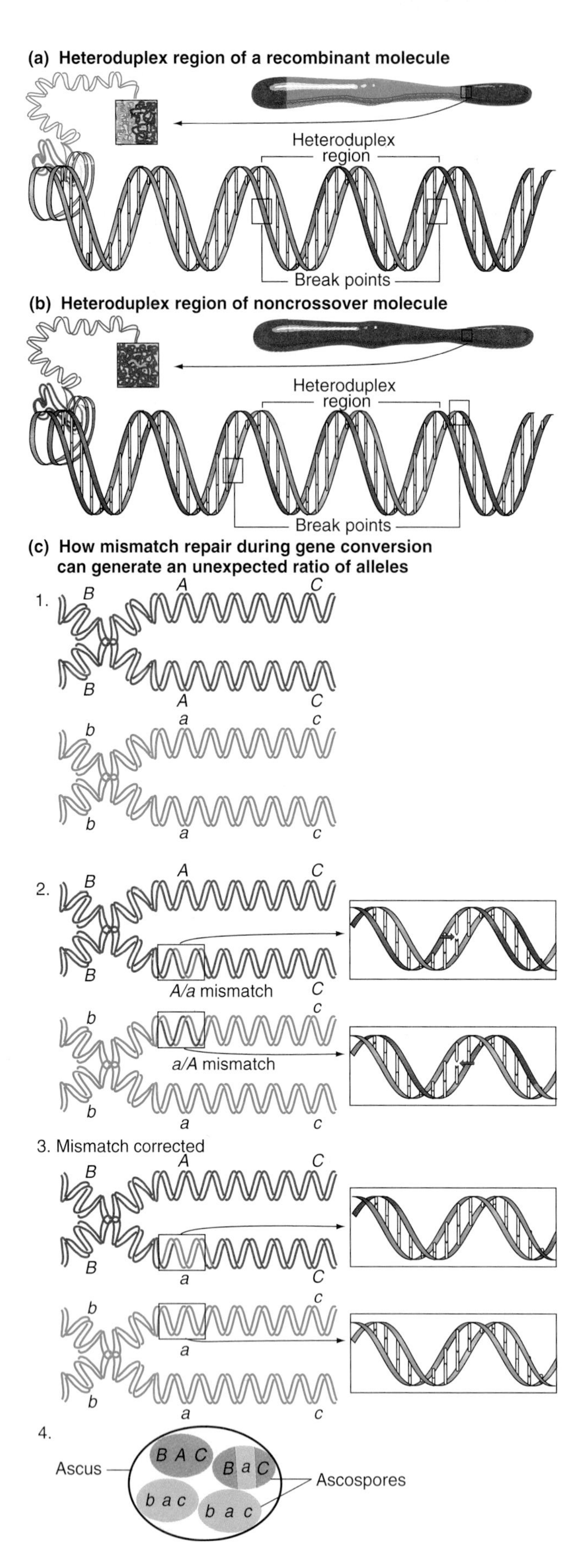

(a) Heteroduplex region of a recombinant molecule

Heteroduplex region

Break points

(b) Heteroduplex region of noncrossover molecule

Heteroduplex region

Break points

(c) How mismatch repair during gene conversion can generate an unexpected ratio of alleles

1.
B   A   C
B
b   a   c
b   a   c

2.
B   A   C
B
A/a mismatch   C
b   c
a/A mismatch
b   a   c

3. Mismatch corrected
B   A   C
B   a   C
b   c
a
b   a   c

4.
Ascus
B A C   *B a C*
b a c   *b a c*
Ascospores

Heteroduplex regions occur at sites of genetic exchange

Figure 6.20

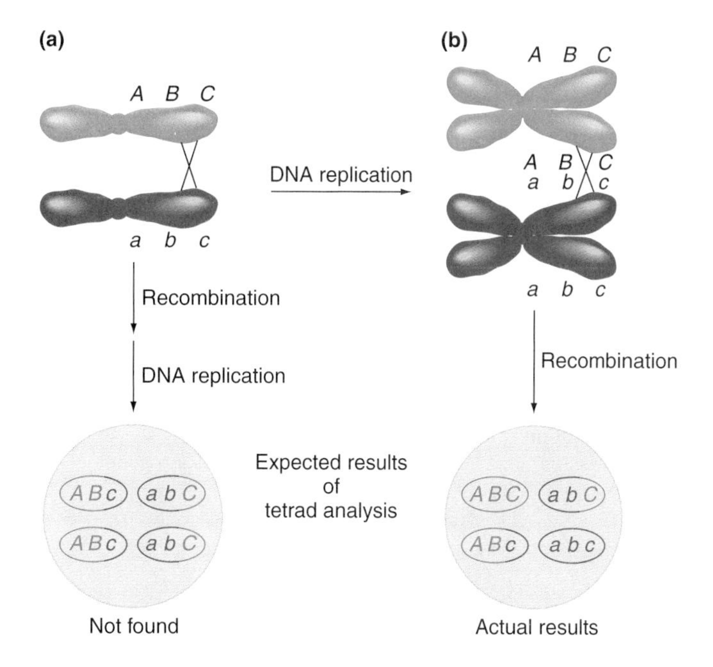

(a)

$A\ B\ C$

DNA replication

$a\ b\ c$

Recombination

DNA replication

ABc abC
ABc abC

Expected results
of
tetrad analysis

Not found

(b)

$A\ B\ C$

$A\ B\ C$
$a\ b\ c$

$a\ b\ c$

Recombination

ABC abC
ABc abc

Actual results

When does crossing-over occur?
Figure 6.21

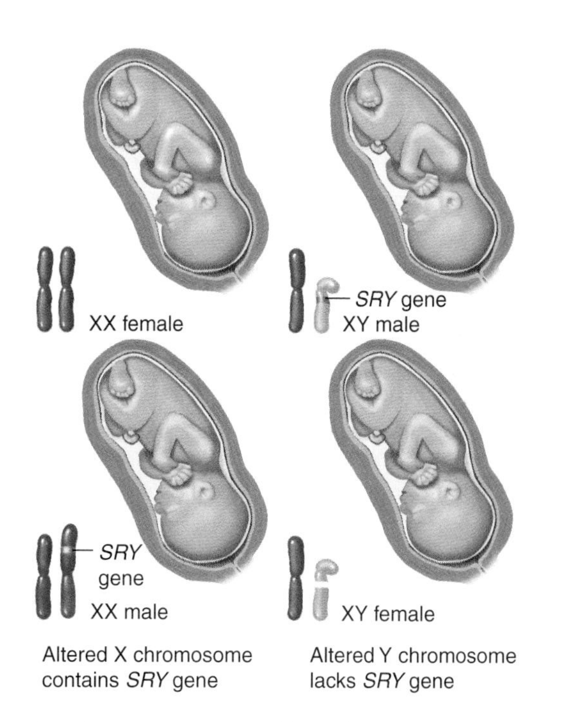

XX female

SRY gene
XY male

SRY
gene
XX male

Altered X chromosome
contains *SRY* gene

XY female

Altered Y chromosome
lacks *SRY* gene

Illegitimate recombination may produce an XY female or an XX male
Figure 6.23

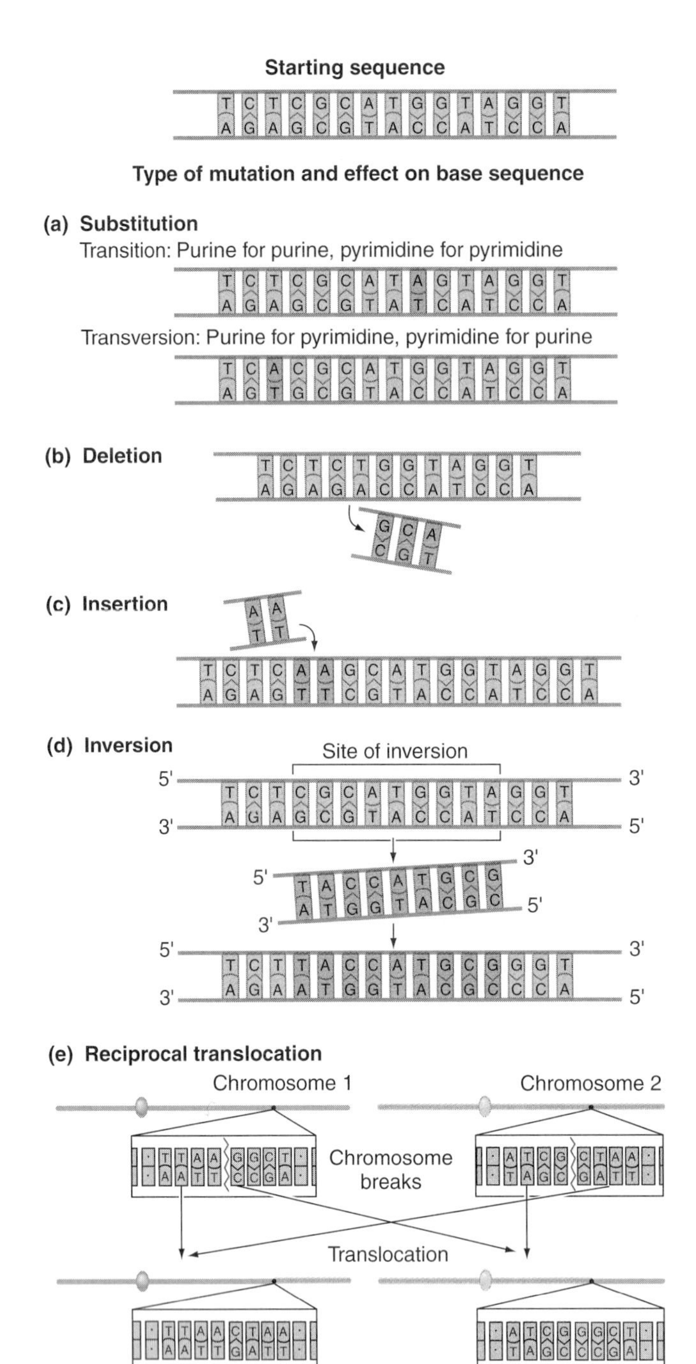

Mutations classified by their effect on DNA
Figure 7.2

(a) Two hypotheses for the origin of bactericide resistance

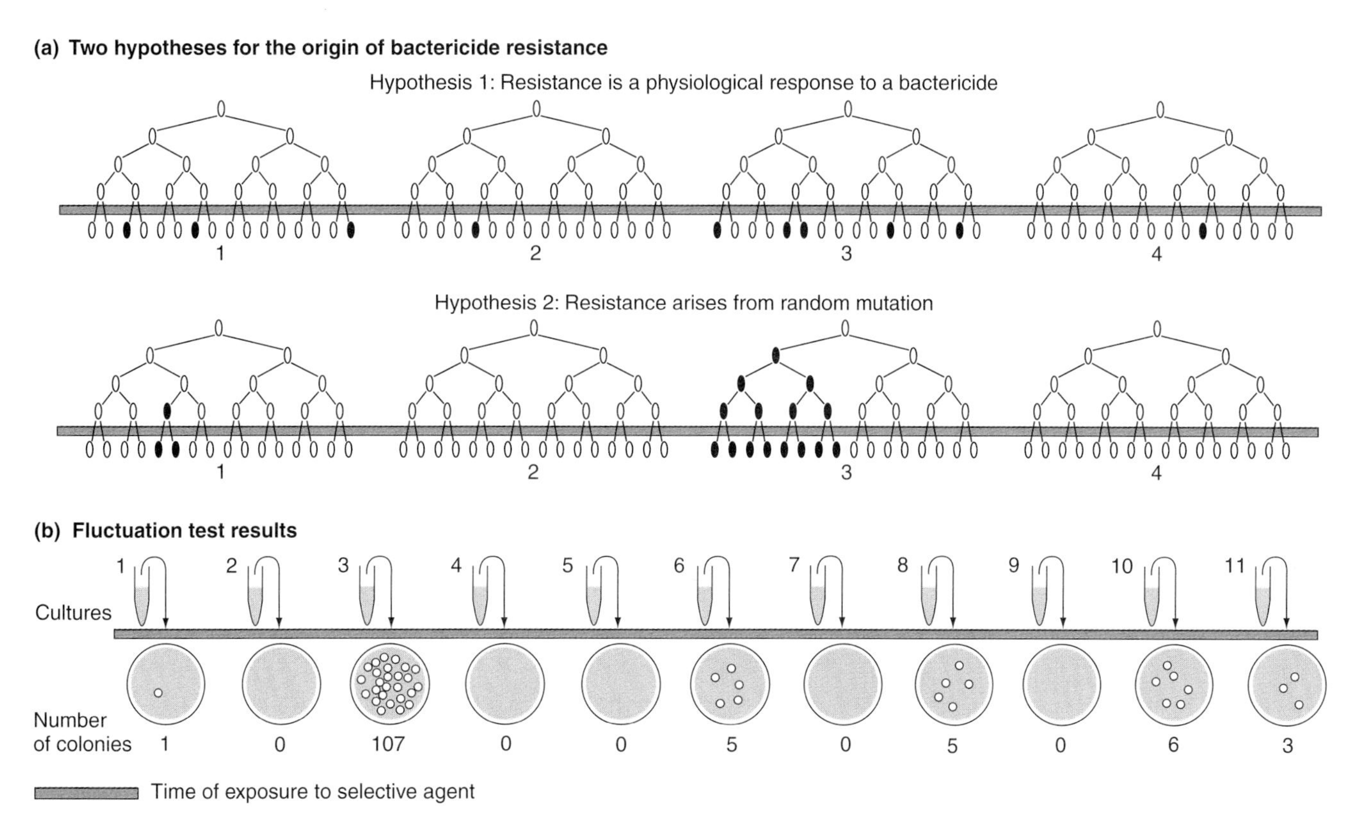

Hypothesis 1: Resistance is a physiological response to a bactericide

Hypothesis 2: Resistance arises from random mutation

(b) Fluctuation test results

Time of exposure to selective agent

The Luria-Delbrück fluctuation experiment
Figure 7.4

(a) The replica plating technique

1. Invert master plate; pressing against velvet surface leaves an imprint of colonies. Save plate.

2. Invert second plate (replica plate); pressing against velvet surface picks up colony imprint.

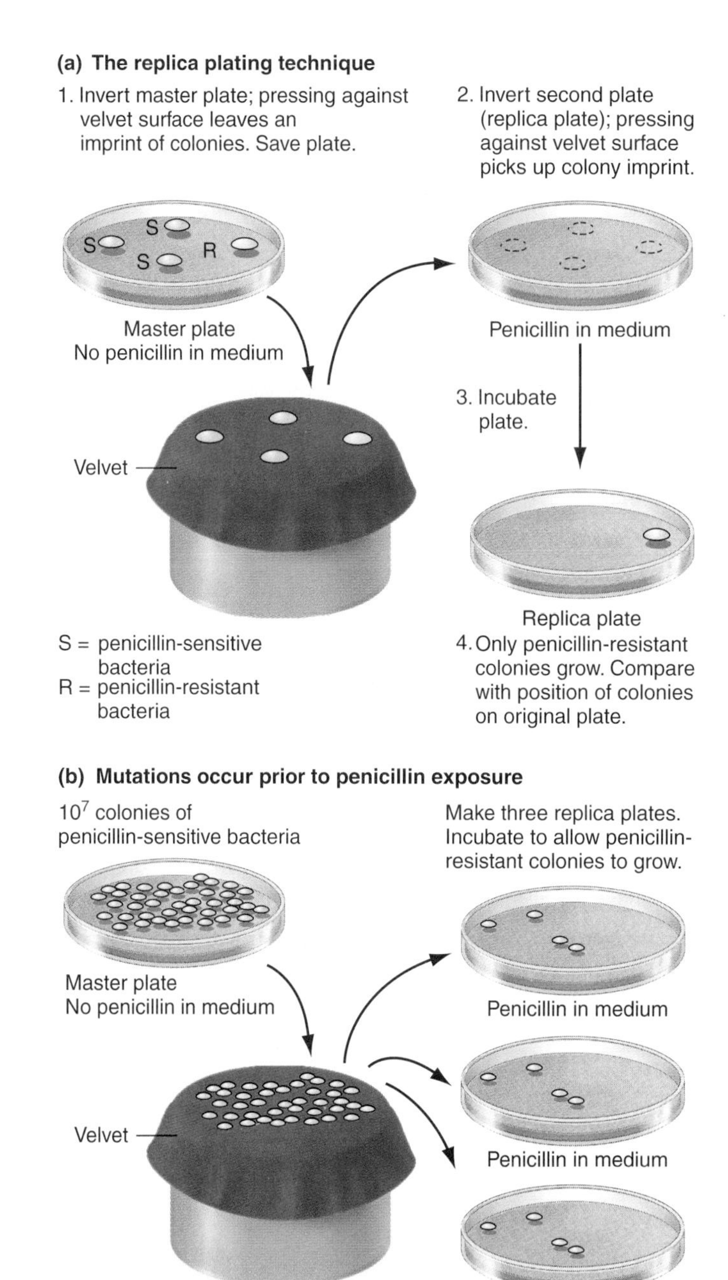

Master plate
No penicillin in medium

Penicillin in medium

Velvet

3. Incubate plate.

S = penicillin-sensitive bacteria
R = penicillin-resistant bacteria

Replica plate
4. Only penicillin-resistant colonies grow. Compare with position of colonies on original plate.

(b) Mutations occur prior to penicillin exposure

10^7 colonies of penicillin-sensitive bacteria

Make three replica plates. Incubate to allow penicillin-resistant colonies to grow.

Master plate
No penicillin in medium

Penicillin in medium

Velvet

Penicillin in medium

Penicillin in medium

Penicillin-resistant colonies grow in the same position on all three plates.

Replica plating verifies that bacterial resistance is the result of preexisting mutations
Figure 7.5

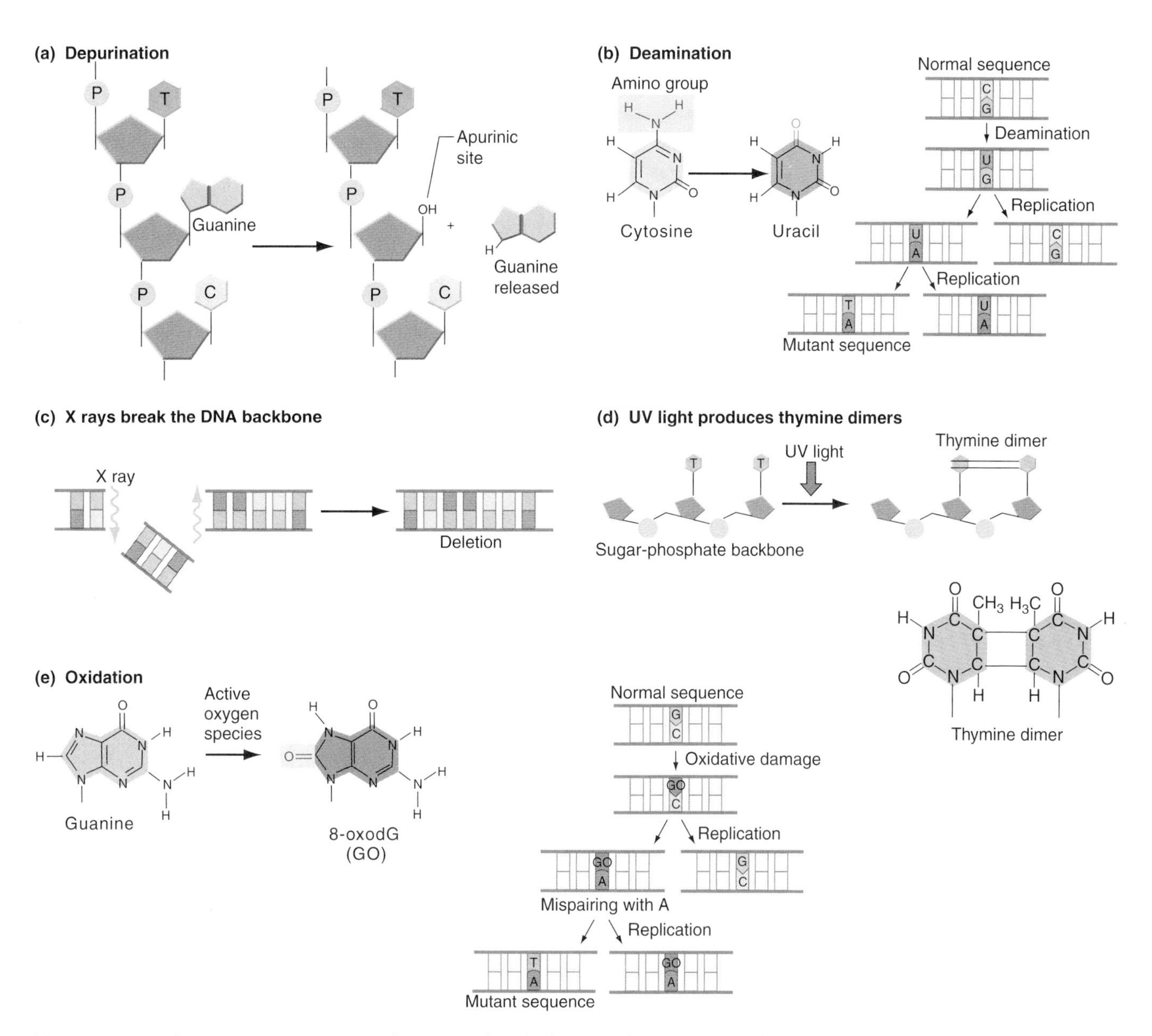

(a) Depurination

Apurinic site

Guanine

Guanine released

(b) Deamination

Amino group

Cytosine

Uracil

Normal sequence

Deamination

Replication

Replication

Mutant sequence

(c) X rays break the DNA backbone

X ray

Deletion

(d) UV light produces thymine dimers

UV light

Thymine dimer

Sugar-phosphate backbone

CH₃ H₃C

Thymine dimer

(e) Oxidation

Active oxygen species

Guanine

8-oxodG (GO)

Normal sequence

Oxidative damage

Replication

Mispairing with A

Replication

Mutant sequence

How natural processes can change the information stored in DNA
Figure 7.6

(a) Excision repair

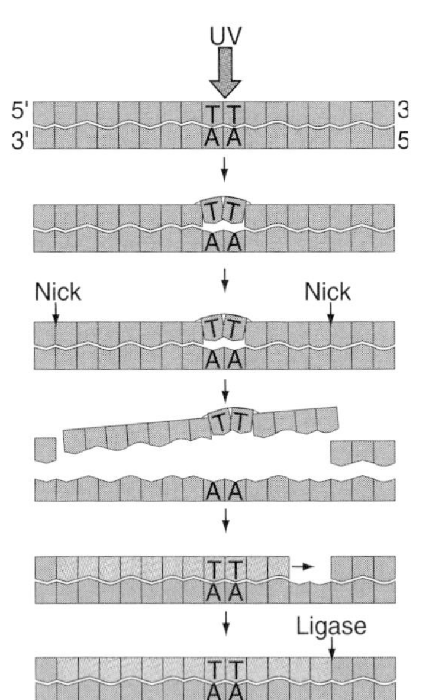

1. Exposure to UV light.

2. Thymine dimer forms.

3. Endonuclease nicks strand containing dimer.

4. Damaged fragment is released from DNA.

5. DNA polymerase fills in the gap with new DNA (yellow).

6. DNA ligase seals the repaired strand.

Excision repair removes damaged DNA and fills in the gap
Figure 7.7

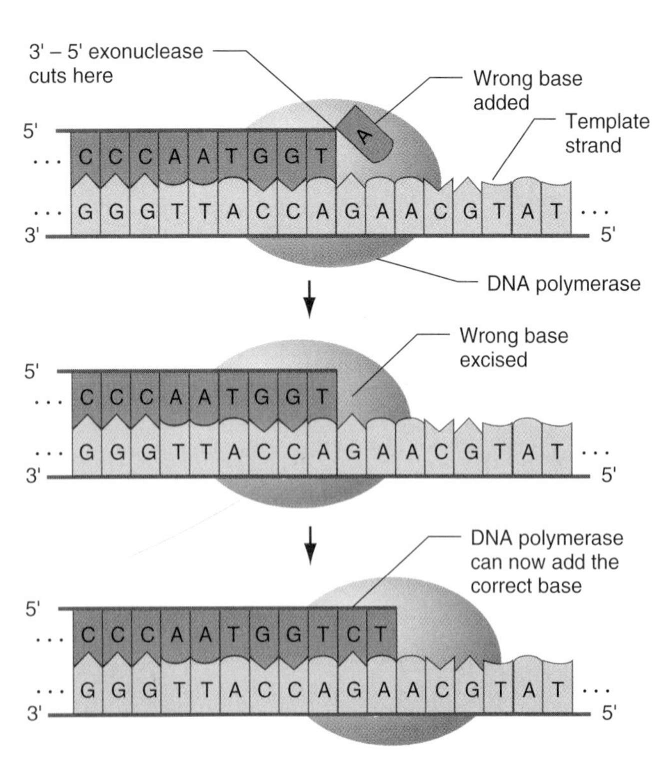

(a) Parental strands are marked with methyl groups.

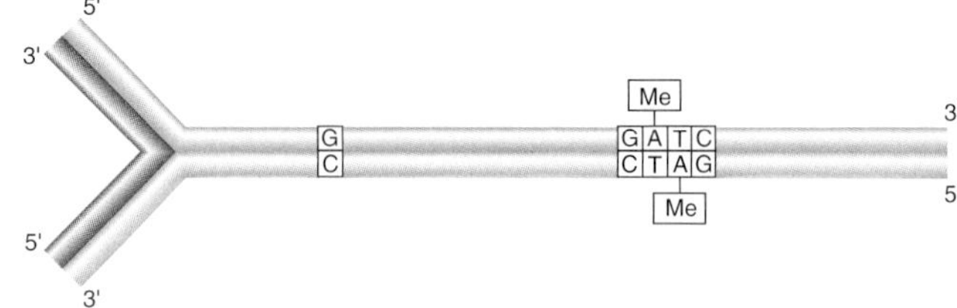

(b) Enzyme system recognizes mismatch in replicated DNA.

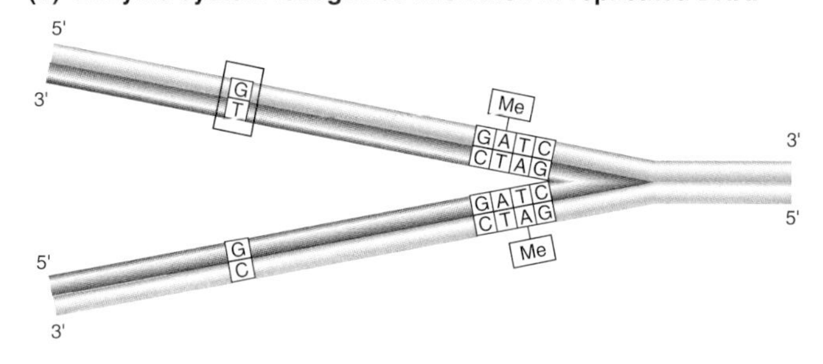

(c) DNA on unmarked new strand is excised.

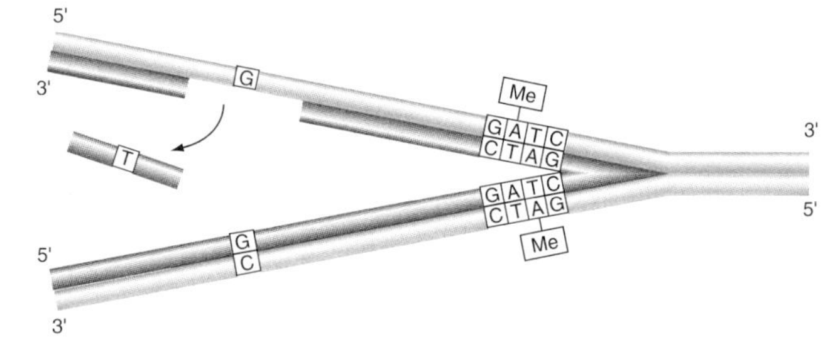

(d) Repair and methylation of newly synthesized DNA strand.

Methyl-directed mismatch repair corrects mistakes in replication
Figure 7.9

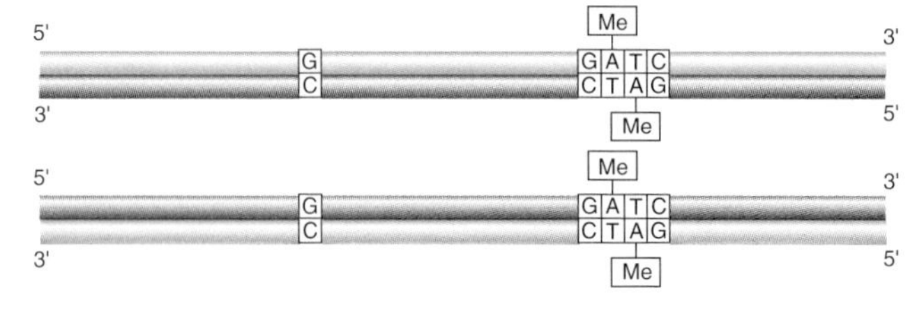

DNA polymerase's proofreading function
Figure 7.8

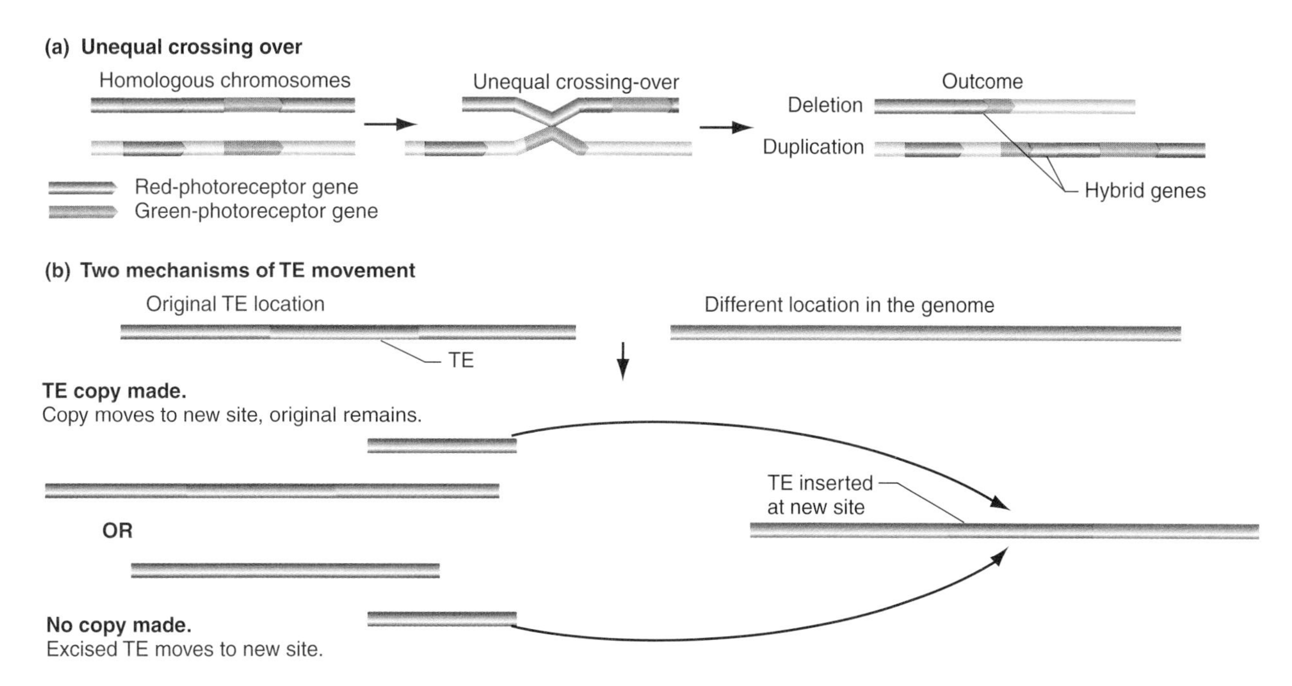

(a) Unequal crossing over

Homologous chromosomes → Unequal crossing-over → Outcome

Deletion

Duplication

═══ Red-photoreceptor gene
═══ Green-photoreceptor gene

Hybrid genes

(b) Two mechanisms of TE movement

Original TE location

Different location in the genome

TE

TE copy made.
Copy moves to new site, original remains.

TE inserted at new site

OR

No copy made.
Excised TE moves to new site.

How unequal crossing-over and the movement of transposable elements (TEs) change DNA's information content
Figure 7.10

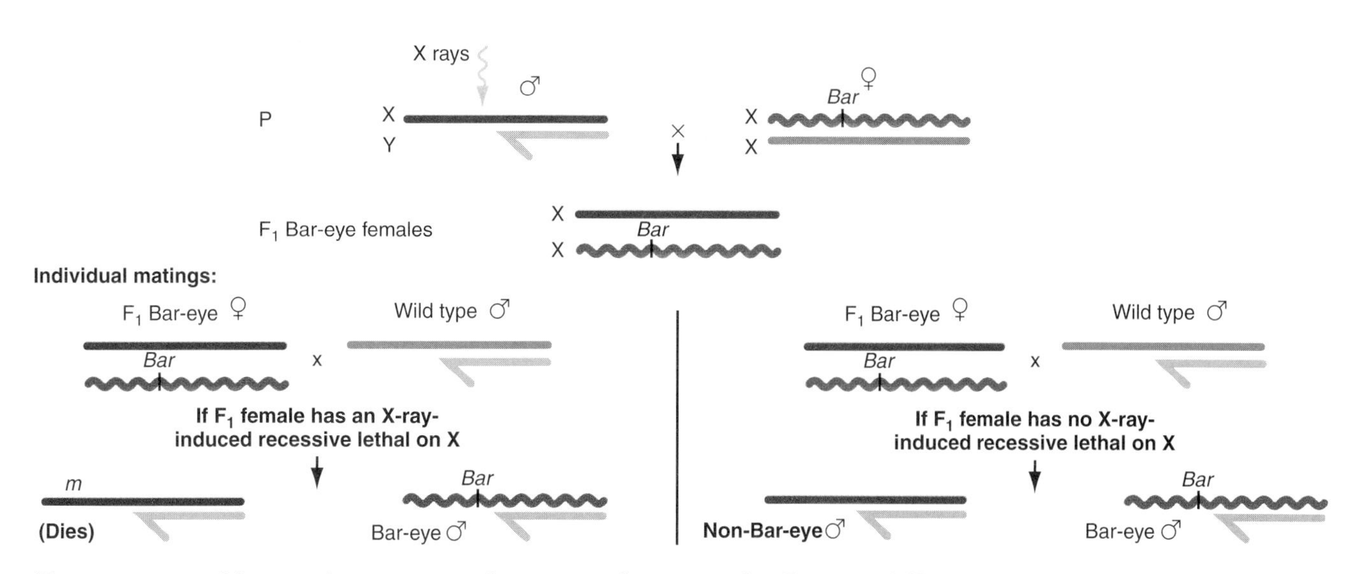

X rays

♂

P   X

        Y

♀

Bar

X

X

F₁ Bar-eye females

X

X

Bar

Individual matings:

F₁ Bar-eye ♀   Wild type ♂

Bar   x

If F₁ female has an X-ray-induced recessive lethal on X

m

(Dies)

Bar

Bar-eye ♂

F₁ Bar-eye ♀   Wild type ♂

Bar   x

If F₁ female has no X-ray-induced recessive lethal on X

Non-Bar-eye ♂

Bar

Bar-eye ♂

Exposure to X rays increases the mutation rate in *Drosophila*
Figure 7.11

Type of mutagen	Chemical action of mutagen
(a) Replace a base: Base analogs have a chemical structure almost identical to that of a DNA base.	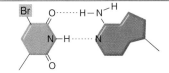 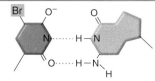 5-Bromouracil–normal state, behaves like thymine Adenine 5-Bromouracil–rare state, behaves like cytosine Guanine 5-Bromouracil: almost identical to thymine. Normally pairs with A; in transient state, pairs with G.

(b) Alter base structure and properties:
Hydroxylating agents: add a hydroxyl (–OH) group

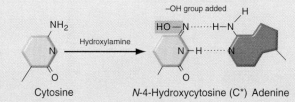

Cytosine   N-4-Hydroxycytosine (C*) Adenine

Hydroxylamine adds –OH to cytosine; with the –OH, hydroxylated C now pairs with A instead of G.

Alkylating agents: add ethyl (–CH₂–CH₃) or methyl (–CH₃) groups

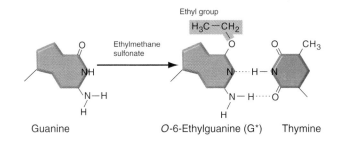

Guanine   O-6-Ethylguanine (G*)   Thymine

Ethylmethane sulfonate adds an ethyl group to guanine or thymine. Modified G pairs with T above, and modified T pairs with G (not shown).

Deaminating agents: remove amine (–NH₂) groups

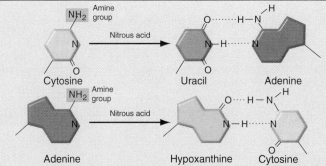

Cytosine   Uracil   Adenine

Adenine   Hypoxanthine   Cytosine

Nitrous acid modifies cytosine to uracil, which pairs with A instead of G; modifies adenine to hypoxanthine, a base that pairs with C instead of T.

(c) Insert between bases:
Intercalating agents

 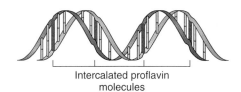

Proflavin   Intercalated proflavin molecules

Proflavin intercalates into the double helix. This disrupts DNA metabolism, eventually resulting in deletion or addition of a base pair.

How mutagens alter DNA
Figure 7.12

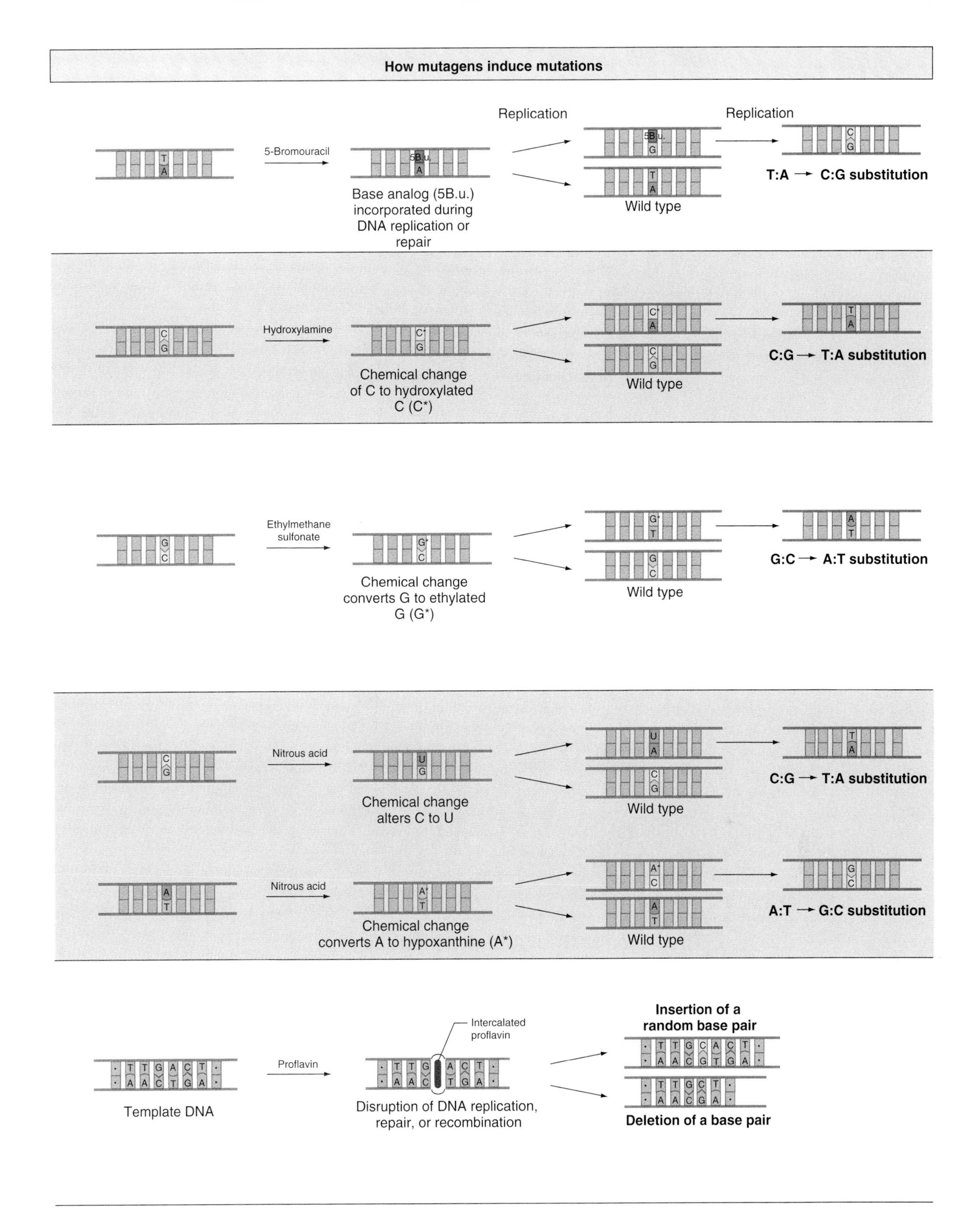

Figure 7.12—*Continued*

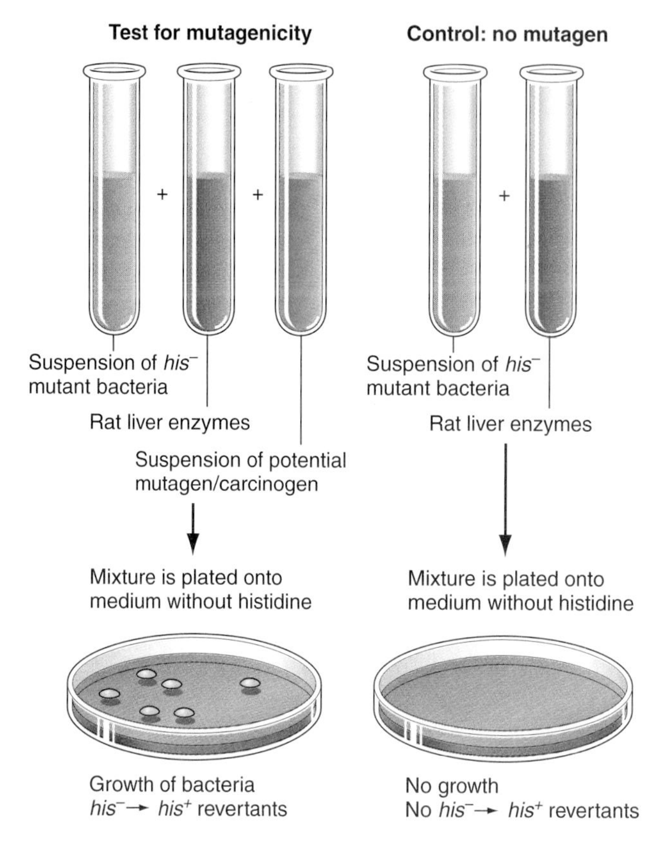

Test for mutagenicity

Control: no mutagen

Suspension of *his⁻* mutant bacteria

Rat liver enzymes

Suspension of potential mutagen/carcinogen

Suspension of *his⁻* mutant bacteria

Rat liver enzymes

Mixture is plated onto medium without histidine

Mixture is plated onto medium without histidine

Growth of bacteria
his⁻ → *his⁺* revertants

No growth
No *his⁻* → *his⁺* revertants

The Ames test identifies potential carcinogens through their mutagenicity
Figure 7.13

(a) Complementation testing

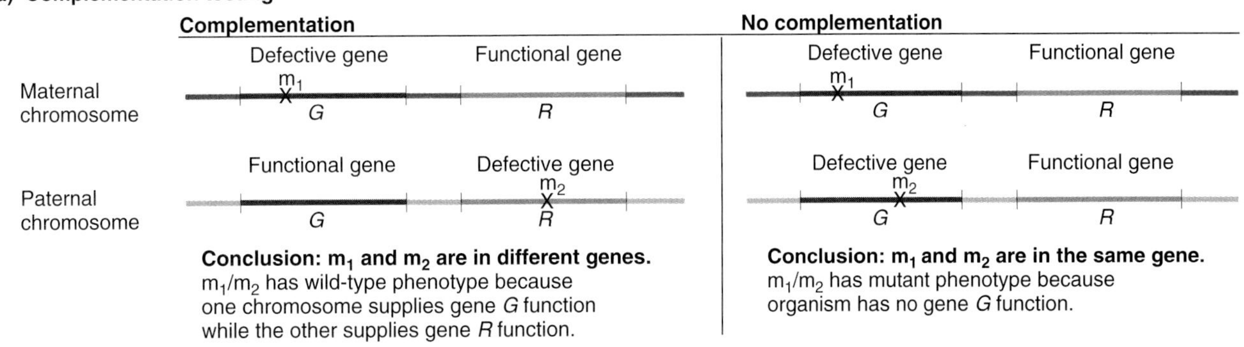

Complementation

Maternal chromosome — Defective gene m_1 / G Functional gene / R

Paternal chromosome — Functional gene / G Defective gene m_2 / R

Conclusion: m_1 and m_2 are in different genes.
m_1/m_2 has wild-type phenotype because one chromosome supplies gene G function while the other supplies gene R function.

No complementation

Maternal chromosome — Defective gene m_1 / G Functional gene / R

Paternal chromosome — Defective gene m_2 / G Functional gene / R

Conclusion: m_1 and m_2 are in the same gene.
m_1/m_2 has mutant phenotype because organism has no gene G function.

(b) A complementation table: X-linked eye color mutations in *Drosophila*

Mutation	white	garnet	ruby	vermilion	cherry	coral	apricot	buff	carnation
white	−	+	+	+	−	−	−	−	+
garnet		−	+	+	+	+	+	+	+
ruby			−	+	+	+	+	+	+
vermilion				−	+	+	+	+	+
cherry					−	−	−	−	+
coral						−	−	−	+
apricot							−	−	+
buff								−	+
carnation									−

(c) Genetic map: X-linked eye color mutations in *Drosophila*

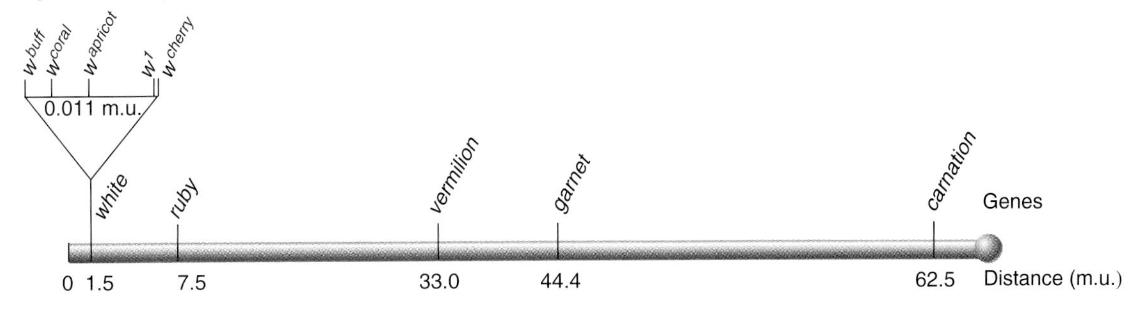

w^{buff} w^{coral} $w^{apricot}$ w^1 w^{cherry}

0.011 m.u.

white   ruby   vermilion   garnet   carnation   Genes

0   1.5   7.5   33.0   44.4   62.5   Distance (m.u.)

Complementation testing of *Drosophila* eye color mutations
Figure 7.15

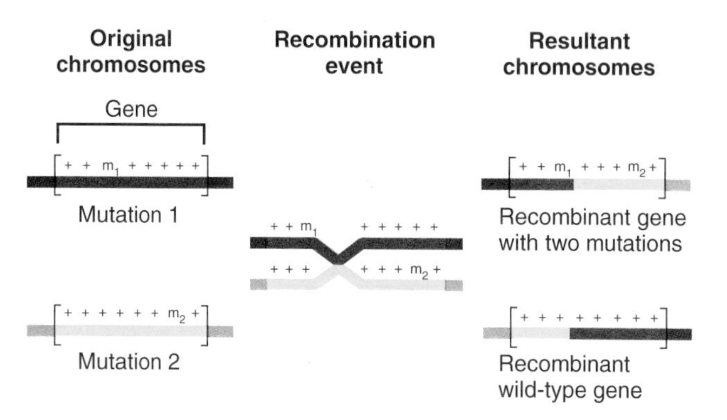

How recombination within a gene could generate a wild-type allele
Figure 7.16

(a) Using deletions for rapid mapping

Point mutation within deletion limits

Point mutation outside deletion limits

m

m

Overlapping deletions

Nonoverlapping deletions

Cannot produce wild-type progeny by recombination

Produce wild-type progeny by recombination

(c) Fine structure of the *rII* region

Each box represents an independent occurrence of a mutation at this site.

Many mutations at a site create a "hot spot."

B cistron
A cistron

(b) Portion of the *rIIA* deletion map at increasing resolutions

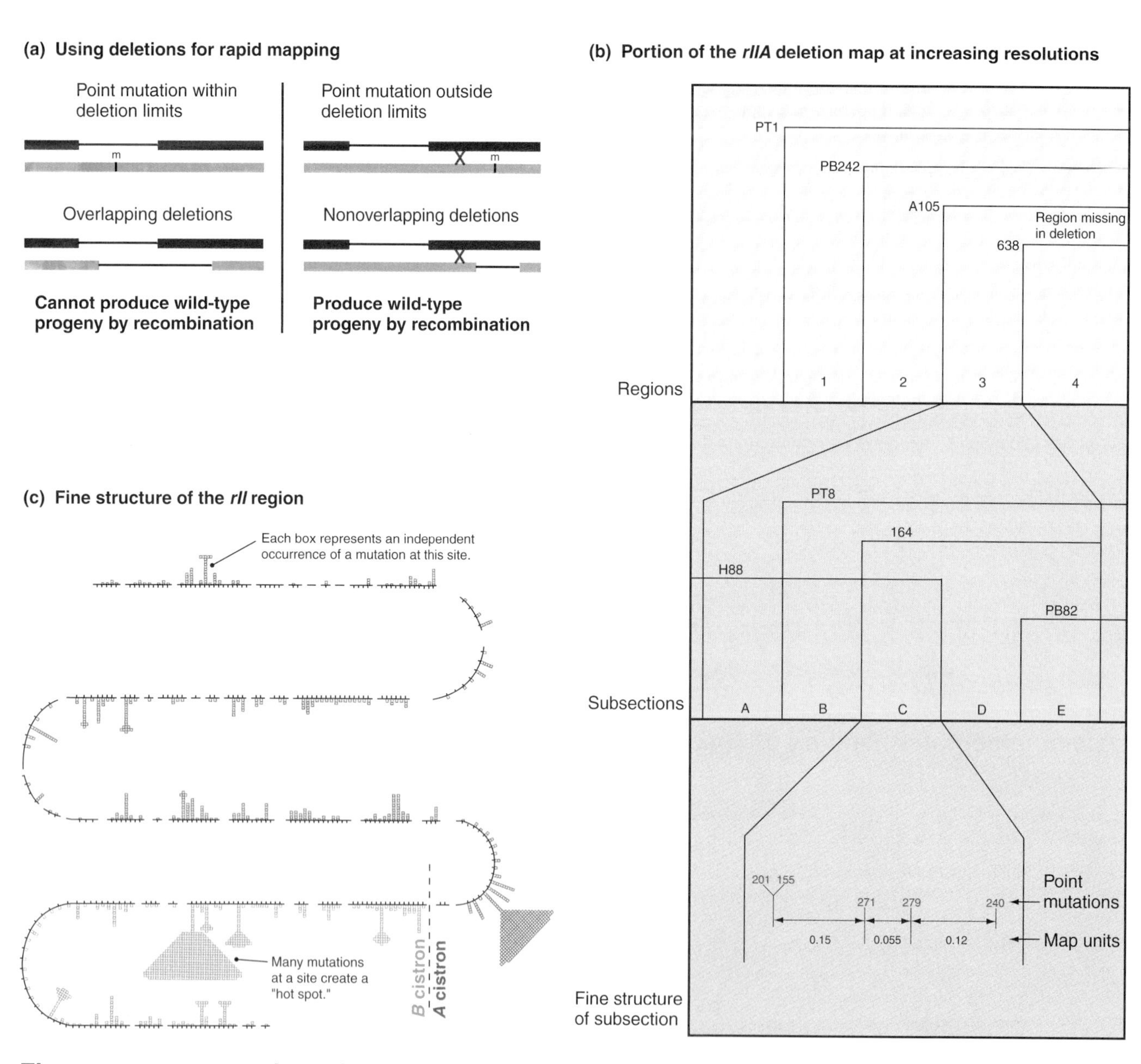

PT1

PB242

A105

Region missing in deletion

638

Regions

1   2   3   4

PT8

164

H88

PB82

Subsections

A   B   C   D   E

201 155

271 279   240 ← Point mutations

0.15   0.055   0.12   ← Map units

Fine structure of subsection

Fine structure mapping of the bacteriophage T4 *rII* genes
Figure 7.18

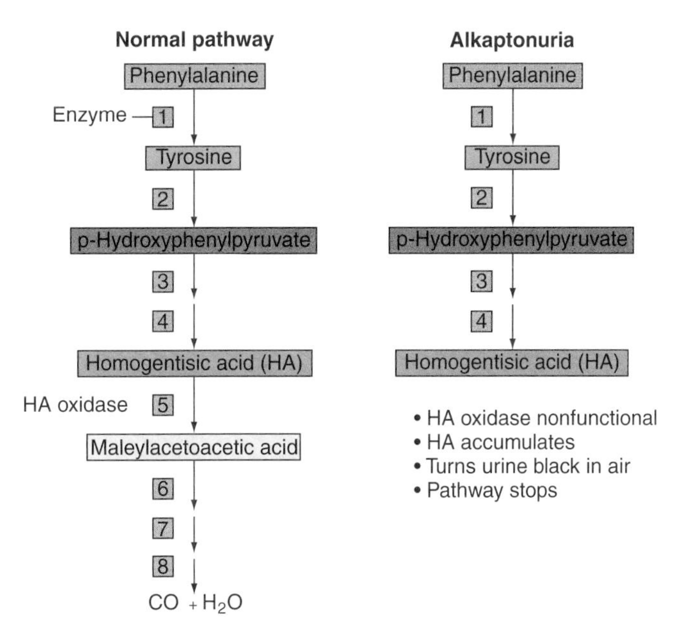

Alkaptonuria: An inborn error of metabolism
Figure 7.19

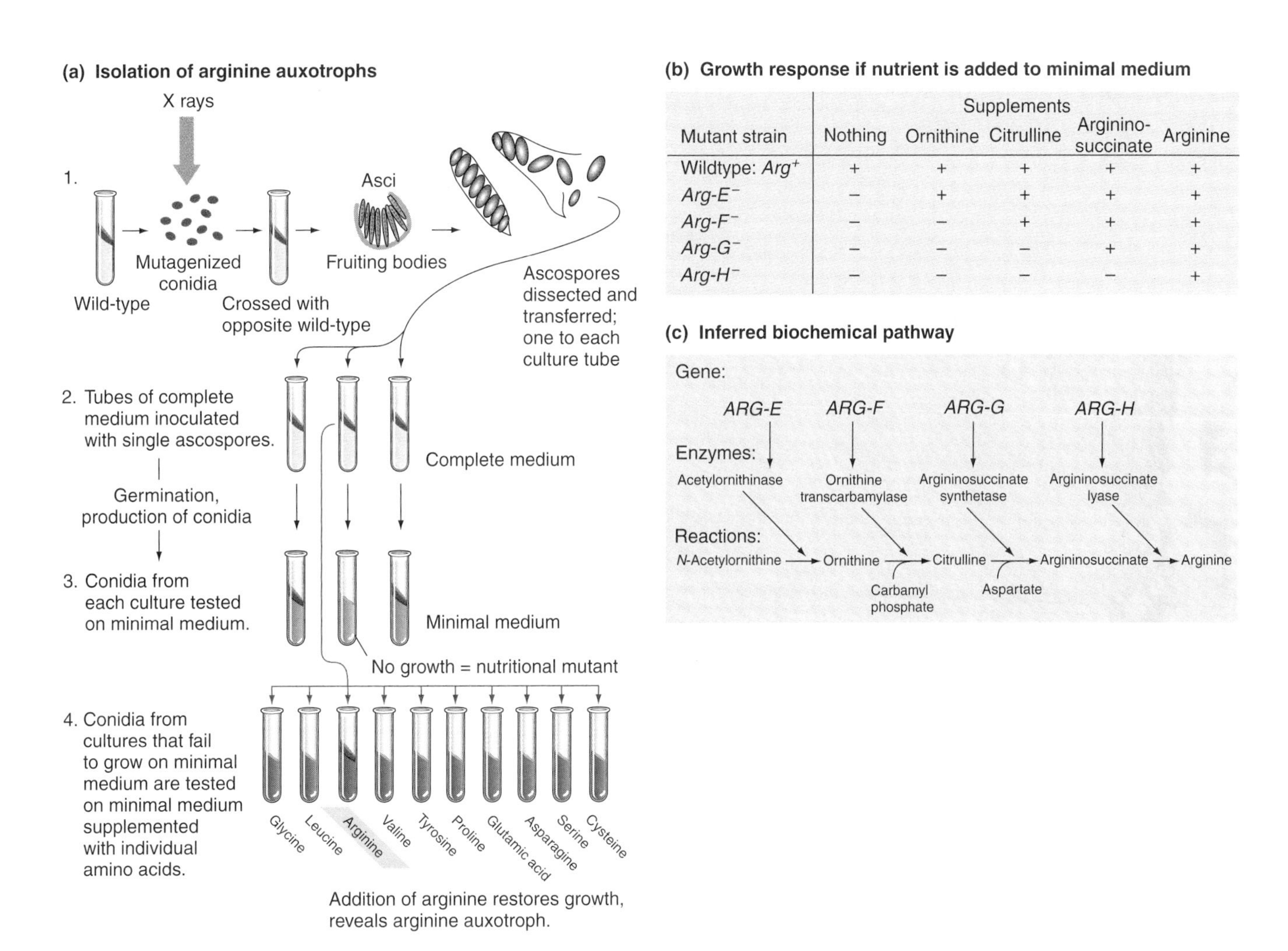

(a) Isolation of arginine auxotrophs

X rays

1.

Wild-type → Mutagenized conidia → Crossed with opposite wild-type → Fruiting bodies → Asci

Ascospores dissected and transferred; one to each culture tube

2. Tubes of complete medium inoculated with single ascospores.

Complete medium

Germination, production of conidia

3. Conidia from each culture tested on minimal medium.

Minimal medium

No growth = nutritional mutant

4. Conidia from cultures that fail to grow on minimal medium are tested on minimal medium supplemented with individual amino acids.

Glycine, Leucine, Arginine, Valine, Tyrosine, Proline, Glutamic acid, Asparagine, Serine, Cysteine

Addition of arginine restores growth, reveals arginine auxotroph.

(b) Growth response if nutrient is added to minimal medium

Mutant strain	Supplements				
	Nothing	Ornithine	Citrulline	Arginino-succinate	Arginine
Wildtype: Arg⁺	+	+	+	+	+
Arg-E⁻	−	+	+	+	+
Arg-F⁻	−	−	+	+	+
Arg-G⁻	−	−	−	+	+
Arg-H⁻	−	−	−	−	+

(c) Inferred biochemical pathway

Gene:

ARG-E ARG-F ARG-G ARG-H

Enzymes:

Acetylornithinase   Ornithine transcarbamylase   Argininosuccinate synthetase   Argininosuccinate lyase

Reactions:

N-Acetylornithine → Ornithine → Citrulline → Argininosuccinate → Arginine

Carbamyl phosphate   Aspartate

Beadle and Tatum's experiment supporting the "one gene, one enzyme" hypothesis
Figure 7.20

(a) Generic amino acid structure

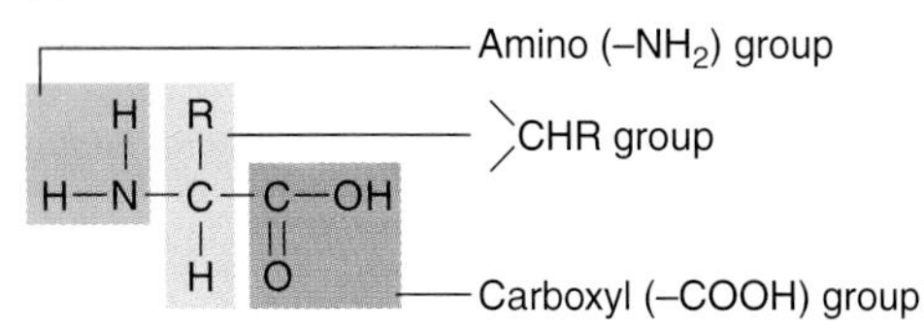

Amino (–NH₂) group
CHR group
Carboxyl (–COOH) group

(c) Peptide bond formation

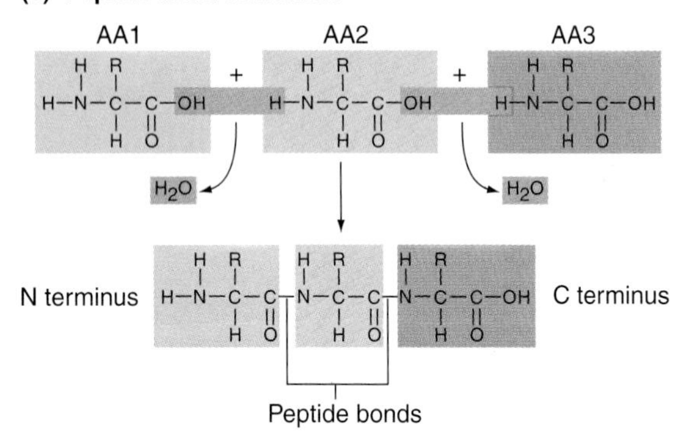

N terminus
C terminus
Peptide bonds

Proteins are chains of amino acids linked by peptide bonds
Figure 7.21

(b) Amino acids with nonpolar R groups

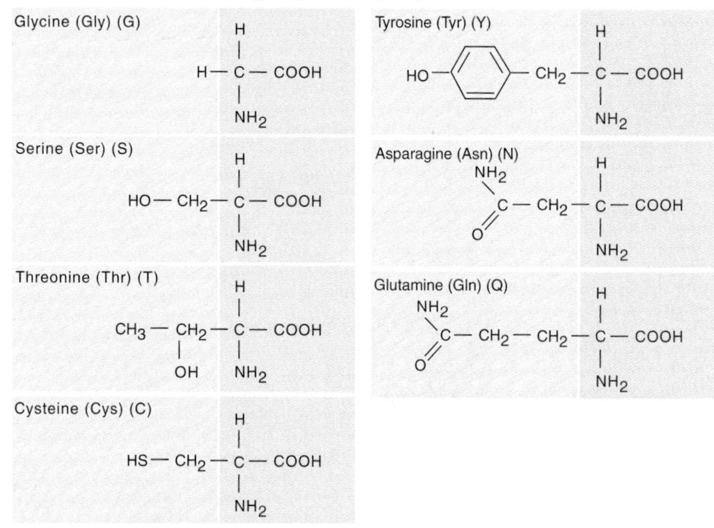

R groups   Backbone   R groups   Backbone

Alanine (Ala) (A)   Proline (Pro) (P)
Valine (Val) (V)   Phenylalanine (Phe) (F)
Leucine (Leu) (L)   Tryptophan (Trp) (W)
Isoleucine (Ile) (I)   Methionine (Met) (M)

Amino acids with uncharged polar R groups

Glycine (Gly) (G)   Tyrosine (Tyr) (Y)
Serine (Ser) (S)   Asparagine (Asn) (N)
Threonine (Thr) (T)   Glutamine (Gln) (Q)
Cysteine (Cys) (C)

Amino acids with basic R groups

Lysine (Lys) (K)   Histidine (His) (H)
Arginine (Arg) (R)

Amino acids with acidic R groups

Aspartic acid (Asp) (D)   Glutamic acid (Glu) (E)

66

The molecular basis of sickle-cell and other anemias

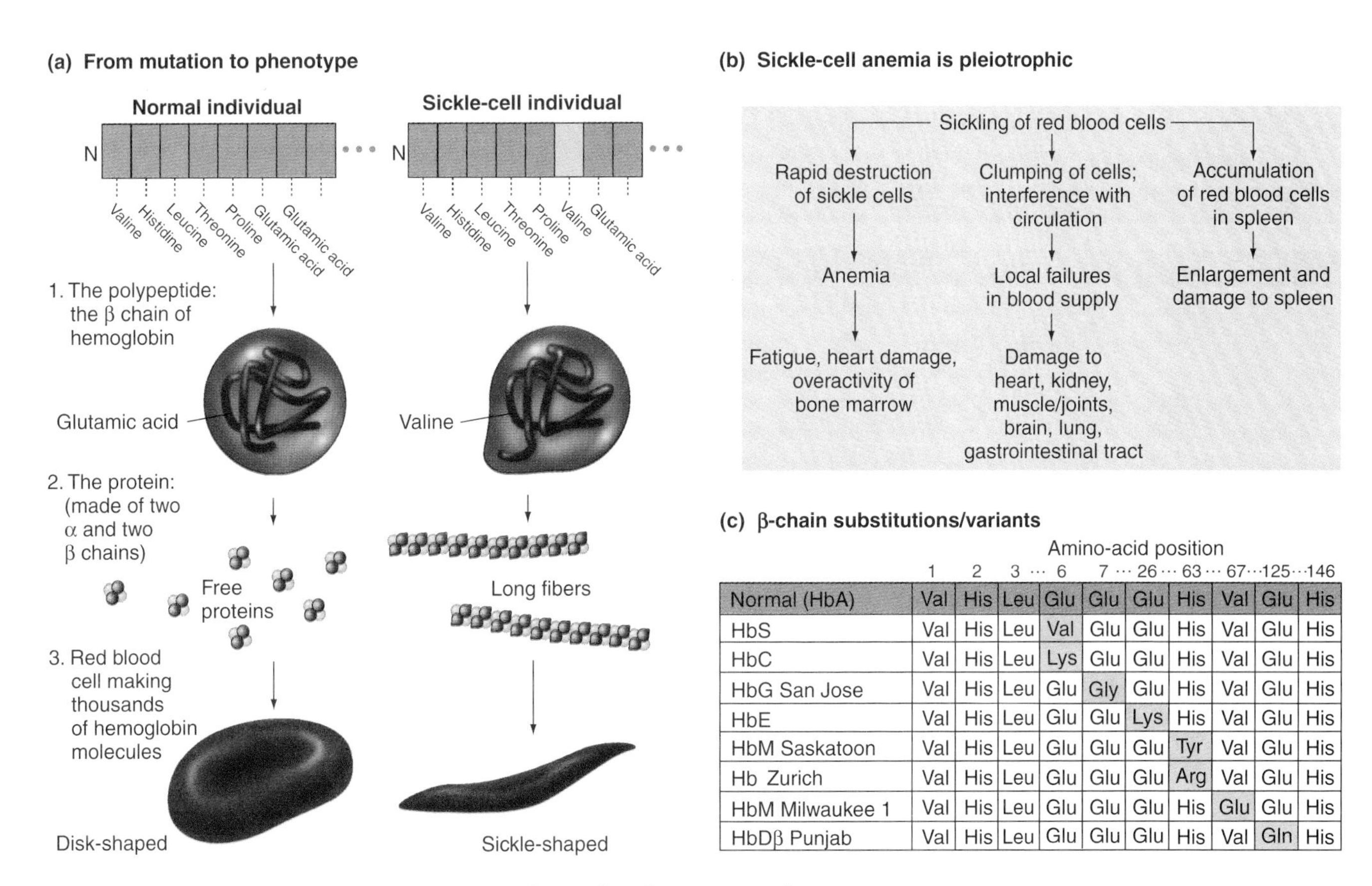

(a) From mutation to phenotype

Normal individual / Sickle-cell individual

N ... Valine, Histidine, Leucine, Threonine, Proline, Glutamic acid, Glutamic acid

1. The polypeptide: the β chain of hemoglobin
   - Glutamic acid / Valine
2. The protein: (made of two α and two β chains) — Free proteins / Long fibers
3. Red blood cell making thousands of hemoglobin molecules

Disk-shaped / Sickle-shaped

(b) Sickle-cell anemia is pleiotrophic

Sickling of red blood cells

- Rapid destruction of sickle cells → Anemia → Fatigue, heart damage, overactivity of bone marrow
- Clumping of cells; interference with circulation → Local failures in blood supply → Damage to heart, kidney, muscle/joints, brain, lung, gastrointestinal tract
- Accumulation of red blood cells in spleen → Enlargement and damage to spleen

(c) β-chain substitutions/variants

	Amino-acid position									
	1	2	3	6	7	26	63	67	125	146
Normal (HbA)	Val	His	Leu	Glu	Glu	Glu	His	Val	Glu	His
HbS	Val	His	Leu	Val	Glu	Glu	His	Val	Glu	His
HbC	Val	His	Leu	Lys	Glu	Glu	His	Val	Glu	His
HbG San Jose	Val	His	Leu	Glu	Gly	Glu	His	Val	Glu	His
HbE	Val	His	Leu	Glu	Glu	Lys	His	Val	Glu	His
HbM Saskatoon	Val	His	Leu	Glu	Glu	Glu	Tyr	Val	Glu	His
Hb Zurich	Val	His	Leu	Glu	Glu	Glu	Arg	Val	Glu	His
HbM Milwaukee 1	Val	His	Leu	Glu	Glu	Glu	His	Glu	Glu	His
HbDβ Punjab	Val	His	Leu	Glu	Glu	Glu	His	Val	Gln	His

The molecular basis of sickle-cell and other anemias
Figure 7.22

Levels of polypeptide structure

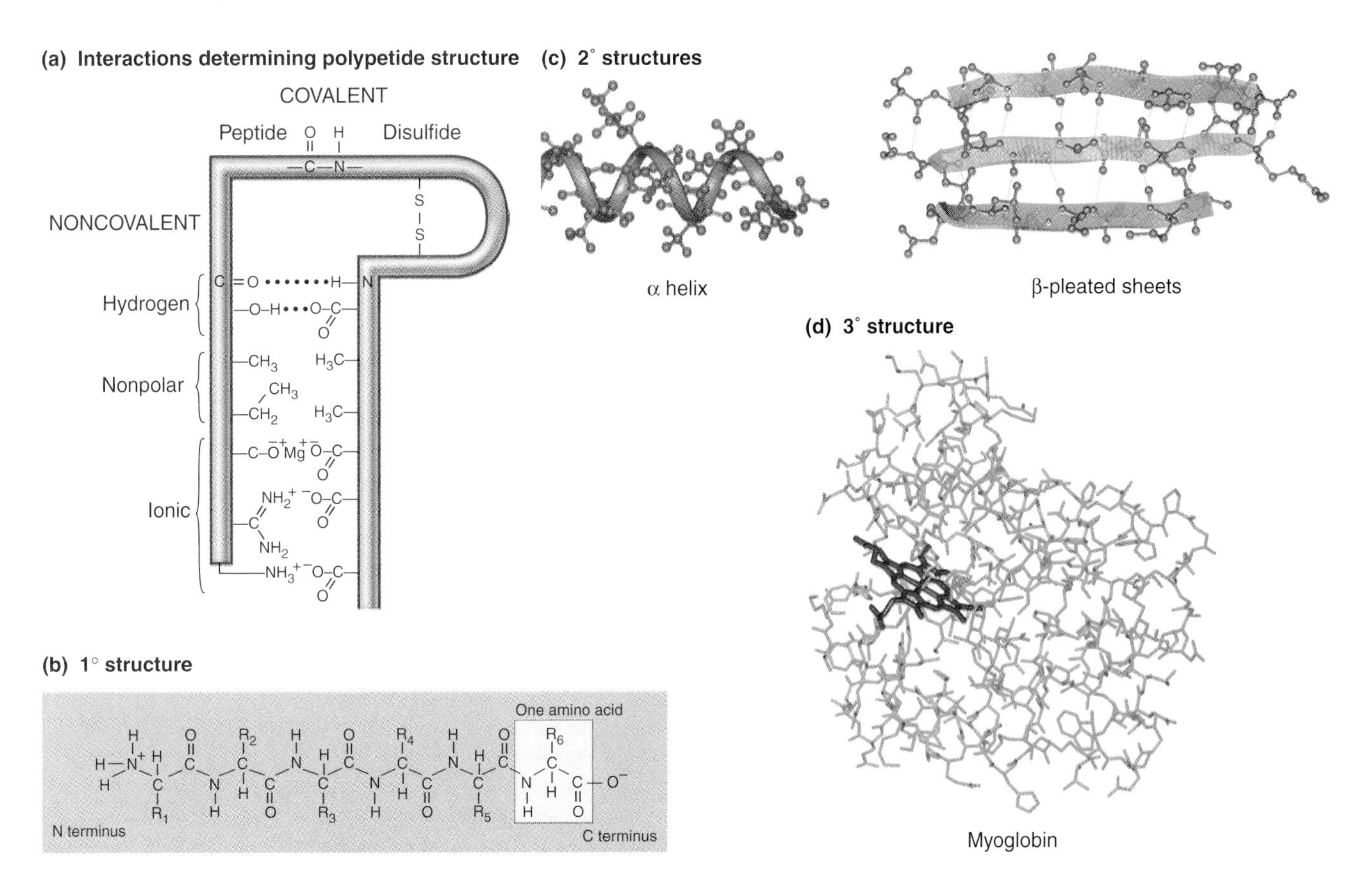

(a) Interactions determining polypeptide structure

COVALENT — Peptide / Disulfide

NONCOVALENT:
- Hydrogen
- Nonpolar
- Ionic

(b) 1° structure

One amino acid

N terminus / C terminus

(c) 2° structures

α helix / β-pleated sheets

(d) 3° structure

Myoglobin

Levels of polypeptide structure
Figure 7.23

(a) A multimer with identical subunits

β2 lens crystallin

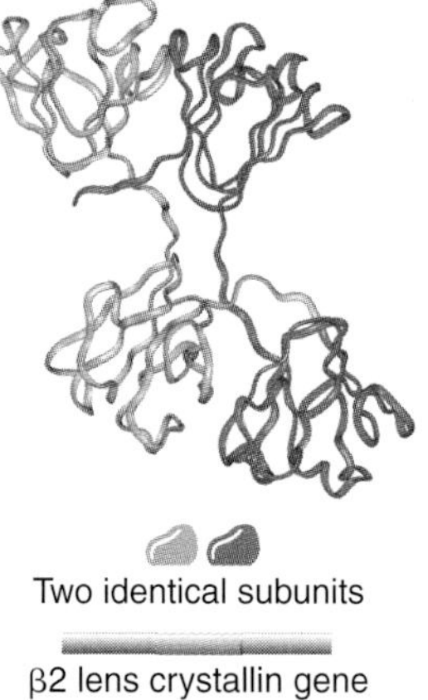

Two identical subunits

β2 lens crystallin gene

(b) A multimer with nonidentical subunits

Hemoglobin

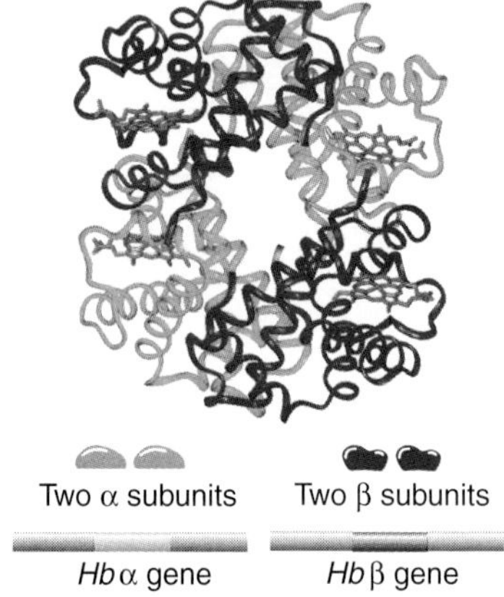

Two α subunits   Two β subunits

*Hb*α gene   *Hb*β gene

(c) One polypeptide in different proteins

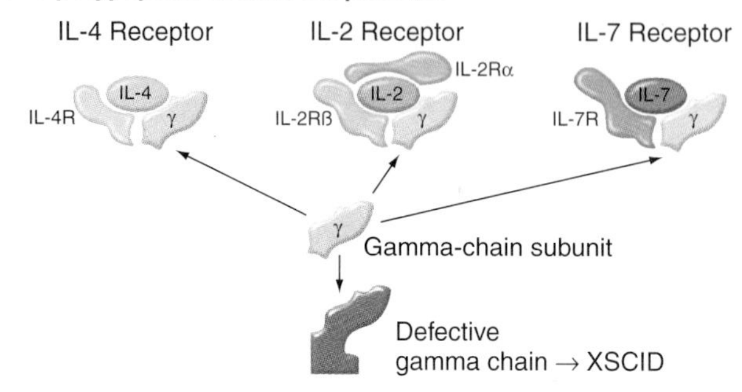

IL-4 Receptor   IL-2 Receptor   IL-7 Receptor

IL-4   IL-2Rα   IL-7

IL-4R   γ   IL-2Rß   γ   IL-7R   γ

γ   Gamma-chain subunit

Defective gamma chain → XSCID

(d) Microtubules: large assemblies of subunits

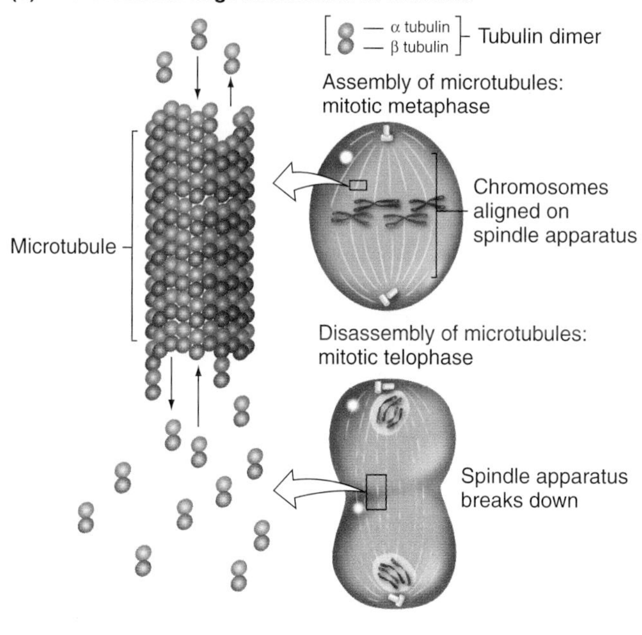

— α tubulin
— β tubulin } Tubulin dimer

Assembly of microtubules: mitotic metaphase

Microtubule

Chromosomes aligned on spindle apparatus

Disassembly of microtubules: mitotic telophase

Spindle apparatus breaks down

Multimeric proteins
Figure 7.24

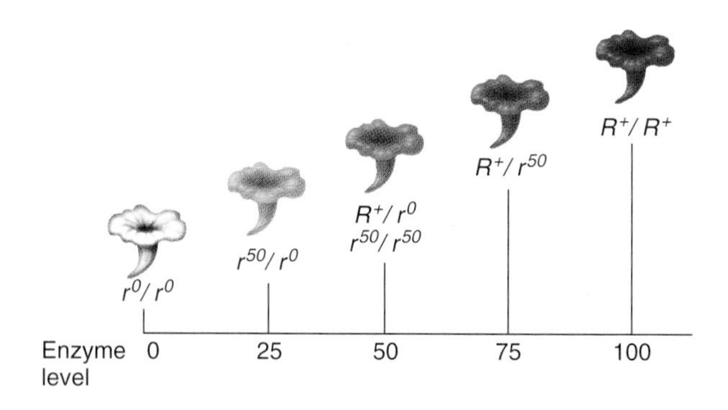

Amount of protein

Threshold for wt eye color

wt/wt   1/1   1/wt   2/2   2/wt

Why some mutant alleles are recessive
Figure 7.25

R^+/R^+

R^+/r^{50}

R^+/r^0
r^{50}/r^{50}

r^{50}/r^0

r^0/r^0

Enzyme   0   25   50   75   100
level

When a phenotype varies continuously with levels of protein function, incomplete dominance results
Figure 7.26

Why some mutant alleles are dominant
Figure 7.27

(b) Dominant negative mutations

Functional enzyme	Nonfunctional enzyme			
				
d+d+d+d+	d+d+d+D	d+d+D d+	d+ D d+d+	D d+d+d+
				
	d+d+D D	d+D D d+	d+ D d+D	D d+D d+
				
	D d+d+D	D D d+d+	d+ D D D	D d+D D
				
	D D d+D	D D D d+	D D D D	

D = dominant mutant subunit
d+ = wild-type subunit

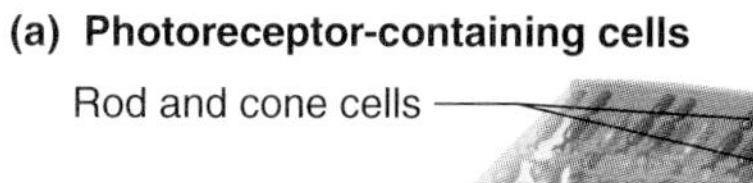

(a) Photoreceptor-containing cells

Rod and cone cells
Pigmented epithelium
Retina surface
Light
Light
Photoreceptor cells
Rod
Cone
Disc membrane
Retinal
Rhodopsin
Membranous disc

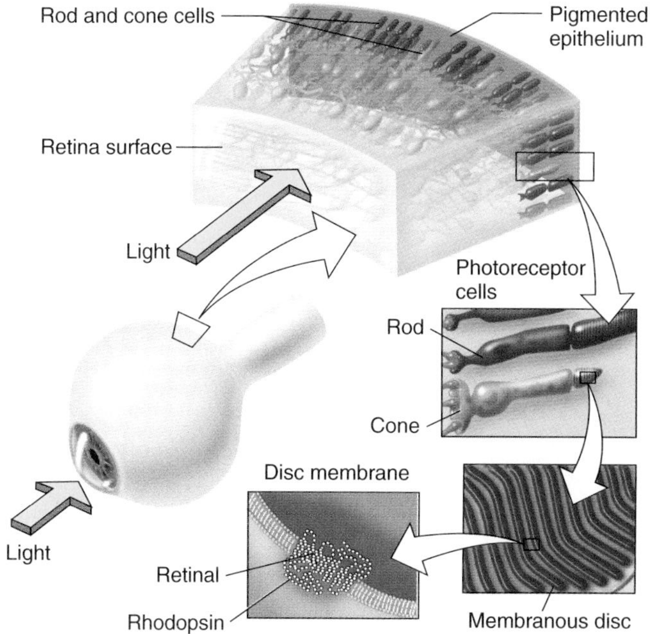

(b) Photoreceptor proteins

Rhodopsin protein
C
N

Blue-receiving protein
C
N

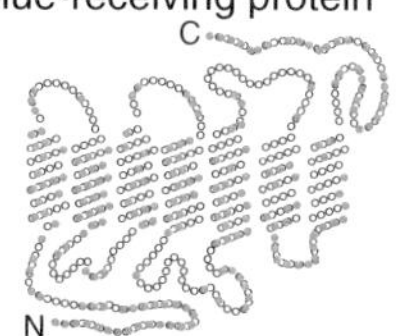

Green-receiving protein
C
N

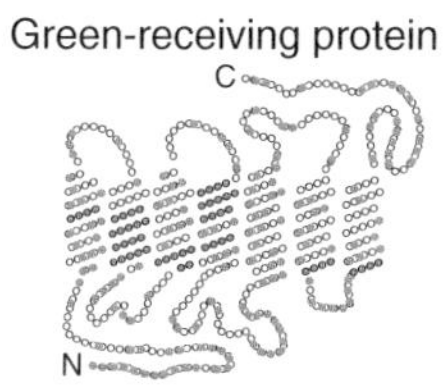

Red-receiving protein
C
N

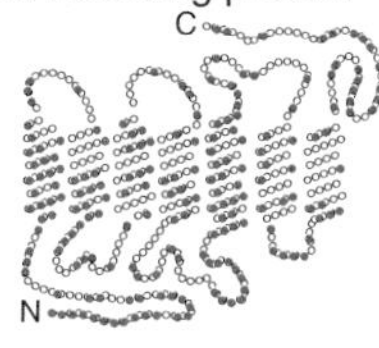

(c) Red/green pigment genes

X chromosomes from normal individuals:

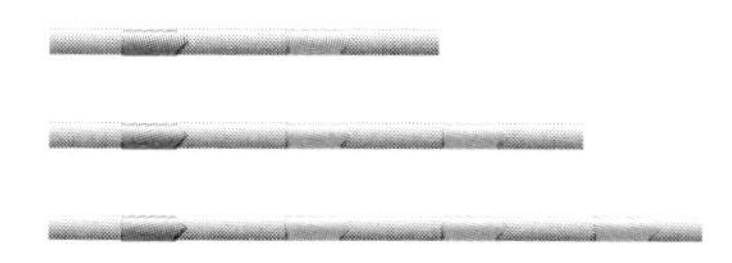

(d) Evolution of visual pigment genes

Primordial gene

Red gene · Green gene · Blue gene · Rhodopsin gene

The cellular and molecular basis of vision
Figure 7.28

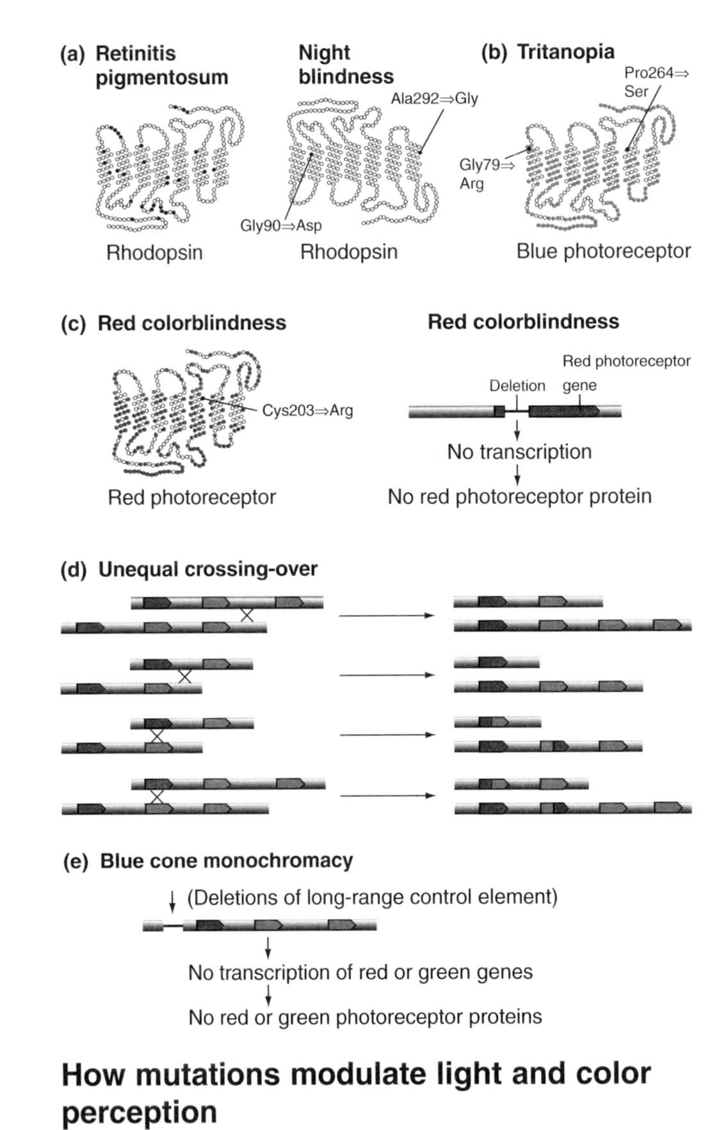

(a) Retinitis pigmentosum

Rhodopsin

Night blindness

Ala292⇒Gly

Gly90⇒Asp

Rhodopsin

(b) Tritanopia

Pro264⇒Ser

Gly79⇒Arg

Blue photoreceptor

(c) Red colorblindness

Cys203⇒Arg

Red photoreceptor

Red colorblindness

Red photoreceptor gene

Deletion

No transcription

No red photoreceptor protein

(d) Unequal crossing-over

(e) Blue cone monochromacy

↓ (Deletions of long-range control element)

↓

No transcription of red or green genes

↓

No red or green photoreceptor proteins

How mutations modulate light and color perception
Figure 7.29

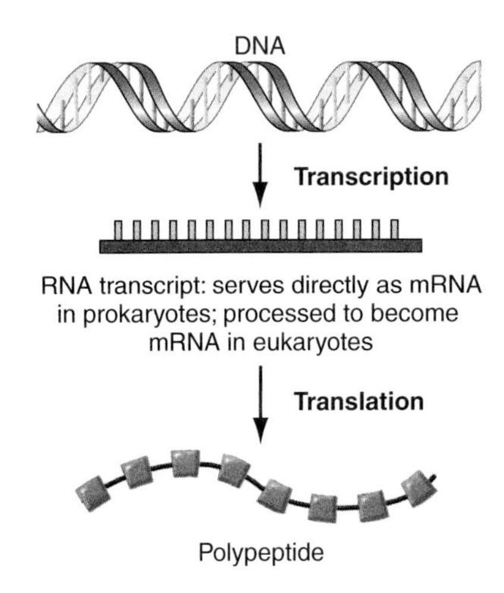

Gene expression: The flow of genetic information from DNA via RNA to protein
Figure 8.2

		Second letter			
		U	C	A	G	
First letter	U	UUU } Phe UUC UUA } Leu UUG	UCU UCC } Ser UCA UCG	UAU } Tyr UAC UAA Stop UAG Stop	UGU } Cys UGC UGA Stop UGG Trp	U C A G
	C	CUU CUC } Leu CUA CUG	CCU CCC } Pro CCA CCG	CAU } His CAC CAA } Gln CAG	CGU CGC } Arg CGA CGG	U C A G
	A	AUU AUC } Ile AUA AUG Met	ACU ACC } Thr ACA ACG	AAU } Asn AAC AAA } Lys AAG	AGU } Ser AGC AGA } Arg AGG	U C A G
	G	GUU GUC } Val GUA GUG	GCU GCC } Ala GCA GCG	GAU } Asp GAC GAA } Glu GAG	GGU GGC } Gly GGA GGG	U C A G

(Third letter runs vertically on right side: U C A G)

The genetic code: 61 codons represent the 20 amino acids, while 3 codons signify stop
Figure 8.3

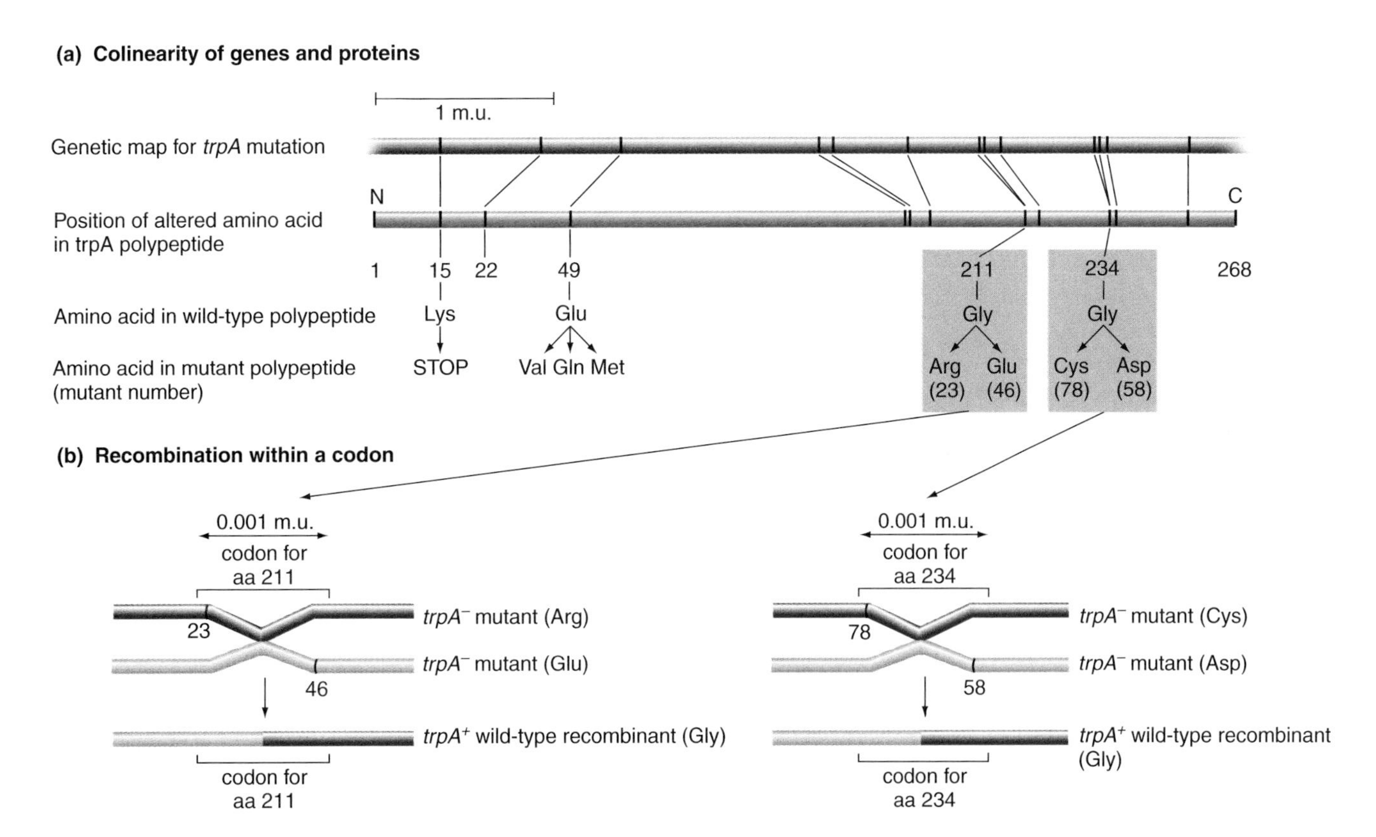

(a) Colinearity of genes and proteins

(b) Recombination within a codon

Mutations in a gene are colinear with the sequence of amino acids in the encoded polypeptide
Figure 8.4

(a) The mutagen proflavin can insert between two base pairs.

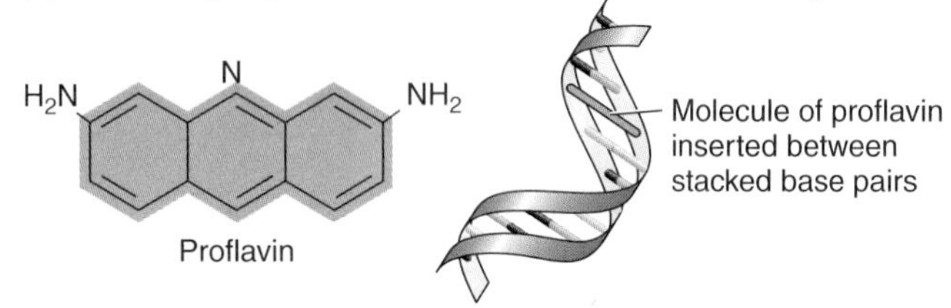

Proflavin

Molecule of proflavin inserted between stacked base pairs

(b) Consequences of exposure to proflavin.

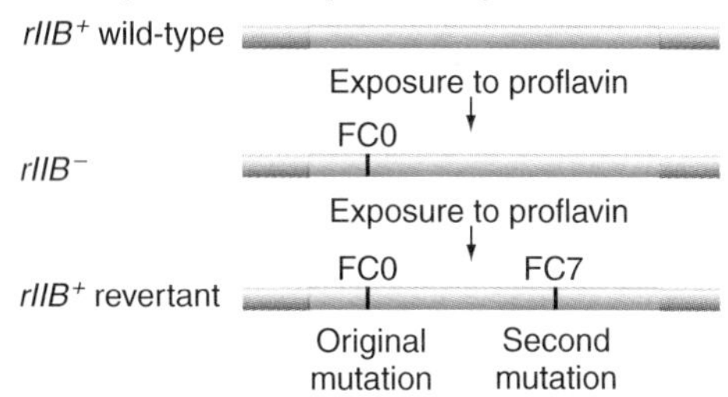

rIIB⁺ wild-type

Exposure to proflavin

FC0

rIIB⁻

Exposure to proflavin

FC0   FC7

rIIB⁺ revertant

Original mutation   Second mutation

(c) *rIIB*⁺ revertant X wild type yields *rIIB*⁻ recombinants.

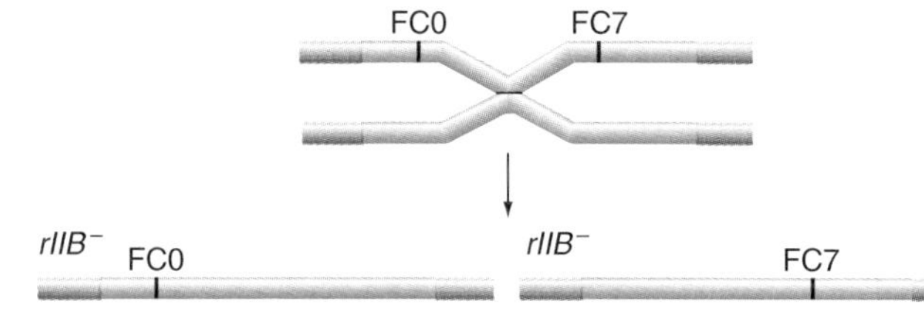

FC0   FC7

rIIB⁻ FC0

rIIB⁻   FC7

(d) Different sets of mutations generate either a mutant or a normal phenotype.

Proflavin-induced mutations (+) insertion (−) deletion	Phenotype
− or +	Mutant
− − or + +	Mutant
− − − − or − − − − − or + + + + or + + + + +	Mutant
− +	Wild type
− − − or − − − − − − or + + + or + + + + + +	Wild type

Studies of frameshift mutations in the bacteriophage T4 *rIIB* gene showed that codons consist of three nucleotides

Figure 8.5

Codons consist of three nucleotides read in a defined reading frame
Figure 8.6

▨ correct triplet
▨ incorrect triplet

(a) Intragenic suppression: 2 mutations of opposite sign.

Single base insertion (+) and single base deletion (−)

```
        C
        ↓                    ↗
ATG AAC AAT GCG CCG GAG GAA GCG GAC
                  ↓
ATG AAC AAT CGC GCCG GAG GA GCG GAC
```

(b) Intragenic suppression: 3 mutations of the same sign.

Three single base deletions (− − −)

```
              ↗       ↗       ↗
ATG AAC AAT GCG CCG GAG GAA GCG GAC
                  ↓
ATG AAC AA  GCG C G G G GAA GCG GAC
```

Three single base insertions (+ + +)

```
             G   T   C
             ↓   ↓   ↓
ATG AAC AAT GCG CCG GAG GAA GCG GAC
                  ↓
ATG AAC AAT GGCGCTCGGCAG GAA GCG GAC
```

(c) Some frameshift mutations.

Single base deletion (−)

```
              ↗
ATG AAC AAT GCG CCG GAG GAA GCG GAC
                  ↓
ATG AAC AA  GCG CCG GAG GAA GCG GAC
```

Single base insertion (+)

```
             G
             ↓
ATG AAC AAT GCG CCG GAG GAA GCG GAC
                  ↓
ATG AAC AAT GGCGCCG GAG GAA GCG GAC
```

(a) Poly-U mRNA encodes polyphenylalanine.

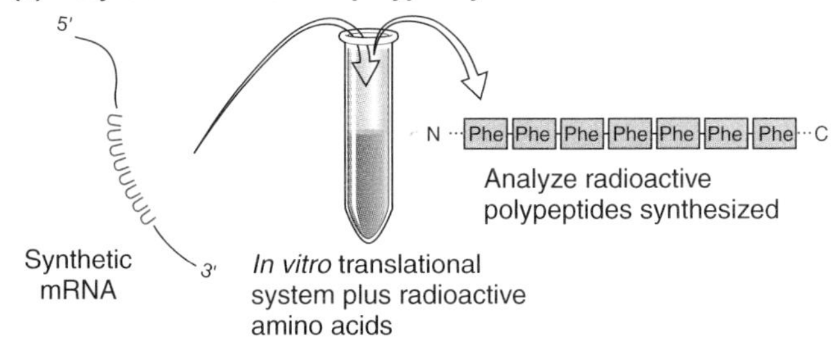

5′

Synthetic mRNA   3′

N ···Phe┤Phe┤Phe┤Phe┤Phe┤Phe┤Phe··· C

Analyze radioactive polypeptides synthesized

In vitro translational system plus radioactive amino acids

(b) Analyzing the coding possibilities.

Synthetic mRNA	Polypeptides synthesized
	Polypeptides with one amino acid
poly-U UUUU ...	Phe-Phe-Phe ...
poly-C CCCC ...	Pro-Pro-Pro ...
poly-A AAAA ...	Lys-Lys-Lys ...
poly-G GGGG ...	Gly-Gly-Gly ...
Repeating dinucleotides	**Polypeptides with alternating amino acids**
poly-UC UCUC ...	Ser-Leu-Ser-Leu ...
poly-AG AGAG ...	Arg-Glu-Arg-Glu ...
poly-UG UGUG ...	Cys-Val-Cys-Val ...
poly-AC ACAC ...	Thr-His-Thr-His ...
Repeating trinucleotides	**Three polypeptides each with one amino acid**
poly-UUC UUCUUCUUC ...	Phe-Phe.... and Ser-Ser.... and Leu-Leu....
poly-AAG AAGAAGAAG ...	Lys-Lys,... and Arg-Arg.... and Glu-Glu....
poly-UUG UUGUUGUUG ...	Leu-Leu.... and Cys-Cys.... and Val-Val....
poly-UAC UACUACUAC ...	Tyr-Tyr.... and Thr-Thr.... and Leu-Leu....
Repeating tetranucleotides	**Polypeptides with repeating units of four amino acids**
poly-UAUC UAUCUAUC ...	Tyr-Leu-Ser-Ile-Tyr-Leu-Ser-Ile...
poly-UUAC UUACUUAC ...	Leu-Leu-Thr-Tyr-Leu-Leu-Thr-Tyr...
poly-GUAA GUAAGUAA ...	none
poly-GAUA GAUAGAUA ...	none

How geneticists used synthetic mRNAs to limit the coding possibilities
Figure 8.7

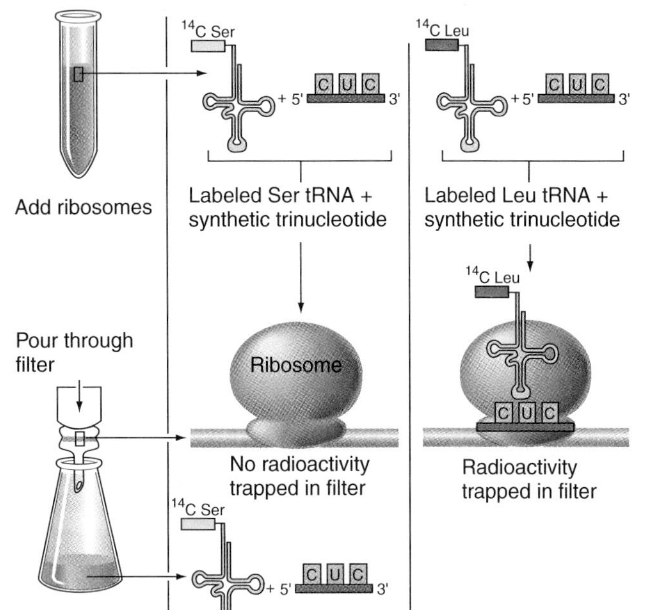

Add ribosomes

Pour through filter

¹⁴C Ser

Labeled Ser tRNA + synthetic trinucleotide

No radioactivity trapped in filter

¹⁴C Leu

Labeled Leu tRNA + synthetic trinucleotide

Radioactivity trapped in filter

Experimental verification of the genetic code
Figure 8.10

Cracking the genetic code with mini-mRNAs
Figure 8.8

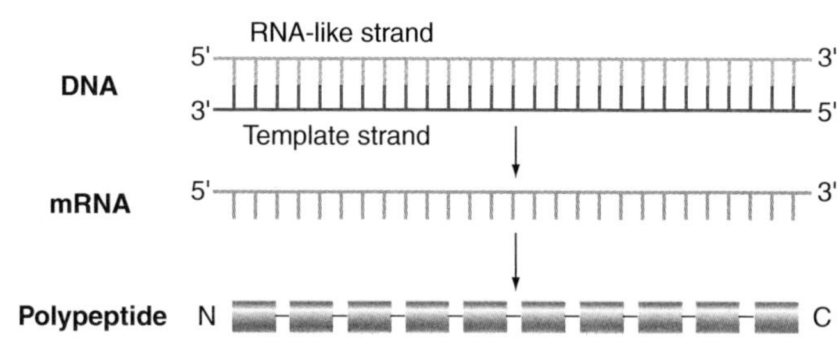

Correlation of polarities in DNA, mRNA, and polypeptide
Figure 8.9

(a) Altered amino acids in *trp⁻* mutations and *trp⁺* revertants

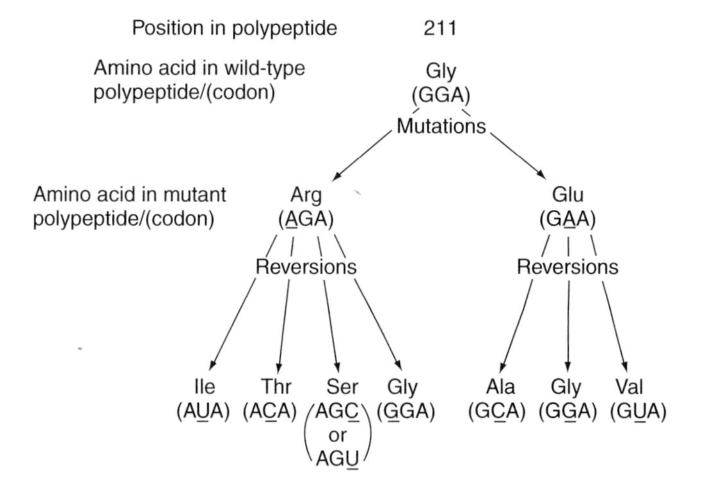

(b) Amino acid alterations that accompany intragenic suppression

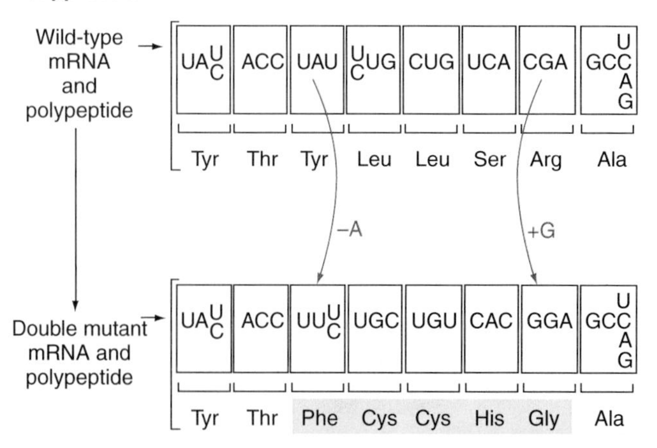

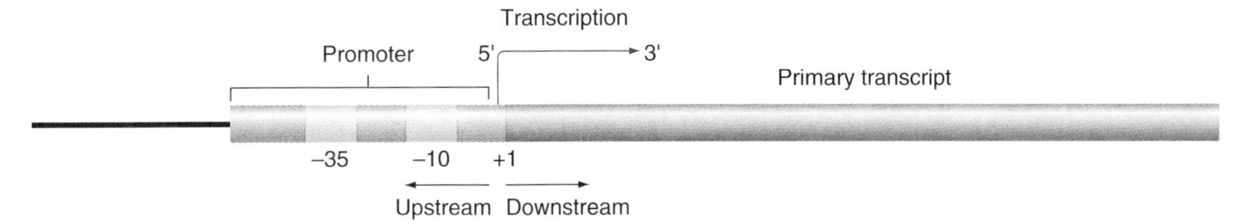

(a) Transcription initiation signals in bacteria

(b) Strong *E. coli* promoters

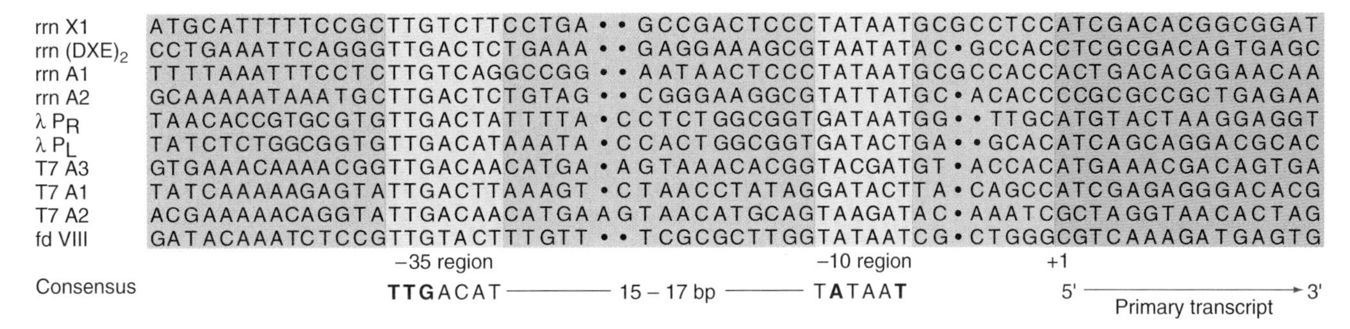

		−35 region		−10 region	+1
rrn X1	ATGCATTTTTCCGCTTGTCTTCCTGA • • GCCGACTCCCTATAATGCGCCTCCATCGACACGGCGGAT
rrn (DXE)₂	CCTGAAATTCAGGGGTTGACTCTGAAA • • GAGGAAAGCGTAATATAC • GCCACCTCGCGACAGTGAGC
rrn A1	TTTTAAATTTCCTCTTGTCAGGCCGG • • AATAACTCCCTATAATGCGCCACCACTGACACGGAACAA
rrn A2	GCAAAAATAAATGCTTGACTCTGTAG • • CGGGAAGGCGTATTATGC • ACACCCCGCGCCGCTGAGAA
λ P_R	TAACACCGTGCGTGTTGACTATTTTA • CCTCTGGCGGTGATAATGG • • TTGCATGTACTAAGGAGGT
λ P_L	TATCTCTGGCGGTGTTGACATAAATA • CCACTGGCGGTGATACTGA • • GCACATCAGCAGGACGCAC
T7 A3	GTGAAACAAAACGGTTGACAACATGA • AGTAAACACGGTACGATGT • ACCACATGAAACGACAGTGA
T7 A1	TATCAAAAAGAGTATTGACTTAAAGT • CTAACCTATAGGATACTTA • CAGCCATCGAGAGGGACACG
T7 A2	ACGAAAAACAGGTATTGACAACATGAAGTAACATGCAGTAAGATAC • AAATCGCTAGGTAACACTAG
fd VIII	GATACAAATCTCCGTTGTACTTTGTT • • TCGCGCTTGGTATAATCG • CTGGGCGTCAAAGATGAGTG

	−35 region	−10 region	+1
Consensus	**TTG**ACAT ——— 15 – 17 bp ——— **T**A**T**AA**T**	5' —————→ 3' Primary transcript

The promoters of 10 different bacterial genes
Figure 8.12

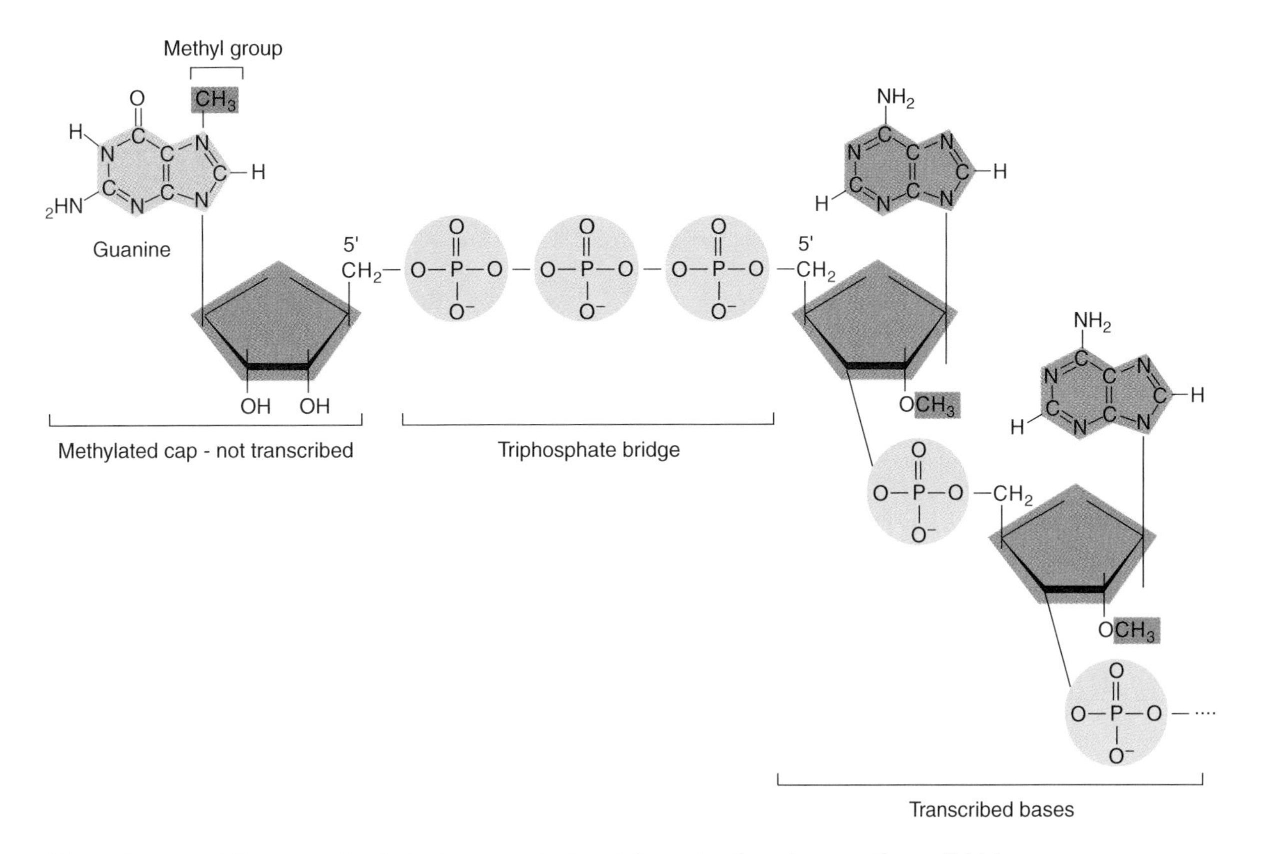

Structure of the methylated cap at the 5′ end of eukaryotic mRNAs
Figure 8.13

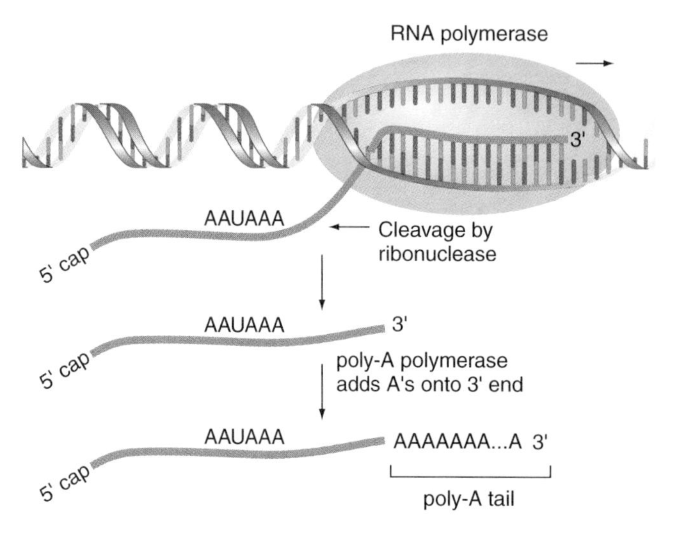

How RNA processing adds a tail to the 3′ end of eukaryotic mRNAs
Figure 8.14

Splicing removes introns from a primary transcript.

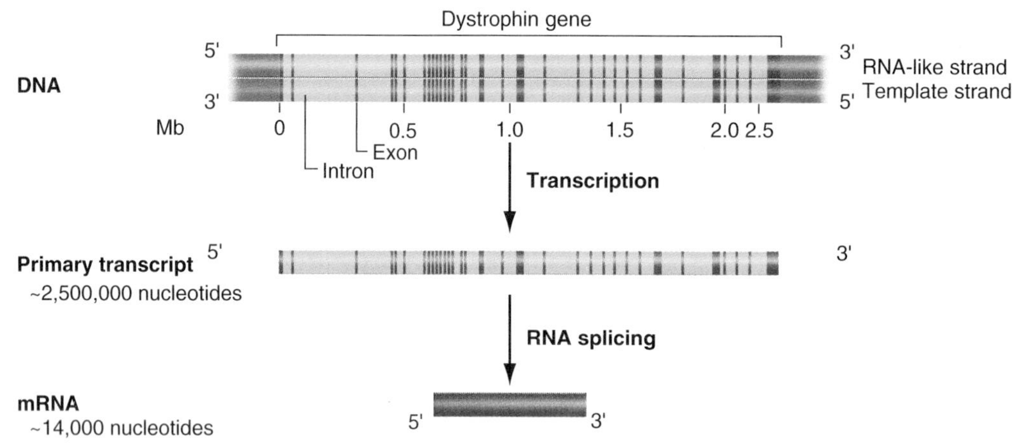

The human dystrophin gene: An extreme example of RNA splicing
Figure 8.15

(a) Short sequences dictate where splicing occurs.

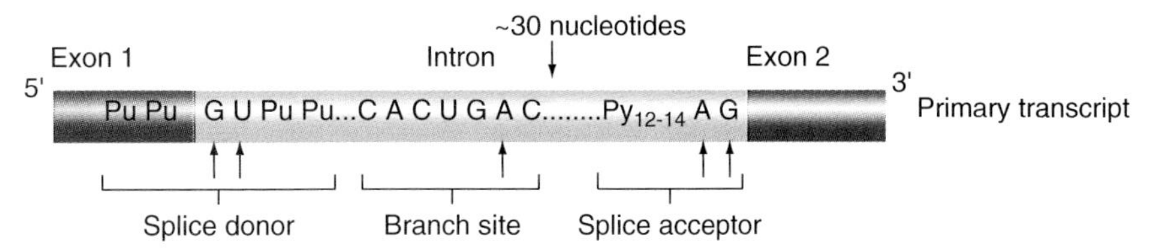

(b) Two sequential cuts remove the intron.

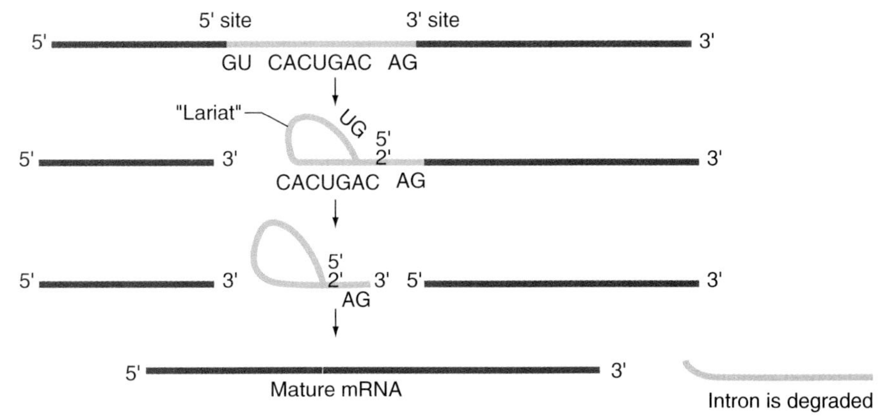

How RNA processing splices out introns and joins adjacent exons
Figure 8.16

(a) Alternative splicing produces two different mRNAs from the same gene.

Spliceosome components

Five snRNAs (small nuclear RNAs) + ~50 proteins

Four snRNPs (small nuclear ribonucleic particles), which assemble into a spliceosome

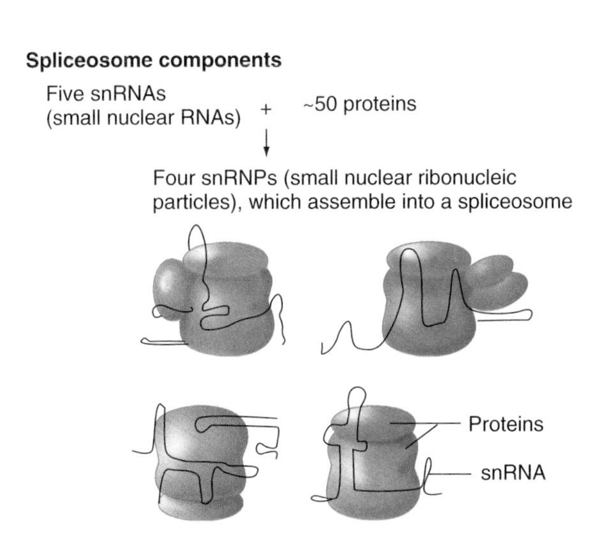

Proteins

snRNA

Splicing is catalyzed by the spliceosome
Figure 8.17

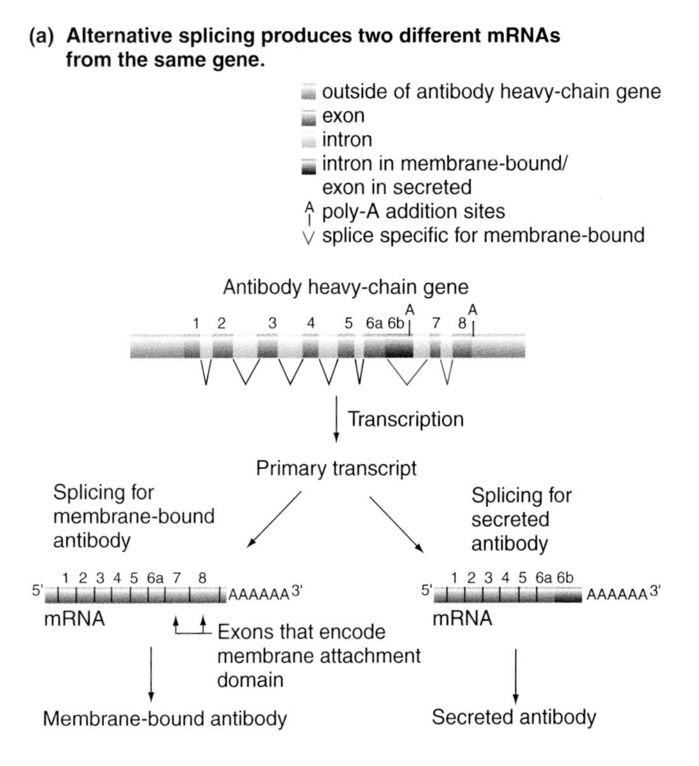

(b) Trans-splicing combines exons from different genes.

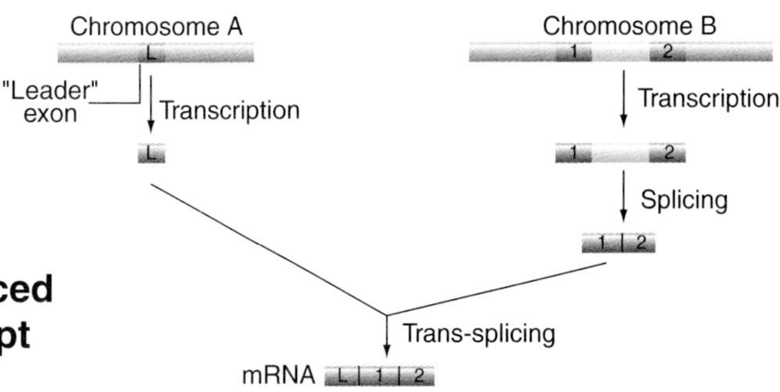

Different mRNAs can be produced from the same primary transcript
Figure 8.18

(a) Some tRNAs contain modified bases.

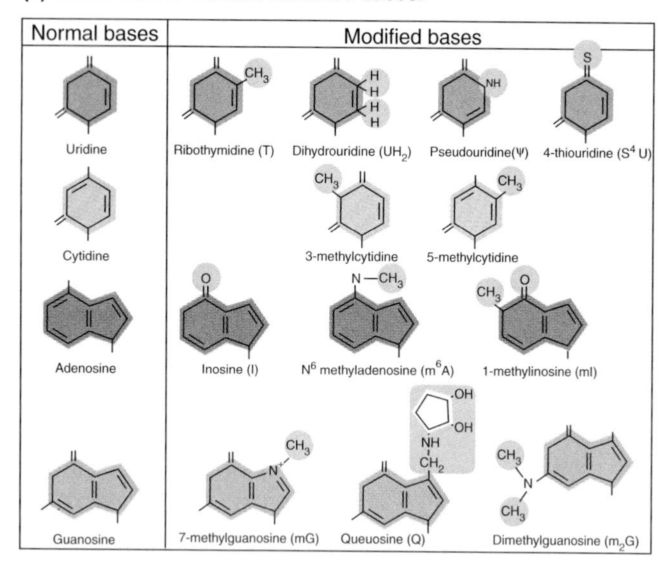

(b) Each tRNA has a primary, secondary, and tertiary structure.

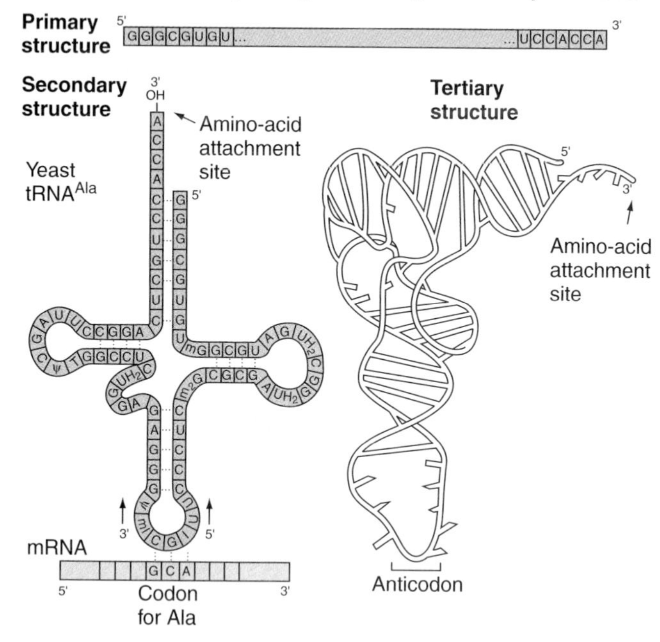

tRNAs mediate the transfer of information from nucleic acid to protein
Figure 8.19

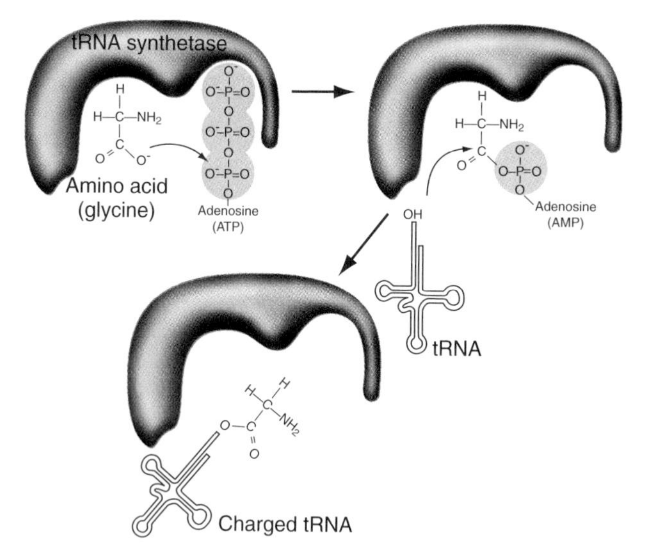

Aminoacyl-tRNA synthetases catalyze the attachment of tRNAs to their corresponding amino acids
Figure 8.20

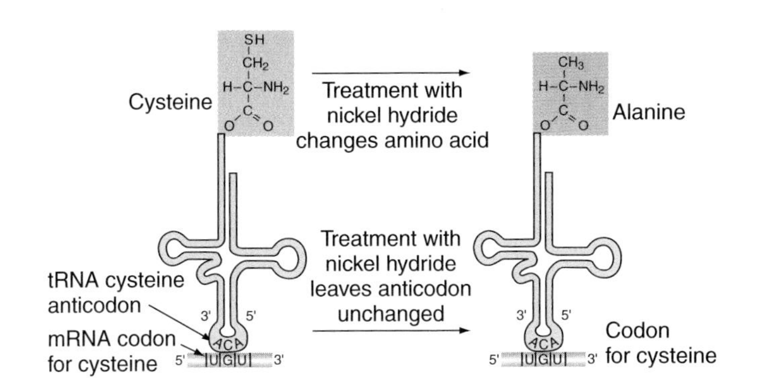

Cysteine

Treatment with nickel hydride changes amino acid

Alanine

tRNA cysteine anticodon

mRNA codon for cysteine

Treatment with nickel hydride leaves anticodon unchanged

Codon for cysteine

Base pairing between an mRNA codon and a tRNA anticodon determines where an amino acid becomes incorporated into a growing polypeptide
Figure 8.21

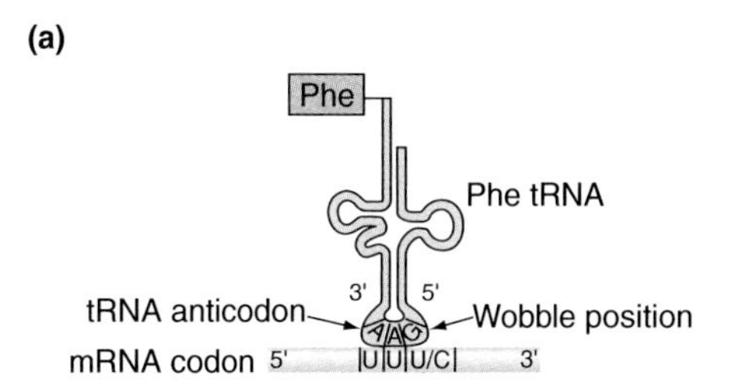

(a)

Phe

Phe tRNA

tRNA anticodon

Wobble position

mRNA codon

Wobble: Some tRNAs recognize more than one codon for the amino acid they carry
Figure 8.22

(b)

Wobble rules		
5' end of anticodon	can pair with	3' end of codon
G		U or C
C		G
A		U
U		A or G
I		U, C, or A

(a) A ribosome has two subunits composed of RNA and protein.

Complete ribosomes	Subunits	Nucleotides	Proteins
Prokaryotic	50S	23S RNA 3000 nucleotides	31
70S	30S	16S RNA 1700 nucleotides / 5S RNA 120 nucleotides	21
Eukaryotic	60S	28S RNA 5000 nucleotides / 5.8S RNA 160 nucleotides / 5S RNA 120 nucleotides	~ 45
80S	40S	18S RNA 2000 nucleotides	~ 33

(b) Different parts of a ribosome have different functions.

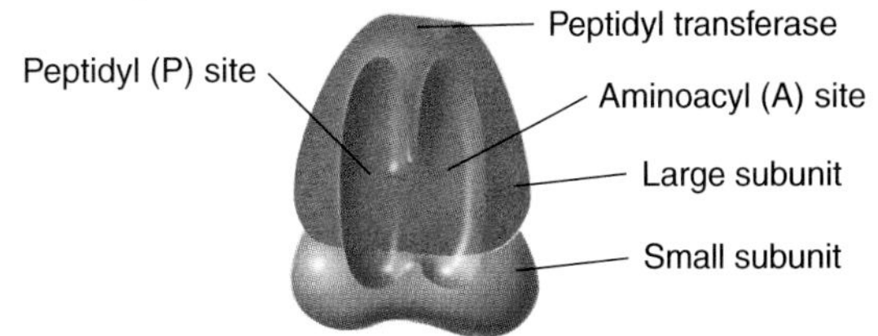

Peptidyl (P) site

Peptidyl transferase

Aminoacyl (A) site

Large subunit

Small subunit

The ribosome: Site of polypeptide synthesis
Figure 8.23

(a) Cleavage may remove an amino acid.

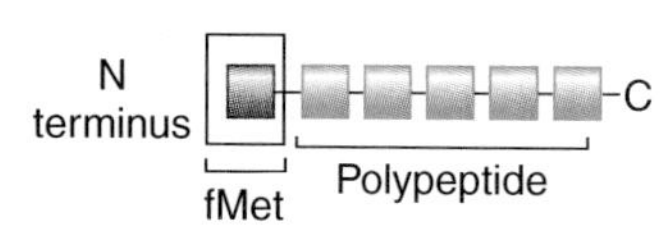

(b) Cleavage may split a polyprotein.

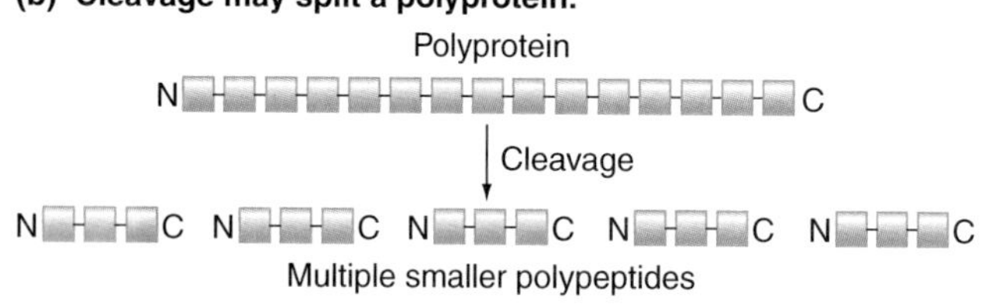

(c) Addition of chemical constituents may modify a protein.

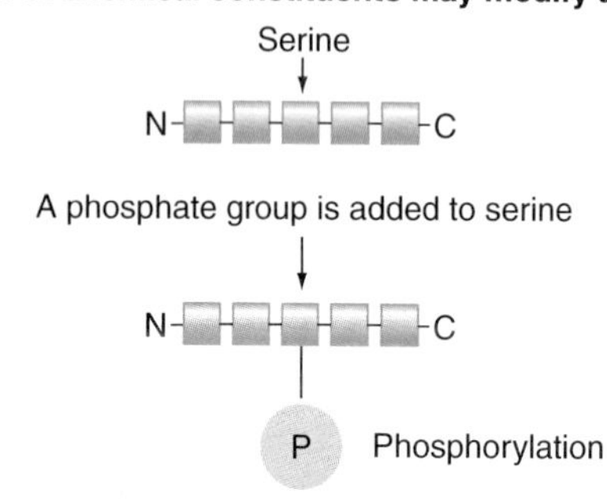

Posttranslational processing can modify polypeptide structure
Figure 8.25

(a) A collagen gene of *C. elegans*

Promoter
ATG (start codon)
TAA (stop codon)
AATAAA addition site for poly (A)

Exon 1 Intron 1 Exon 2 Intron 2 Exon 3 200 bp

Direction of transcription ⟶

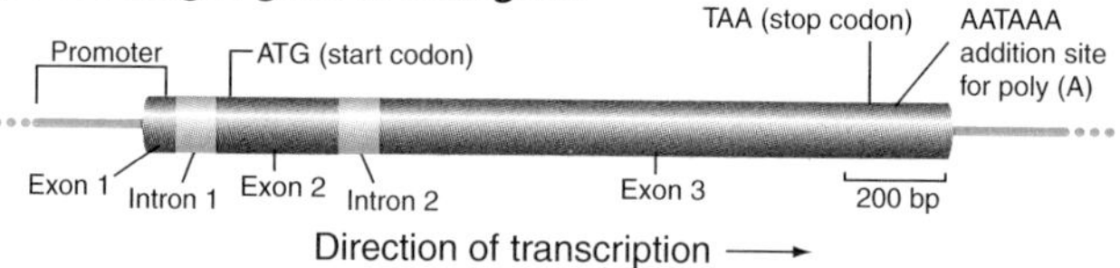

(b) Sequence of a *C. elegans* collagen gene, mRNA, and polypeptide

RNA-like strand 5'... ACAACACTAGGTATAAAGCGGAAGTGGTGGCTTTAAAATCACTTGGCTTCTAAAGTCCAGTGACAGGTAAG

Template strand 3'...TGTTGTGATCCATATTTCGCCTTCACCACCGAAATTTTAGTGAACCGAAGATTTCAGGTCACTGTCCATTC

mRNA   5' cap–CACUUGGCUUCUAAAGUCCAGUGACAG

Polypeptide

GTTCTCGTTACTTCCGTCTCGATTACTAAGATTTGATTACTTTTAGAAAAATGACCGAAGATCCAAAGCAGATTGCCCAGGAGACTGAG
CAAGAGCAATGAAGGCAGAGCTAATGATTCTAAACTAATGAAAATCTTTTTACTGGCTTCTAGGTTTCGTCTAACGGGTCCTCTGACTC...
                              AAAAAUGACCGAAGAUCCAAAGCAGAUUGCCCAGGAGACUGAG...
                              Met Thr Glu Asp Pro Lys Gln Ile Ala Gln Glu Thr Glu...

GTTGAATTCTGCCAACACAGATCAAATGGACTTTGGGATGAGTATAAGAGAGTATGTTTTTTTGTTGAATAATTTTAATTTTAGTTAAATGTTT
CAACTTAAGACGGTTGTGTCTAGTTTACCTGAAACCCTACTCATATTCTCTCATACAAAAAAAACAACTTATTAAAATTAAAATCAATTTACAAA
GUUGAAUUCUGCCAACACAGAUCAAAUGGACUUUGGGAUGAGUAUAAGAGA
Val Glu Phe Cys Gln His Arg Ser Asn Gly Leu Trp Asp Glu Tyr Lys Arg

GATTTCAGTTCCAAGGAGTTTCTGGAGTTGAAGGACGTATCAAGAGAGACGCATATCACCGTAGCCTCGGAGTTTCTGGTGCTTCCCGC
CTAAAGTCAAGGTTCCTCAAAGACCTCAACTTCCTGCATGAGTTCTCTCTGCGTATAGTGGCATCGGAGCCTCAAAGACCACGAAGGGCG
UUCCAAGGAGUUUCUGGAGUUGAAGGACGUAUCAAGAGAGACGCAUAUCACCGUAGCCUCGGAGUUUCUGGUGCUUCCCGC
Phe Gln Gly Val Ser Gly Val Glu Gly Arg Ile Lys Arg Asp Ala Tyr His Arg Ser Leu Gly Val Ser Gly Ala Ser Arg

AAGGCTCGTCGTCAATCTTATGGAAATGACGCTGCTGTCGGAGGATTCGGTGGATCATCTGGAGGATCATGCTGCTCATGCGGATCT...
TTCCGAGCAGCAGTTAGAATACCTTTACTGCGACGACAGCCTCCTAAGCGCACCTAGTAGACCTCCTAGTACGACGAGTACGCCTAGA...
AAGGCUCGUCGUCAAUCUUAUGGAAAUGACGCUGCUGUCGGAGGAUUCGGUGGAUCAUCUGGAGGAUCAUGCUGCUCAUGCGGAUCU...
Lys Ala Arg Arg Gln Ser Tyr Gly Asn Asp Ala Ala Val Gly Gly Phe Gly Gly Ser Ser Gly Gly Ser Cys Cys Ser Gly Ser...

CCAGGACAAGCTGGAGCACCAGGACAAGATGGAGAGAGTGGATCCGAGGGAGCTTGCGATCACTGCCCACCACCACGTACCGGCTCCA
GGTCCTGTTCGACCTCGTGGTCCTGTTCTACCTCTCTCACCTAGGCTCCCTCGAACGCTAGTGACGGGTGGTGGTGCATGGCGAGGT
CCAGGACAAGCUGGAGCACCAGGACAAGAUGGAGAGAGUGGAUCCGAGGGAGCUUGCGAUCACUGCCCACCACCACGUACCGGCUCCA
Pro Gly Gln Ala Gly Ala Pro Gly Gln Asp Gly Glu Ser Gly Ser Glu Gly Ala Cys Asp His Cys Pro Pro Pro Arg Thr Ala Pro

GGATATTAAGCGCTTCAATGACATCTCATTTGATTATCTCTGCTTTATCTCATTTGTATGTTTTGTGTATGAAAAACGAACACACTTAGAATAG
CCTATAATTCGCGAAGTTACTGTAGAGTAAACTAATAGAGCAGAAATAGAGTAAACATACAAAACACTACTTTTTGCTTGTGTGAATCTTATC
GGAUAUUAAGCGCUUCAAUGACAUCUCAUUUGAUUAUCUCUGCUUUAUCUCAUUUGUAUGUUUUGUGUAUGAAAAACGAACACACUUAGAAUAG
Gly Tyr Stop

TGGAATAAATGATTTCATTACAAATTTGAAATTGAATAAAGACAAATGTGAAATGAAAGTATAAAAGAAAATGAGAGAC...3'
ACCTTATTTACTAAAGTAATGTTTAAAACTTTAACTTATTCTGTTTACACTTTACTTTTCATATTTTCTTTTACTCTCTG...5'
UGGAAUAAAUGAUUUCAUUACAAAUUUGAAAUUGAAAUUUGAAAUGAAAGUAUAAAAGAAAAUGAGAGAC...
         (poly (A))

Expression of a *C. elegans* gene for collagen
Figure 8.26

(a) Types of mutation in a gene's coding sequence

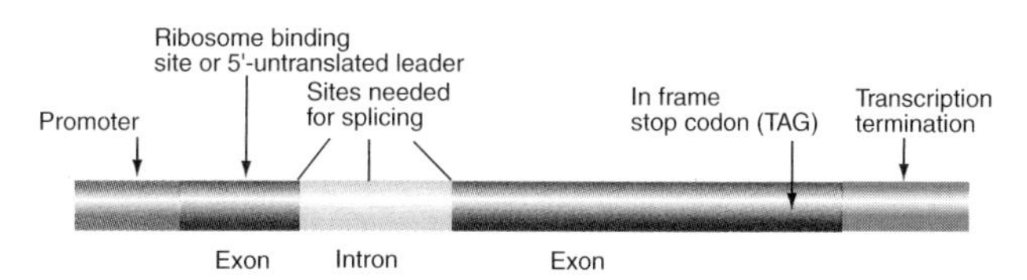

Wild-type mRNA   5' GCU GGA GCA CCA GGA CAA GAU GGA 3'
Wild-type polypeptide N Ala Gly Ala Pro Gly Gln Asp Gly C

Silent mutation   GCU GGA GCC CCA GGA CAA GAU GGA
                          Ala Gly Ala Pro Gly Gln Asp Gly

Missense mutation   GCU GGA GCA CCA AGA CAA GAU GGA
                          Ala Gly Ala Pro Arg Gln Asp Gly

Nonsense mutation   GCU GGA GCA CCA GGA UAA GAU GGA
                          Ala Gly Ala Pro Gly Stop

Frameshift mutation   GCU GGA GCC ACC AGG ACA AGA UGG A
                          Ala Gly Ala Thr Arg Thr Arg Trp

(b) Sites of mutation outside the coding sequence that can disrupt gene expression

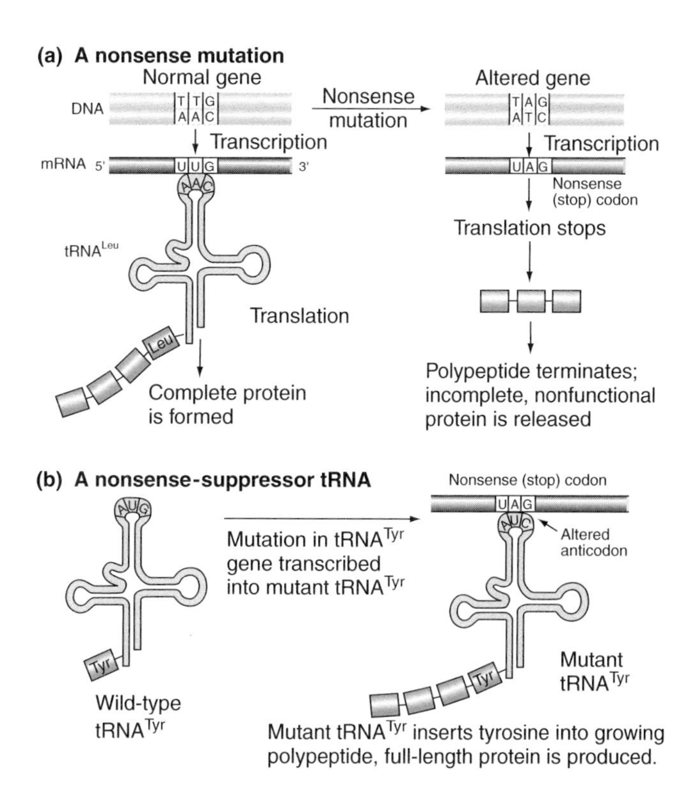

Promoter

Ribosome binding site or 5'-untranslated leader

Sites needed for splicing

In frame stop codon (TAG)

Transcription termination

Exon   Intron   Exon

How mutations in a gene can affect its expression
Figure 8.27

(a) A nonsense mutation

Normal gene

DNA   T T G / A A C

Nonsense mutation

Altered gene

T A G / A T C

Transcription

mRNA 5' U U G 3'

tRNA^Leu

Translation

Leu

Complete protein is formed

Transcription

U A G

Nonsense (stop) codon

Translation stops

Polypeptide terminates; incomplete, nonfunctional protein is released

(b) A nonsense-suppressor tRNA

Nonsense (stop) codon

U A G

AUG

Mutation in tRNA^Tyr gene transcribed into mutant tRNA^Tyr

AUC

Altered anticodon

Tyr

Wild-type tRNA^Tyr

Mutant tRNA^Tyr

Tyr

Mutant tRNA^Tyr inserts tyrosine into growing polypeptide, full-length protein is produced.

Nonsense suppression
Figure 8.28

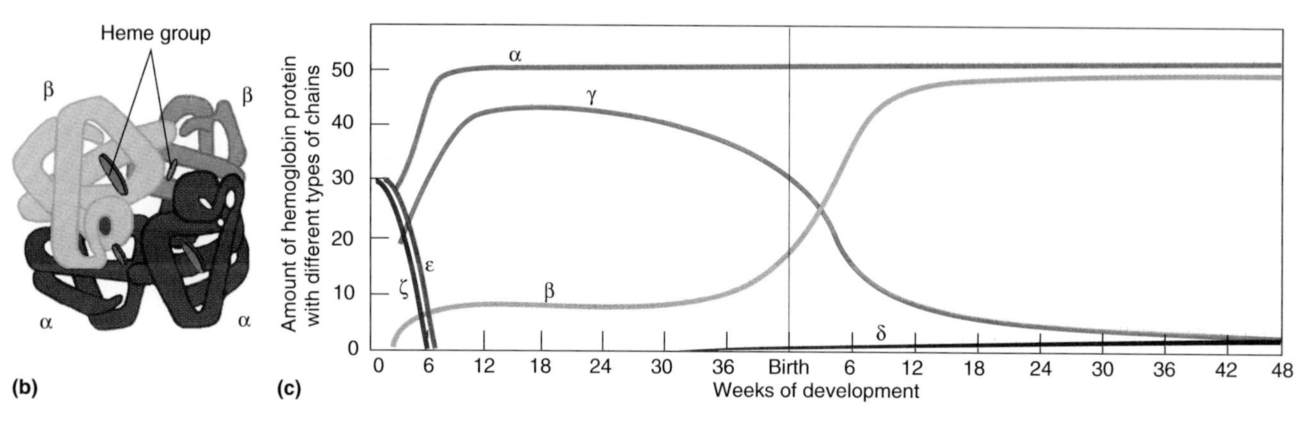

Hemoglobin is composed of four polypeptide chains that change during development
Figure 9.1

(a) 1. Probability that a four-base recognition site will be found in a genome =

$$1/4 \times 1/4 \times 1/4 \times 1/4 = 1/256$$

2. Probability that a six-base recognition site will be found =

$$1/4 \times 1/4 \times 1/4 \times 1/4 \times 1/4 \times 1/4 = 1/4,096$$

(b) Intact Human DNA

1. Four-base

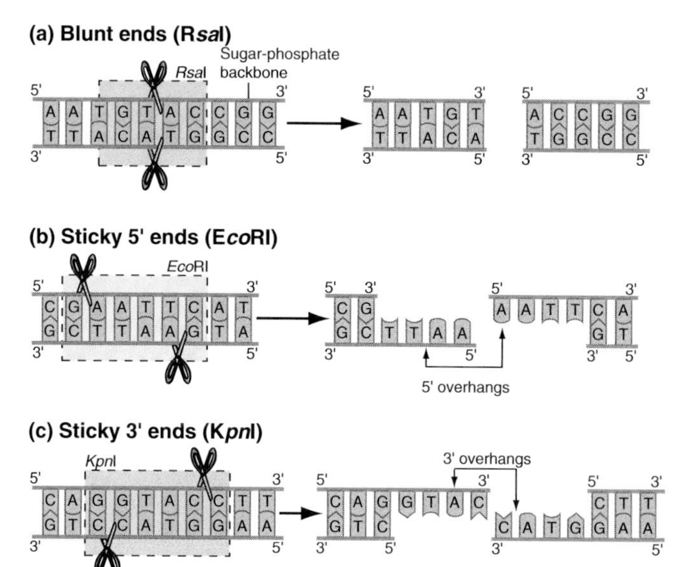

*Rsa*I sites

2. Six-base

*Eco*RI sites

3. Eight-base

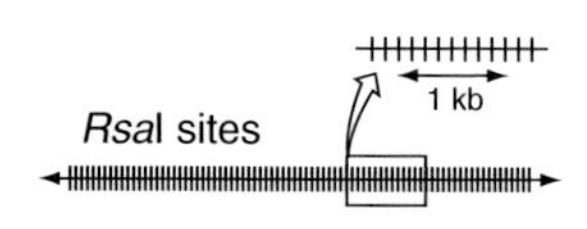

*Not*I sites

The number of base pairs in a restriction enzyme
Figure 9.3

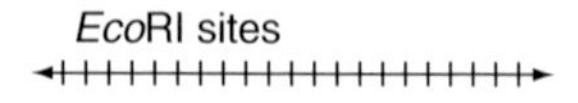

Restriction enzymes cut DNA molecules at specific locations to produce restriction fragments with either blunt or sticky ends
Figure 9.2

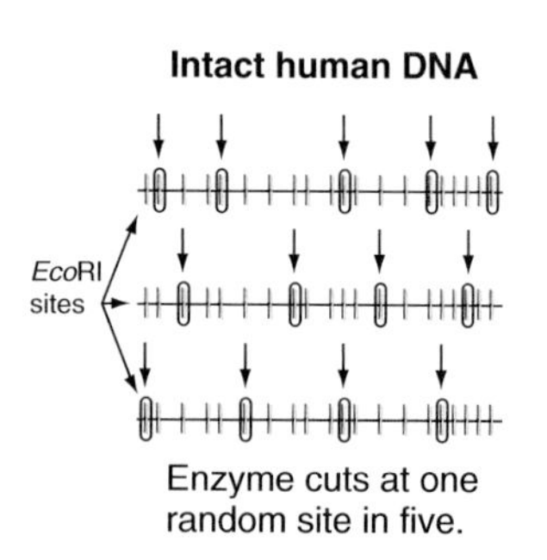

Intact human DNA

*Eco*RI sites

Enzyme cuts at one random site in five.

Partial digests
Figure 9.4

How to infer a restriction map from the sizes of restriction fragments produced by two restriction enzymes
Figure 9.6

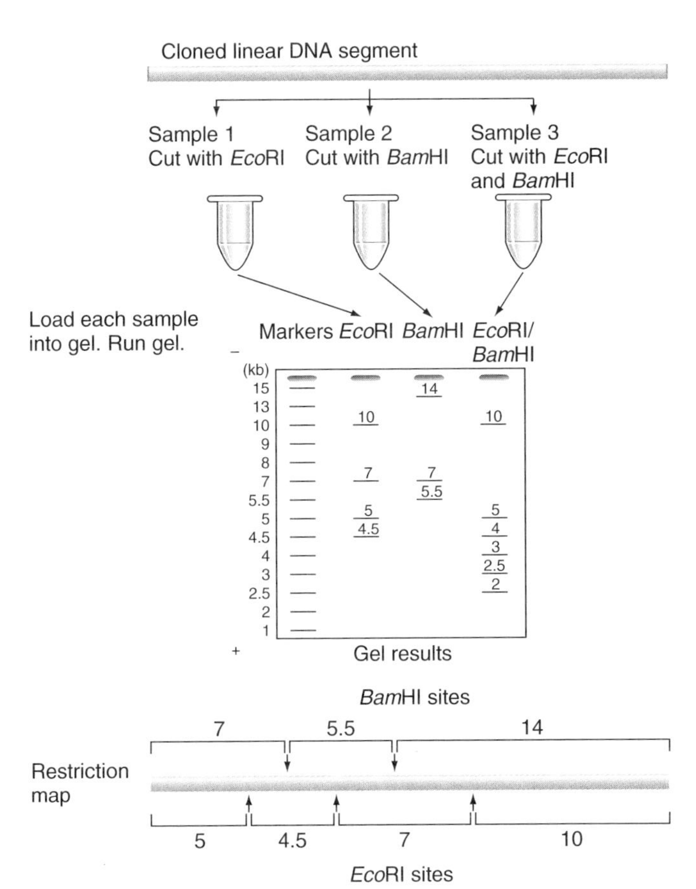

Cloned linear DNA segment

Sample 1
Cut with *Eco*RI

Sample 2
Cut with *Bam*HI

Sample 3
Cut with *Eco*RI
and *Bam*HI

Load each sample
into gel. Run gel.

Markers *Eco*RI *Bam*HI *Eco*RI/
*Bam*HI

(kb)
15
13
10
9
8
7
5.5
5
4.5
4
3
2.5
2
1

14

10

7
5.5

7

5
4.5

10

5
4
3
2.5
2

Gel results

*Bam*HI sites

7   5.5   14

Restriction
map

5   4.5   7   10

*Eco*RI sites

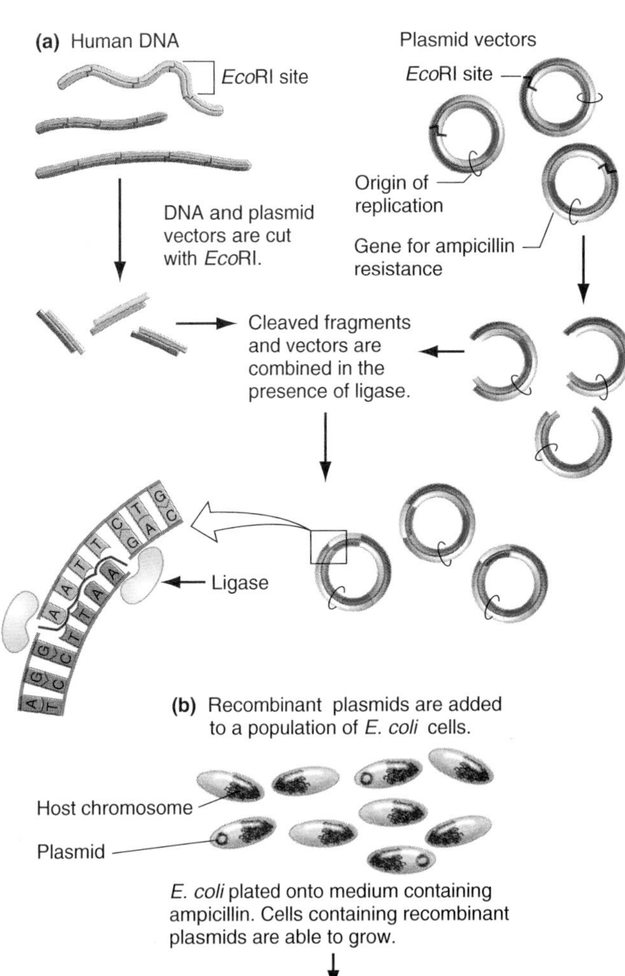

(a) Human DNA

*Eco*RI site

Plasmid vectors

*Eco*RI site

Origin of
replication

Gene for ampicillin
resistance

DNA and plasmid
vectors are cut
with *Eco*RI.

Cleaved fragments
and vectors are
combined in the
presence of ligase.

Ligase

(b) Recombinant plasmids are added
to a population of *E. coli* cells.

Host chromosome

Plasmid

E. coli plated onto medium containing
ampicillin. Cells containing recombinant
plasmids are able to grow.

Creating recombinant DNA molecules with plasmid vectors
Figure 9.7

83

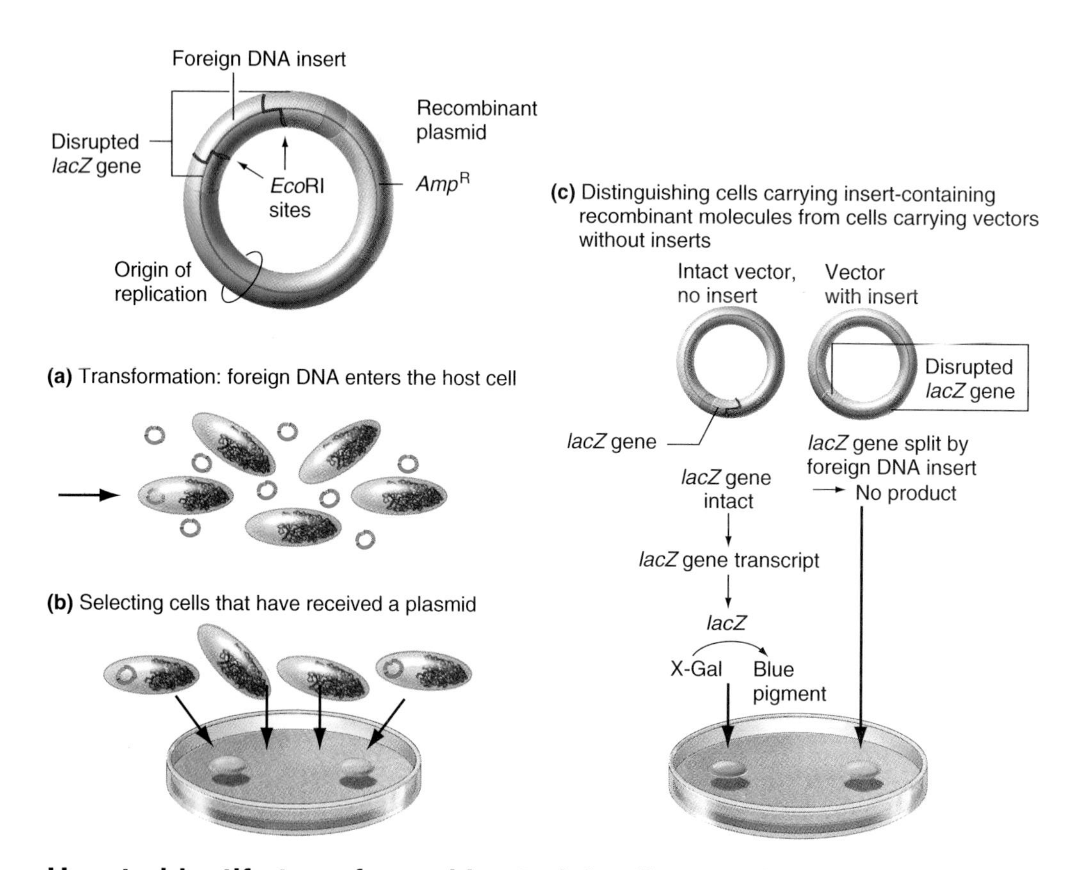

Foreign DNA insert

Disrupted
lacZ gene

*Eco*RI
sites

Recombinant
plasmid

*Amp*ᴿ

Origin of
replication

(a) Transformation: foreign DNA enters the host cell

(b) Selecting cells that have received a plasmid

(c) Distinguishing cells carrying insert-containing
recombinant molecules from cells carrying vectors
without inserts

Intact vector,
no insert

Vector
with insert

Disrupted
lacZ gene

lacZ gene

lacZ gene
intact

lacZ gene split by
foreign DNA insert
→ No product

lacZ gene transcript

lacZ

X-Gal ⟶ Blue
pigment

**How to identify transformed bacterial cells containing plasmids
with DNA inserts**
Figure 9.8

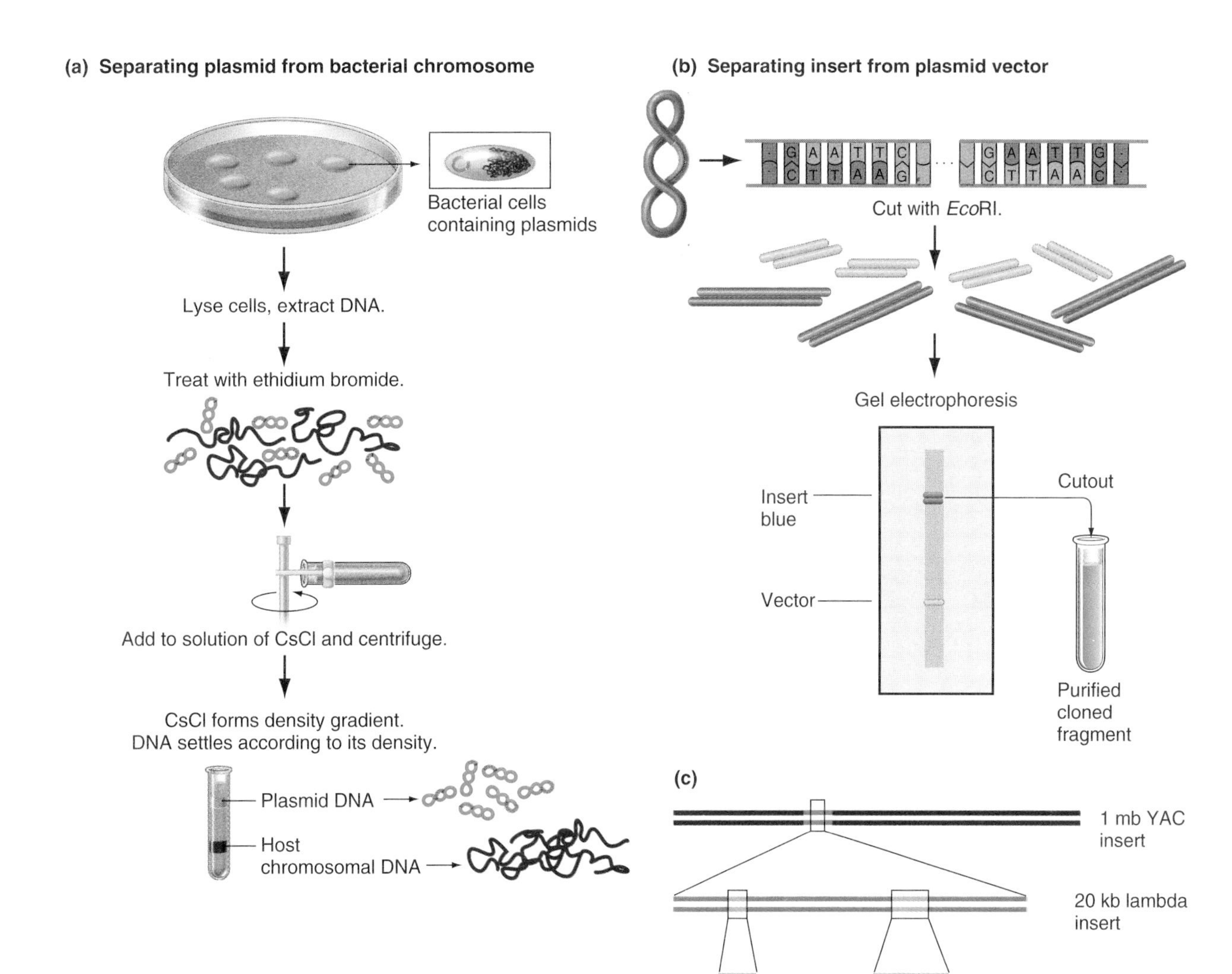

(a) Separating plasmid from bacterial chromosome

Bacterial cells containing plasmids

Lyse cells, extract DNA.

Treat with ethidium bromide.

Add to solution of CsCl and centrifuge.

CsCl forms density gradient.
DNA settles according to its density.

Plasmid DNA

Host chromosomal DNA

(b) Separating insert from plasmid vector

Cut with *Eco*RI.

Gel electrophoresis

Insert blue

Cutout

Vector

Purified cloned fragment

(c)

1 mb YAC insert

20 kb lambda insert

1 kb Plasmid insert

2 kb Plasmid insert

Purifying cloned DNA
Figure 9.9

(a) Red blood cell precursors

Release mRNA from cytoplasm and purify.

| | | | | | | | | | | | | | |A|A|A|A| mRNA
5'   3'

5'   3'
| | | | | | | | | | | | |A|A|A|A| mRNA

5'   3'
| | | | | | | | | | | | | | | |A|A|A|A| mRNA

(b) Add oligo(dT) primer. Treat with reverse transcriptase in presence of four nucleotides.

1.
                                          Primer
                                          |T|T|T| 5'
5' |A|A|A|A| mRNA
                                              3'

2.   5' Growing
                                          cDNA
                              |T|T|T|
5' |A|A|A|A| mRNA
                              3'
            Reverse transcriptase

3. 3'
                                          5' cDNA
                                          mRNA
5'   3'

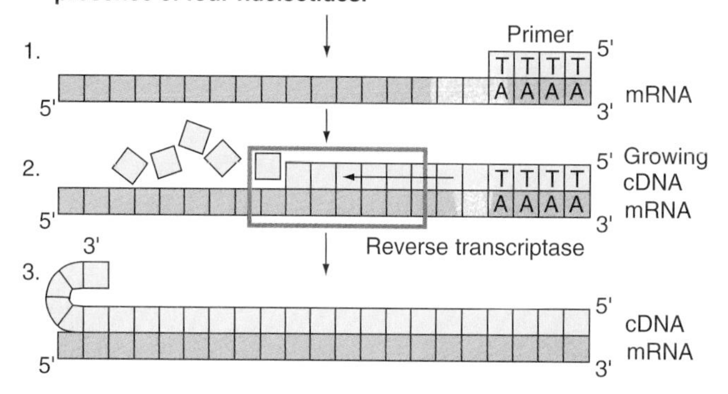

(c) Denature cDNA-mRNA hybrids and digest mRNA with RNase. cDNA acts as template for synthesis of second cDNA strand in the presence of four nucleotides and DNA polymerase.

                              Growing second strand
                                                  5'
                                                  cDNA

S1 nuclease   DNA
cuts hairpin loop   polymerase

5'   3' cDNA
                                              double
3'   5' helix

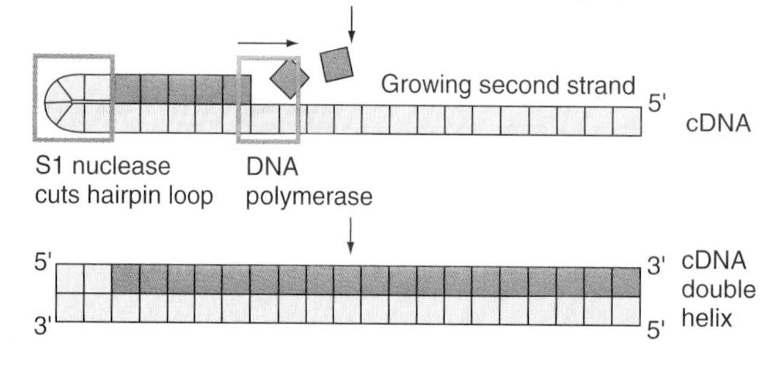

(d) Insert cDNA into vector.

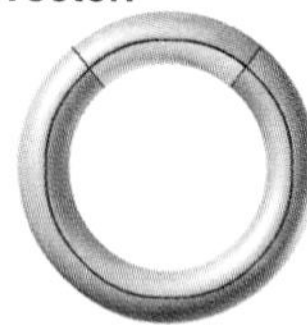

Converting RNA transcripts to cDNA
Figure 9.10

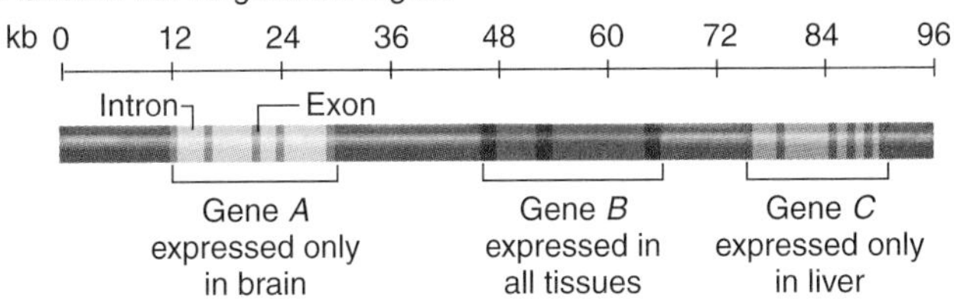

Random 100 kb genomic region

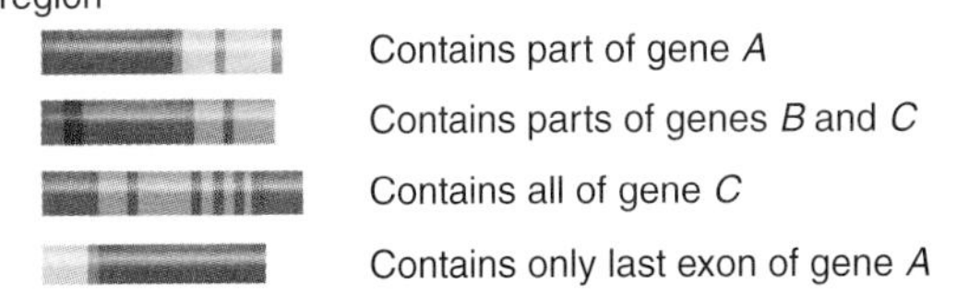

Clones from a genomic library with 20 kb inserts that are homologous to this region

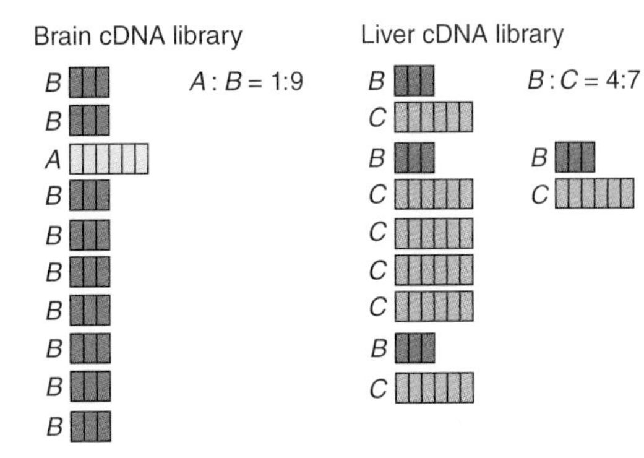

Contains part of gene *A*

Contains parts of genes *B* and *C*

Contains all of gene *C*

Contains only last exon of gene *A*

Clones from cDNA libraries

Brain cDNA library   Liver cDNA library

A comparison of genomic and cDNA libraries
Figure 9.11

(a) An expression vector allows production of specific polypeptide

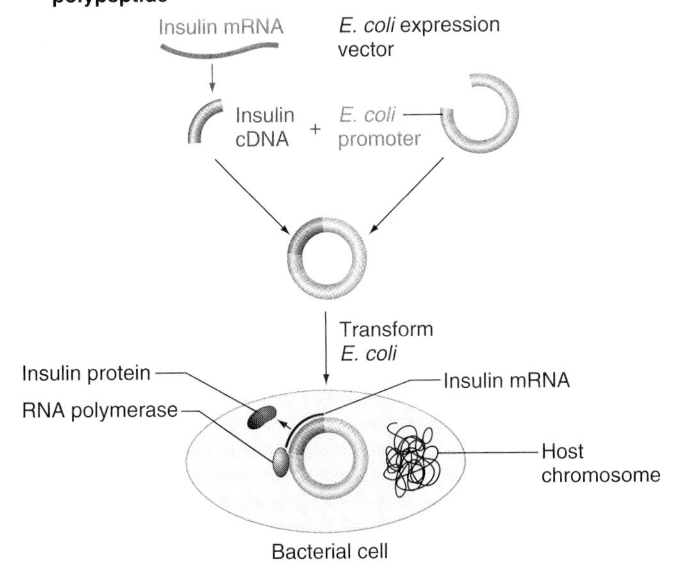

(b) Screening for insulin gene expression

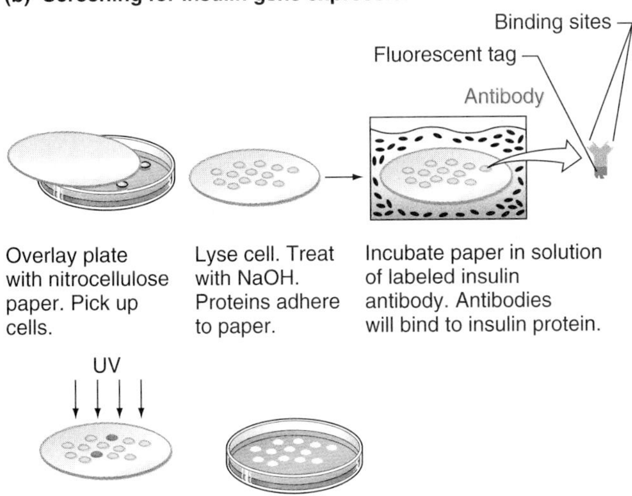

Overlay plate with nitrocellulose paper. Pick up cells.

Lyse cell. Treat with NaOH. Proteins adhere to paper.

Incubate paper in solution of labeled insulin antibody. Antibodies will bind to insulin protein.

Wash filter. Expose to UV light and identify flourescent spots. Compare with original plate in order to find bacterial clone containing human insulin gene.

An expression vector can be used to produce a desired polypeptide
Figure 9.12

Screening a library of clones by hybridization to a labeled probe
Figure 9.13

Master plate containing genomic library of mouse clones.

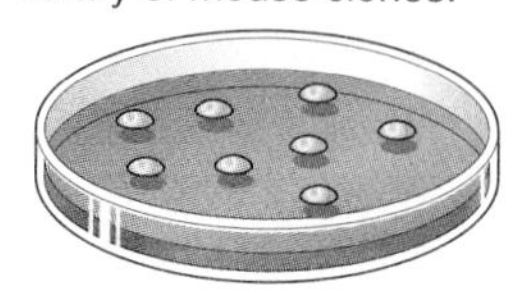

Overlay a nitrocellulose disk to make a replica of the plate.

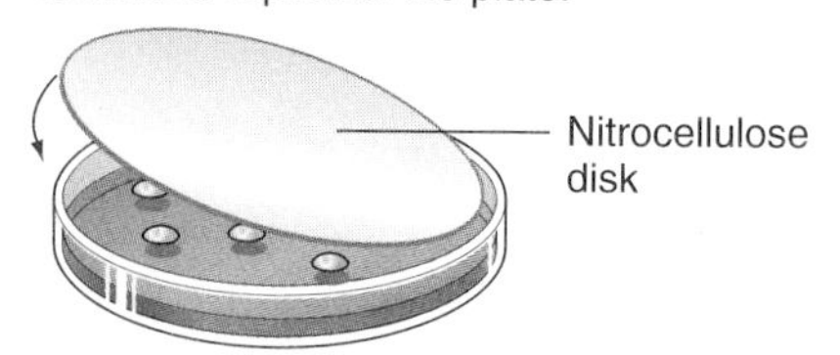

Nitrocellulose disk

Remove disk from plate and lyse cells on it and denature DNA with NaOH. Bake and treat with UV light to bind DNA strands to disk.

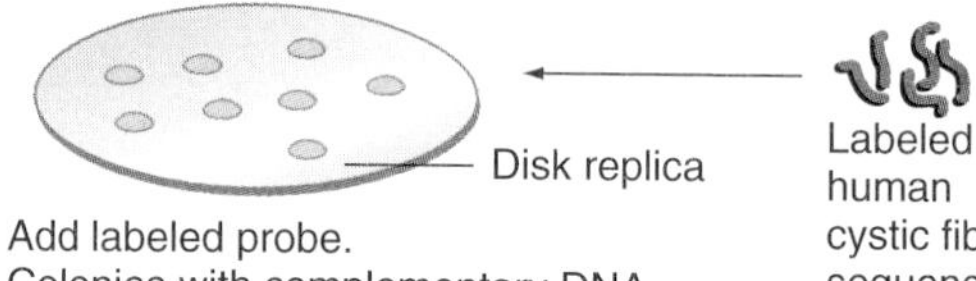

Disk replica

Labeled human cystic fibrosis sequences

Add labeled probe.
Colonies with complementary DNA sequences hybridize to probe and restrain it.

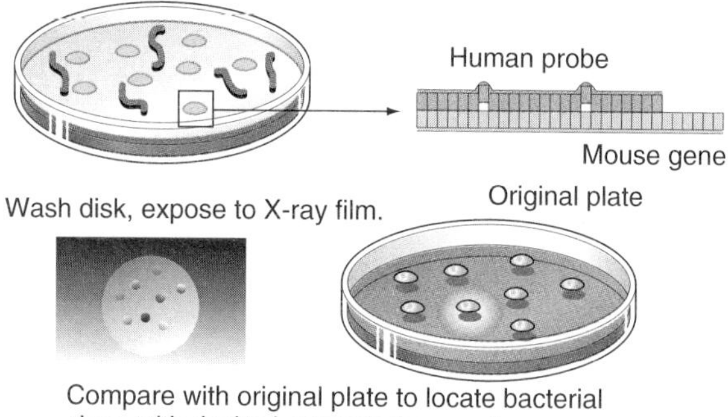

Human probe

Mouse gene

Wash disk, expose to X-ray film.

Original plate

Compare with original plate to locate bacterial clone with desired genomic fragment.

(a) Automated sequencing

1. Generate nested array of fragments; each with a fluorescent label corresponding to the terminating 3' base.

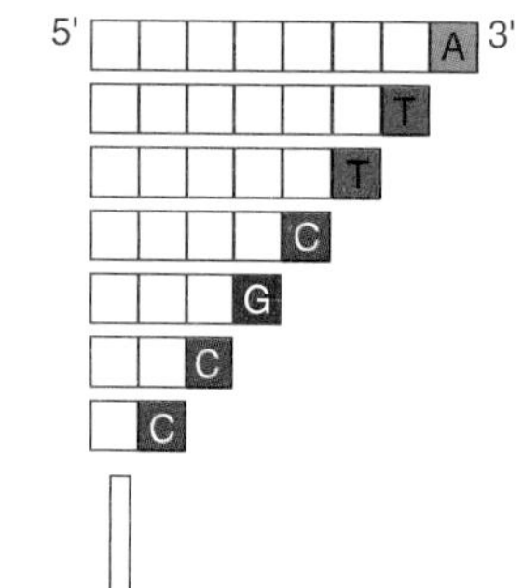

2. Fragments separated by electrophoresis in a single vertical gel lane.

3. As migrating fragments pass through the scanning laser, they fluoresce. A fluorescent detector records the color order of the passing bands. That order is translated into sequence data by a computer.

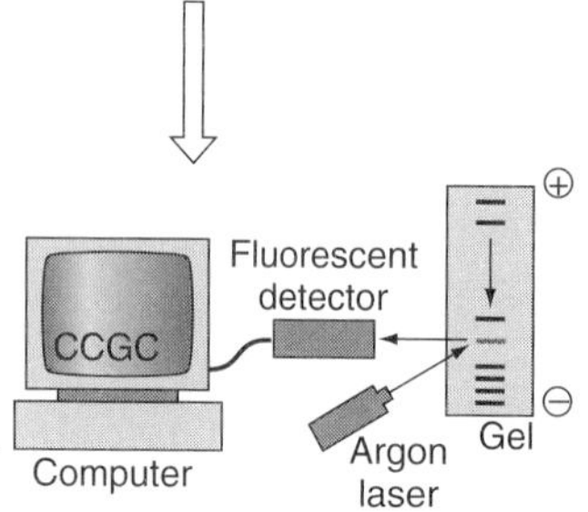

Computer

CCGC

Fluorescent detector

Argon laser

Gel

(c)

CTNGCTTTGGAGAAAGGCTCCATTGNCAATCAAGACACACAGAGGTGTCCTCTTTTTCCCCTGGTCAGCGNCCAGGTACATNGCACCAAGGCTGCGTAGTGAACTTGNCACCAGNCCATGGAC
CTatGCTTTGGAGAAAGGCTCCATTGgCAATCAAGACACACAGAGGTGTCCTCTTTTTCcCCTGGTCAGCGaCCAGGTACATgGCACCAAGGCTGCGTAGTGAACTTGcCACCAGcCCATGGAC

Automated sequencing
Figure 9.18

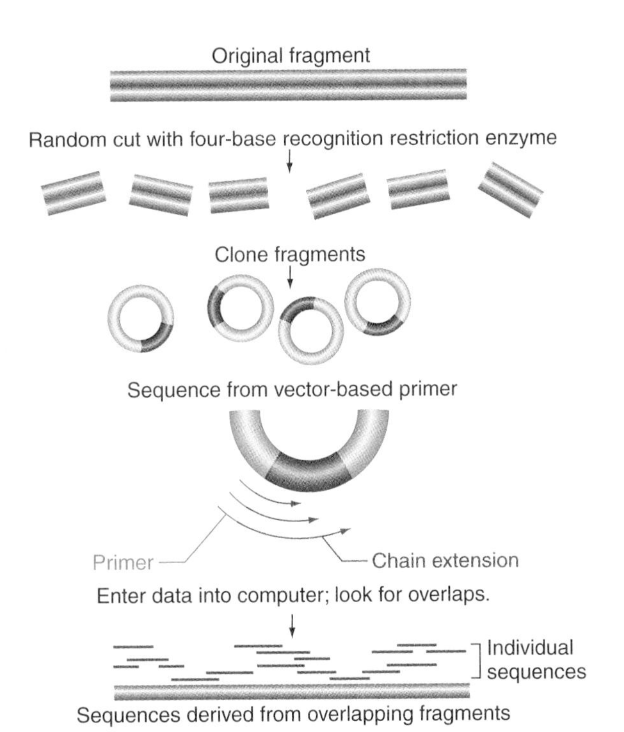

Sequencing a long DNA molecule
Figure 9.19

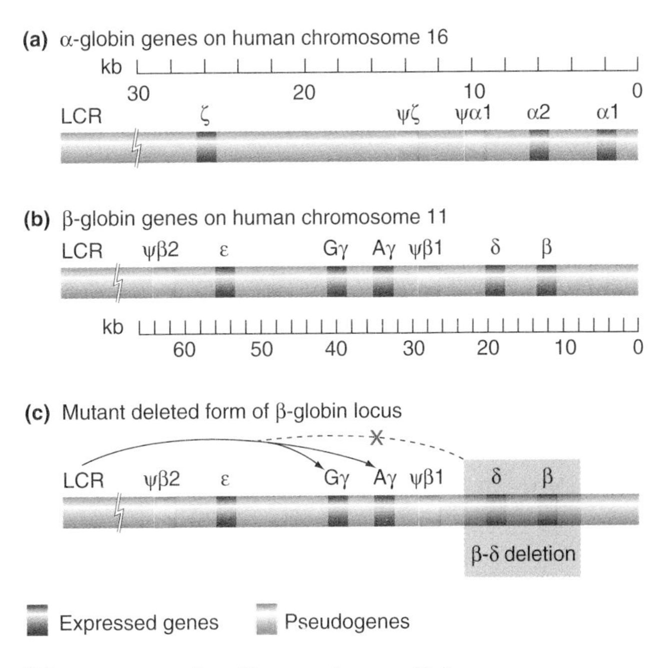

(a) α-globin genes on human chromosome 16

kb | | | | | | | | | | | | | | | |
30   20   10   0
LCR   ζ   ψζ   ψα1   α2   α1

(b) β-globin genes on human chromosome 11

LCR   ψβ2   ε   Gγ   Aγ   ψβ1   δ   β

kb | | | | | | | | | | | | | | | | | | | | | | | | | | | | | | | | | | |
60   50   40   30   20   10   0

(c) Mutant deleted form of β-globin locus

LCR   ψβ2   ε   Gγ   Aγ   ψβ1   δ   β

β-δ deletion

■ Expressed genes   ▨ Pseudogenes

The genes for the polypeptide components of human hemoglobin are located in two genomic clusters on two different chromosomes
Figure 9.20

(a.1) Major types of structural variants causing hemolytic anemias

Name	Molecular basis of mutation	Change in polypeptide	Pathophysiological effect of mutation	Inheritance
HbS	Single nucleotide substitution	β 6 Glu ↓ Val	Deoxygenated HbS polymerizes → sickle cells → vascular occlusion and hemolysis	Autosomal Recessive
HbC	Single nucleotide substitution	β 6 Glu ↓ Lys	Oxygenated HbC tends to crystallize → less deformable cells → mild hemolysis; the disease in HbS:HbC compounds is like mild sickle-cell anemia	Autosomal Recessive
Hb Hammer-smith	Single nucleotide substitution	β 42 Phe ↓ Ser	An unstable Hb → Hb precipitation → hemolysis; also low O_2 affinity	Autosomal Dominant

(a.2) Basis of sickle-cell anemia

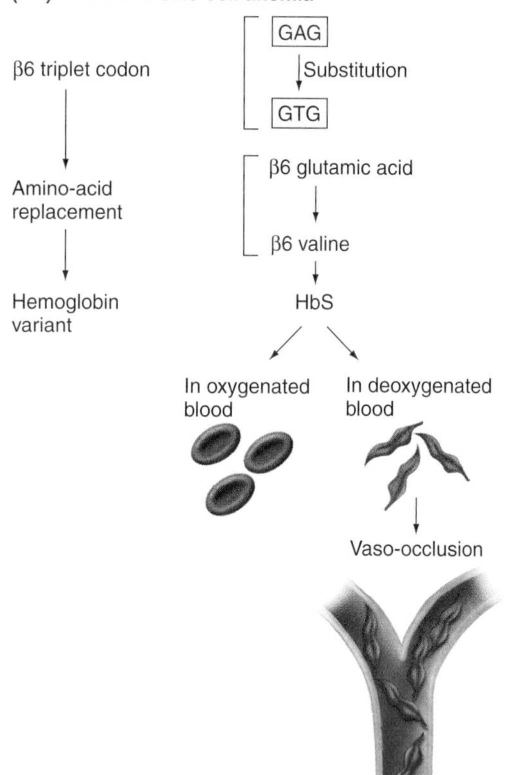

(b.1) Clinical results of various α-thalassemia genotypes

Clinical condition	Genotype		Number of functional α genes	α-chain production
Normal	ζ α2 α1	αα/αα	4	100%
Silent carrier		αα/α–	3	75%
Heterozygous α-thalassemia—mild anemia	or	α–/α– or αα/– –	2	50%
HbH (β₄) disease—moderately severe anemia		α–/– –	1	25%
Homozygous α-thalassemia—lethal		– –/– –	0	0%

(b.2) A β-thalassemia patient makes only α globin, not β globin.

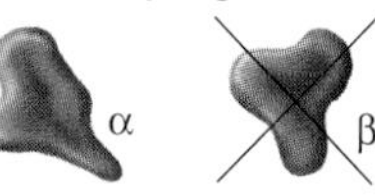

Four α subunits combine to make abnormal hemoglobin.

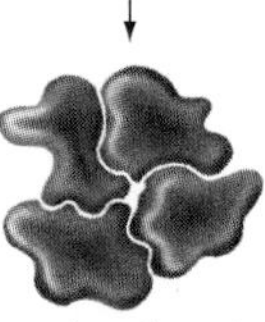

Abnormal hemoglobin molecules clump together, altering shape of red blood cells. Abnormal cells carry reduced amounts of oxygen.

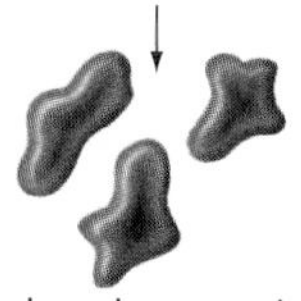

To compensate for reduced oxygen level, medullary cavities of bones enlarge to produce more red blood cells.

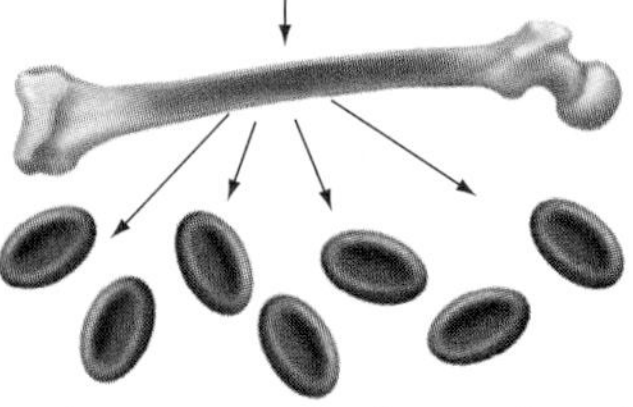

Spleen enlarges to remove excessive number of abnormal red blood cells.

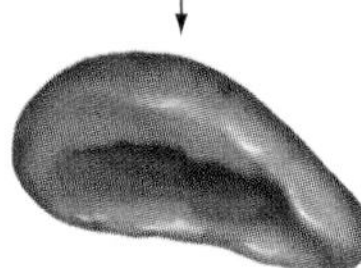

Too few red blood cells in circulation result in anemia.

Mutations in the DNA for hemoglobin produce two classes of disease
Figure 9.21

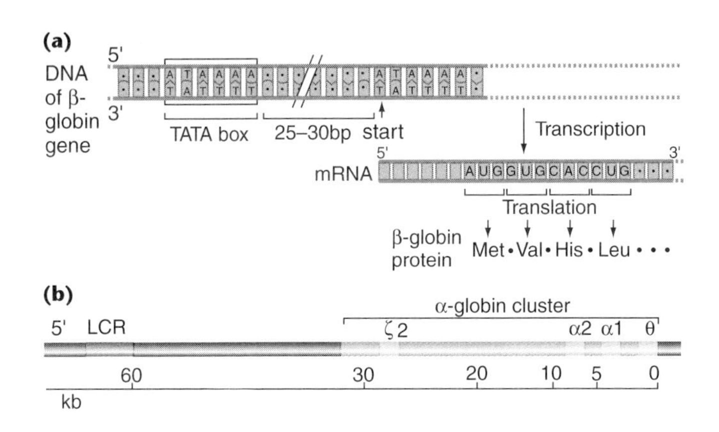

(a)

DNA of β-globin gene

TATA box   25–30bp start

Transcription

mRNA

Translation

β-globin protein   Met · Val · His · Leu · · ·

(b)

α-globin cluster

5' LCR   ζ2   α2 α1 θ

60   30   20   10 5   0

kb

Mutations in globin regulatory regions can affect globin gene activity and result in thalassemia
Figure 9.22

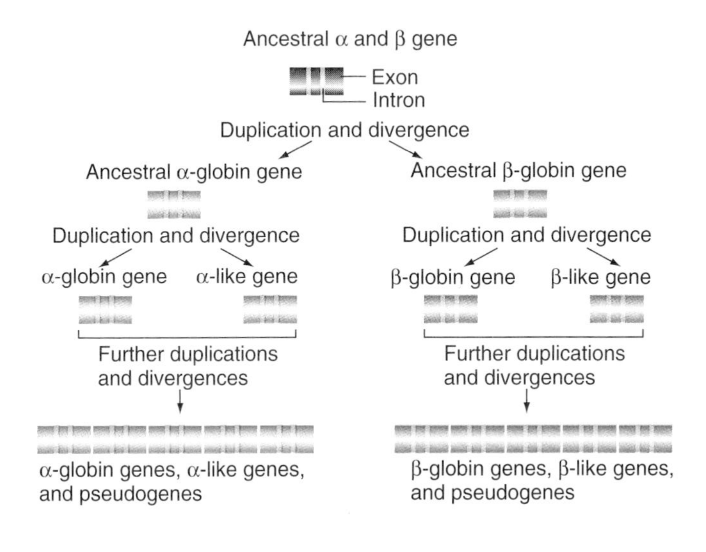

Ancestral α and β gene

Exon
Intron

Duplication and divergence

Ancestral α-globin gene   Ancestral β-globin gene

Duplication and divergence   Duplication and divergence

α-globin gene   α-like gene   β-globin gene   β-like gene

Further duplications   Further duplications
and divergences   and divergences

α-globin genes, α-like genes,   β-globin genes, β-like genes,
and pseudogenes   and pseudogenes

Evolution of the globin gene family
Figure 9.23

Single nucleotide polymorphism (SNP)   ...GCAA T TCCCGATT...
                                        ...GCAA G TCCCGATT...

Simple sequence repeat (SSR)   ...GCATTATATATATATC...
                                        ...GCATTATAT[   ]C...

Two common types of polymorphisms employed for genetic mapping
Figure 10.3

(a) PCR amplification

PCR primer 1

Individual 1   GCAGCAAT(AT)$_{25}$ GGTAAAAC
Individual 2   Unknown
Individual 3   GCAGCAAT(AT)$_{5}$ GGTAAAAC

                                      PCR primer 2

(b) Size separation

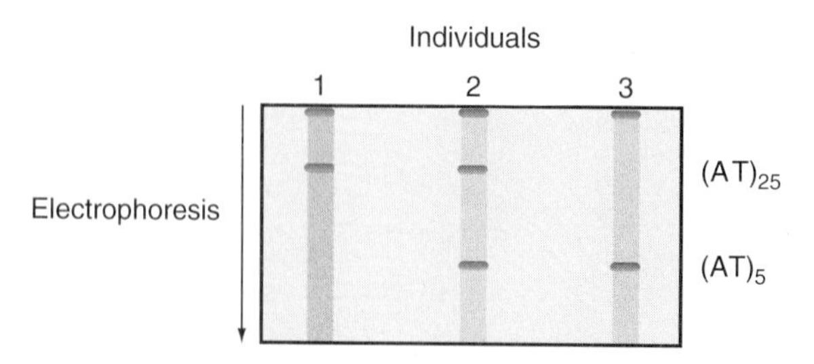

Individuals

            1   2   3

Electrophoresis   (AT)$_{25}$

                                             (AT)$_{5}$

The two-stage assay for simple sequence repeats (SSRs)
Figure 10.4

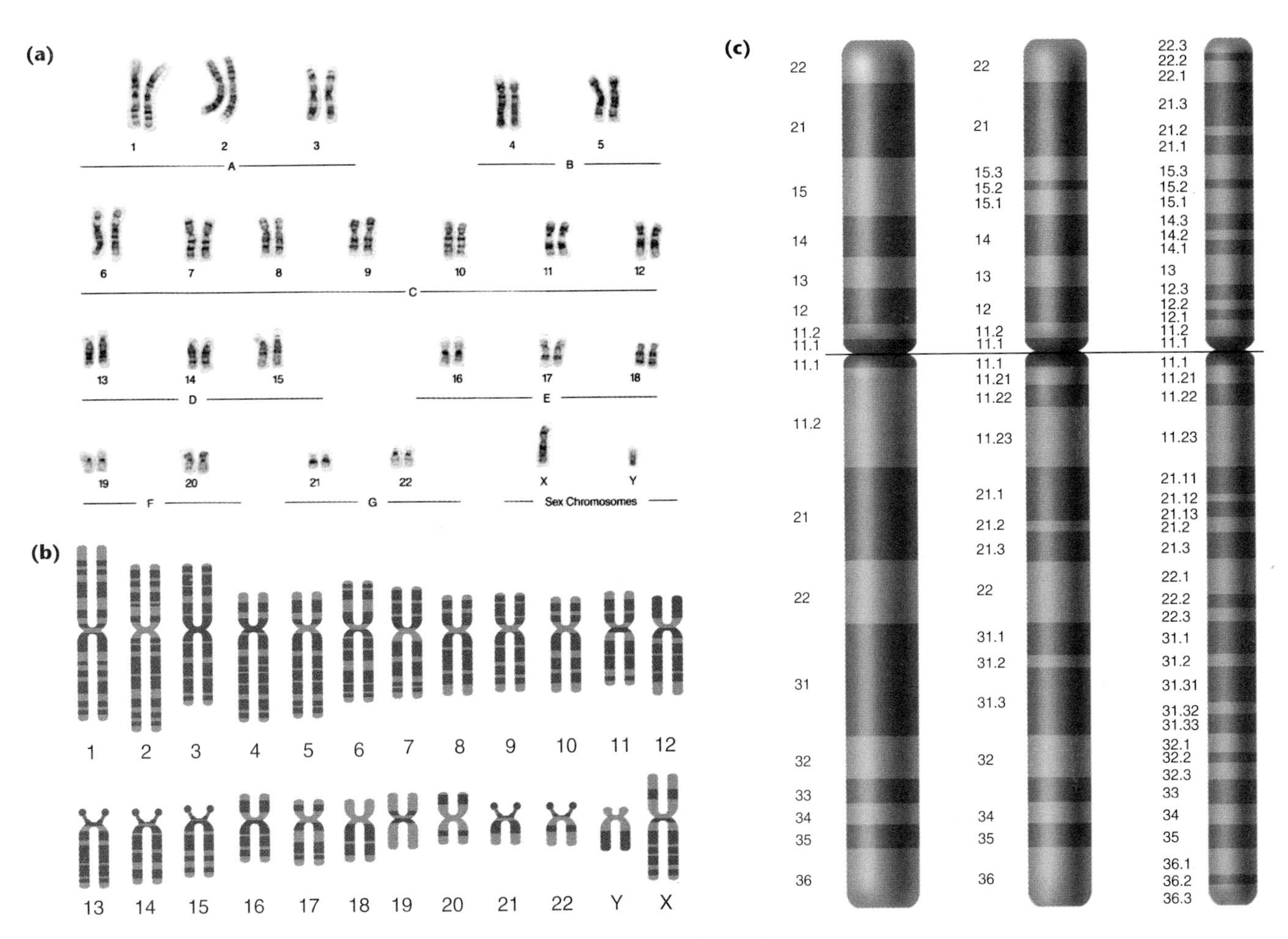

The human karyotype: Banding distinguishes the chromosomes
Figure 10.5
© Lee Silver, Princeton University

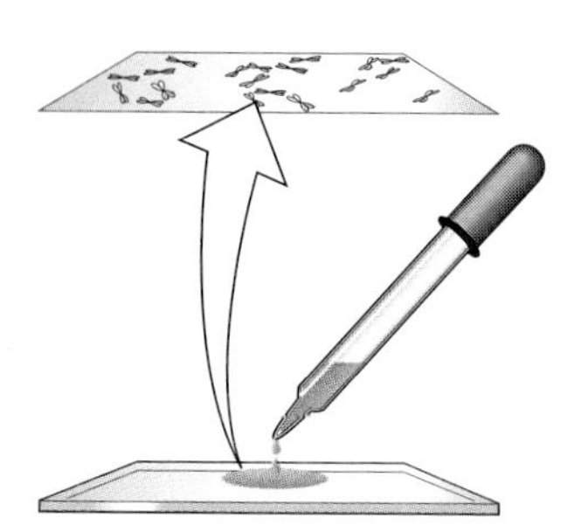

1. Drop cells onto a glass slide.

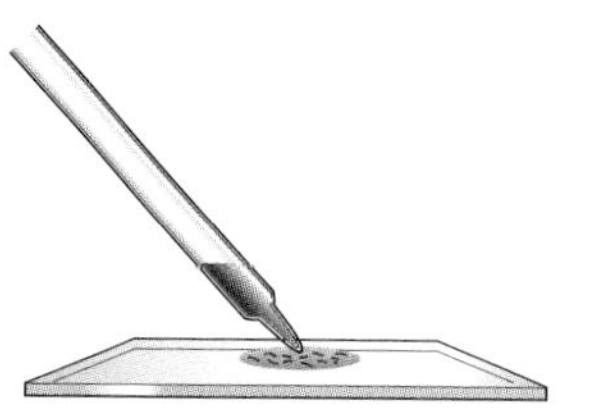

2. Gently denature DNA by treating briefly with DNase.

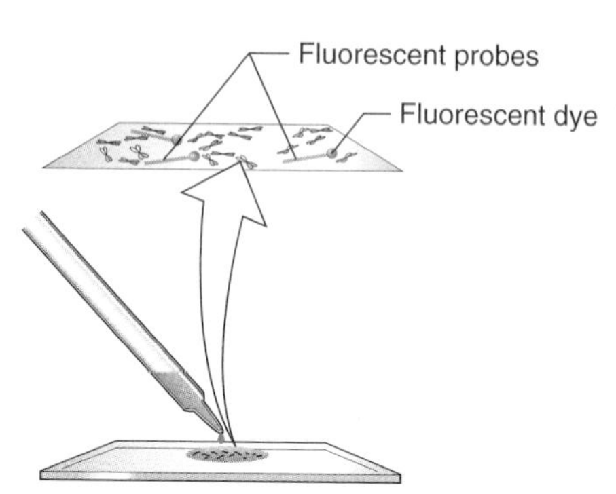

Fluorescent probes

Fluorescent dye

3. Add hybridization probes labeled with fluorescent dye and wash away unhybridized probe.

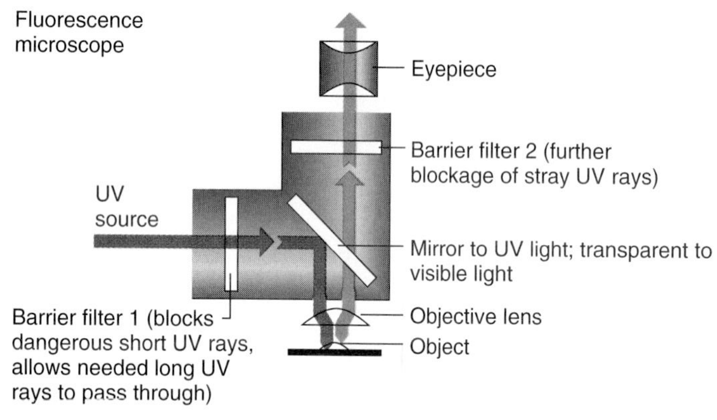

Fluorescence microscope

Eyepiece

Barrier filter 2 (further blockage of stray UV rays)

UV source

Mirror to UV light; transparent to visible light

Barrier filter 1 (blocks dangerous short UV rays, allows needed long UV rays to pass through)

Objective lens

Object

4. Expose to ultraviolet (UV) light.
   Take picture of fluorescent chromosomes.

The FISH protocol
Figure 10.6

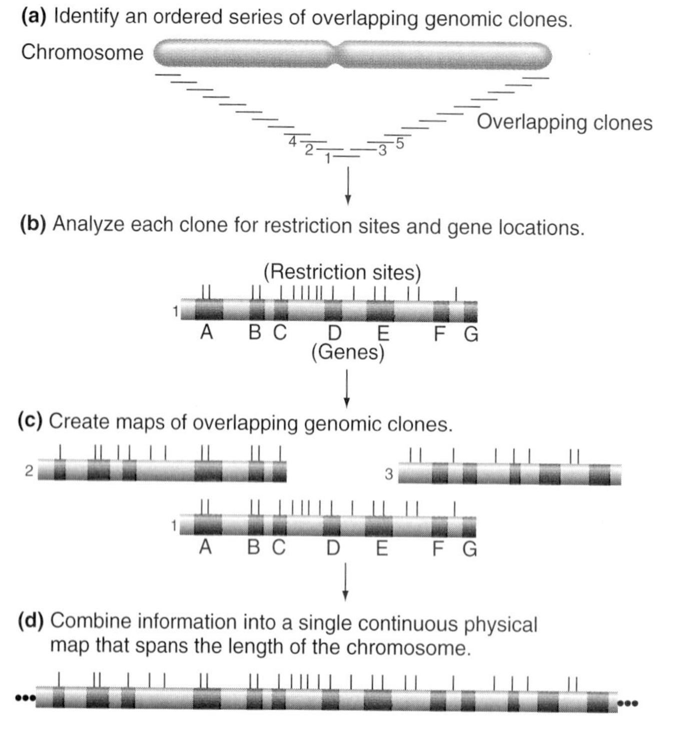

(a) Identify an ordered series of overlapping genomic clones.

Chromosome

Overlapping clones

(b) Analyze each clone for restriction sites and gene locations.

(Restriction sites)

A B C D E F G
(Genes)

(c) Create maps of overlapping genomic clones.

A B C D E F G

(d) Combine information into a single continuous physical map that spans the length of the chromosome.

Building a whole chromosome physical map
Figure 10.7

Assigning chromosomal positions: A top-down approach

Use M1 as hybridization probe to screen YAC library.

Use M2 as hybridization probe to screen YAC library.

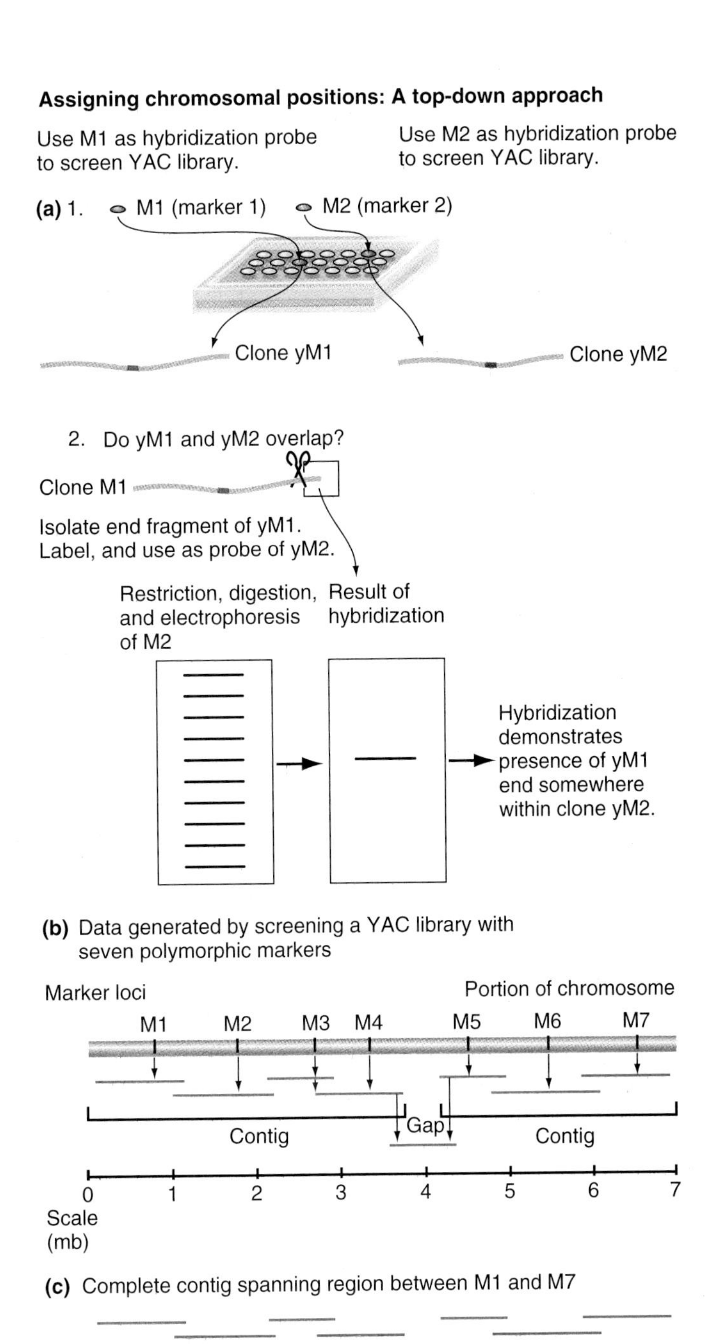

(a) 1.

M1 (marker 1)   M2 (marker 2)

Clone yM1

Clone yM2

2. Do yM1 and yM2 overlap?

Clone M1

Isolate end fragment of yM1.
Label, and use as probe of yM2.

Restriction, digestion, and electrophoresis of M2

Result of hybridization

Hybridization demonstrates presence of yM1 end somewhere within clone yM2.

(b) Data generated by screening a YAC library with seven polymorphic markers

Marker loci

Portion of chromosome

M1   M2   M3   M4   M5   M6   M7

Contig

Gap

Contig

Scale (mb)

0   1   2   3   4   5   6   7

(c) Complete contig spanning region between M1 and M7

M1   M2   M3 M4   M5   M6   M7

Using a high-density linkage map to build an overlapping set of genomic clones
Figure 10.8

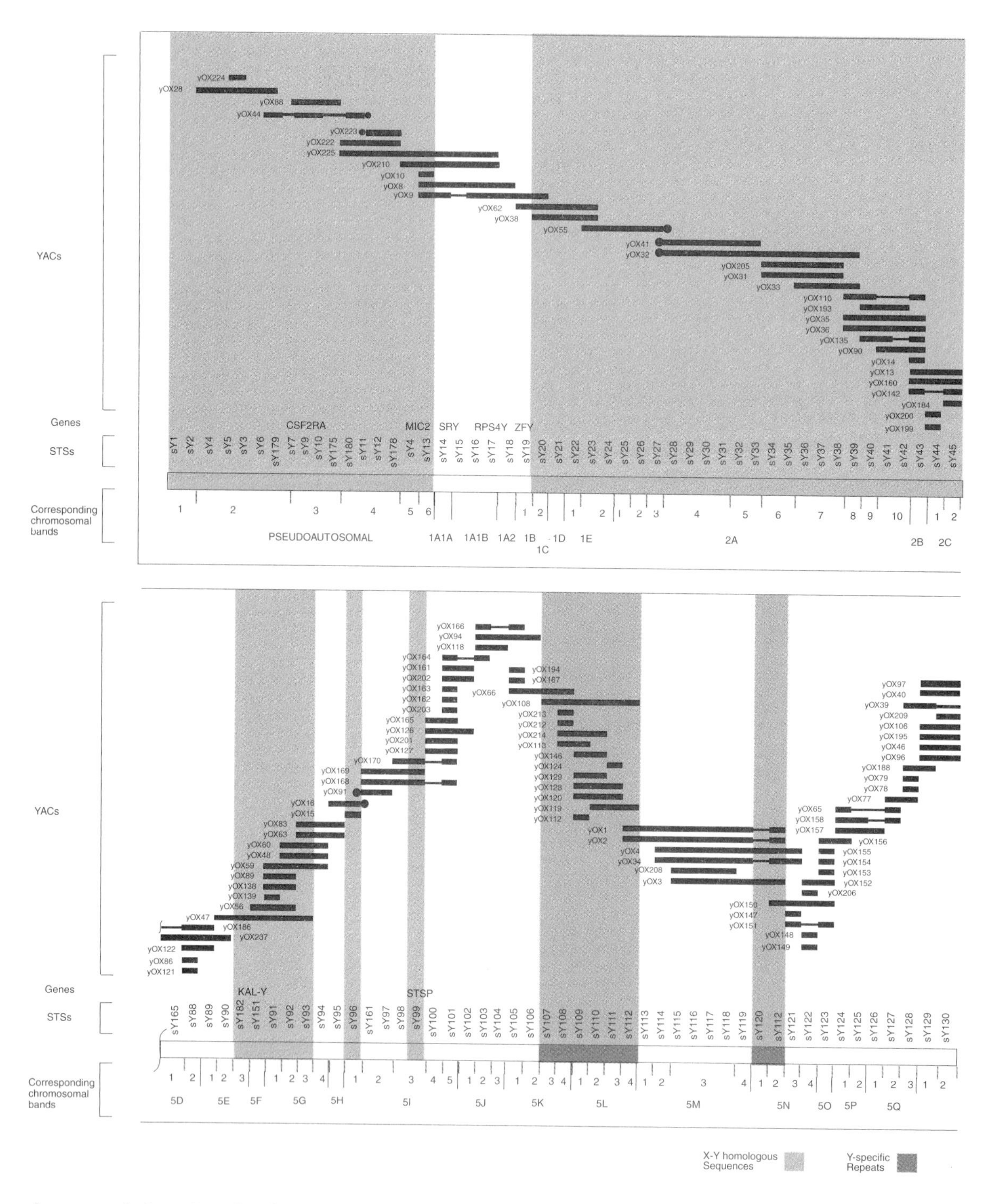

A complete physical map of the human Y chromosome
Figure 10.9

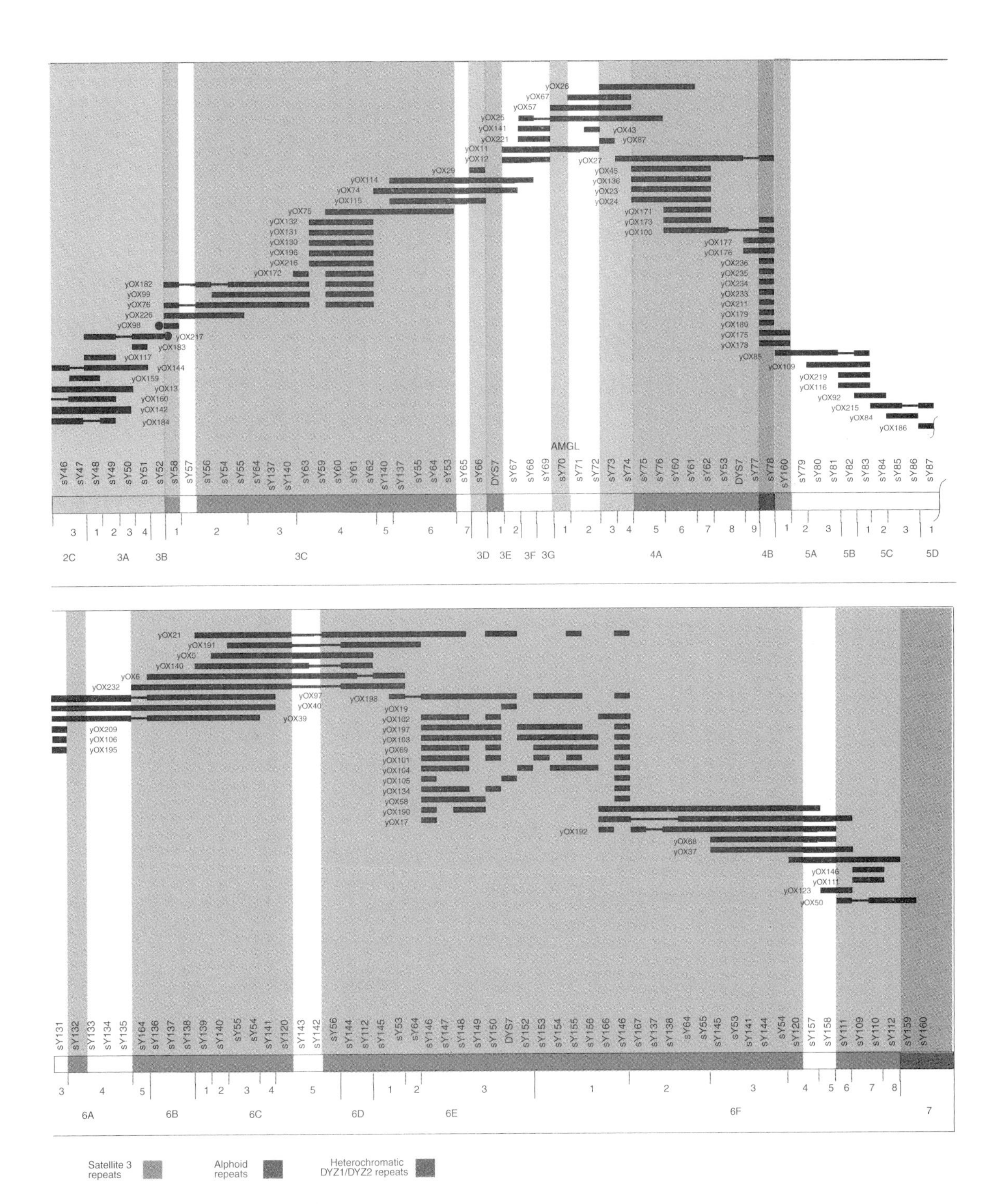

Figure 10.9 (Continued)

Bottom-up approach

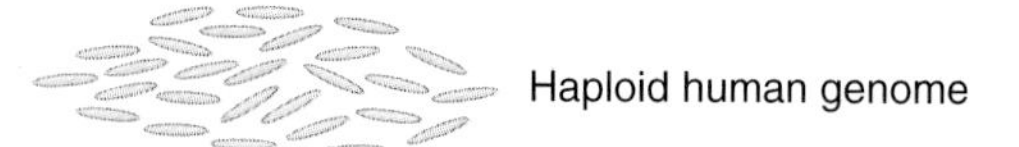

Haploid human genome

(a) Digest chromosomes with restriction enzyme.
Clone fragments into a genomic cosmid library.

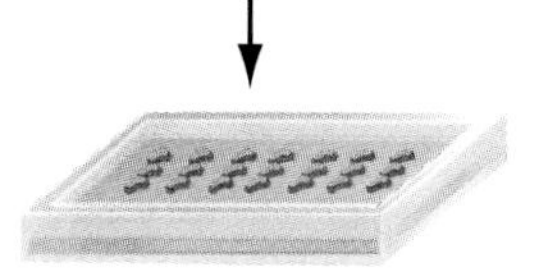

(b) Fingerprint each cosmid for the pattern of restriction
fragment sites.

↓ – Restriction sites

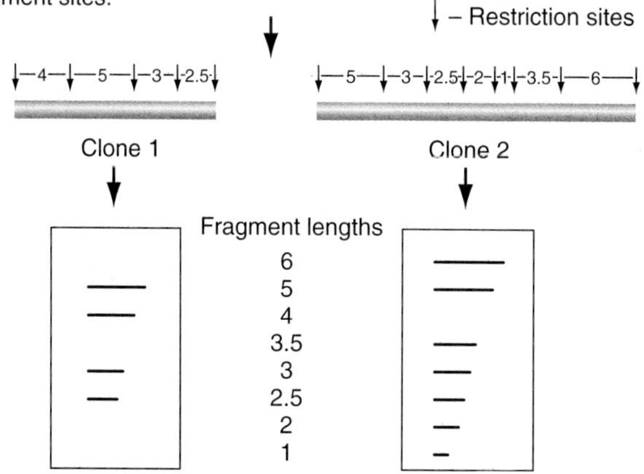

(c) Feed fingerprint data into computer programmed to look
for overlaps.

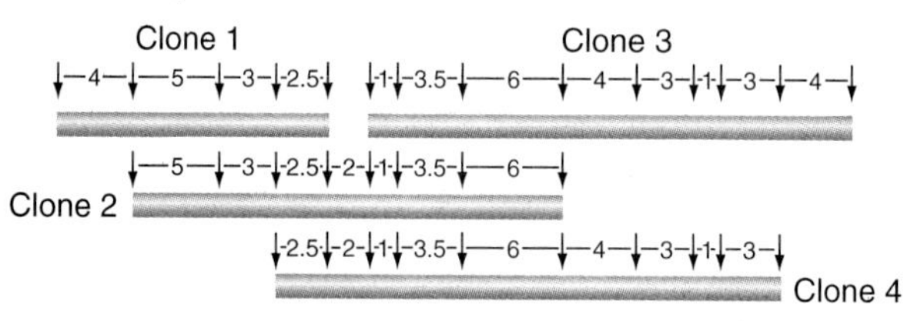

(d) Arrange overlapping cosmid inserts into 24 chromosomal contigs.

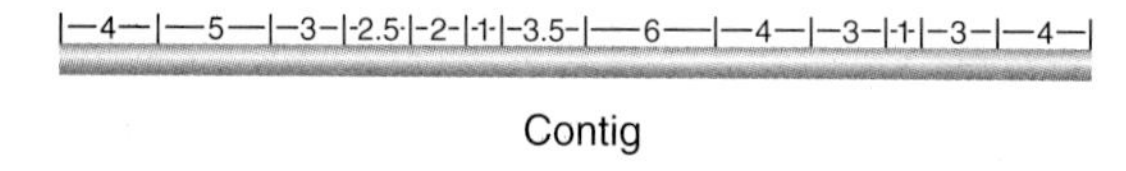

Contig

Building a whole chromosome physical map of overlapping genomic clones without linkage information
Figure 10.10

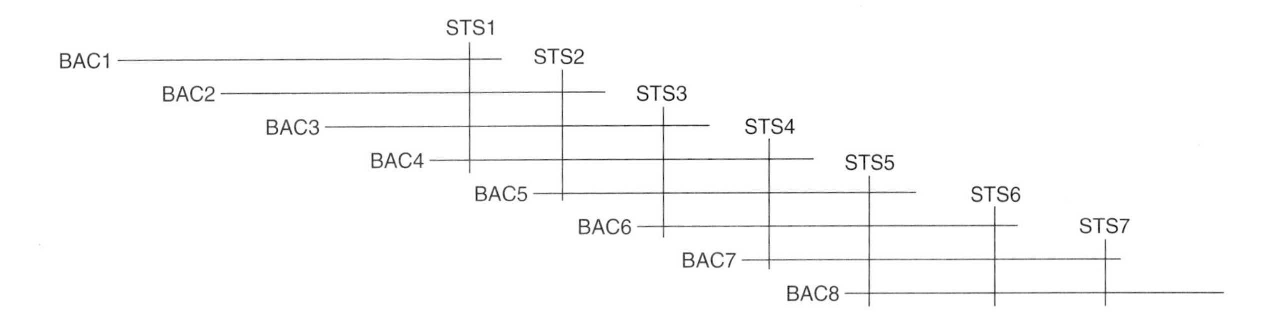

A hypothetical physical map generated by the analysis of sequence tagged sites (STS)
Figure 10.11

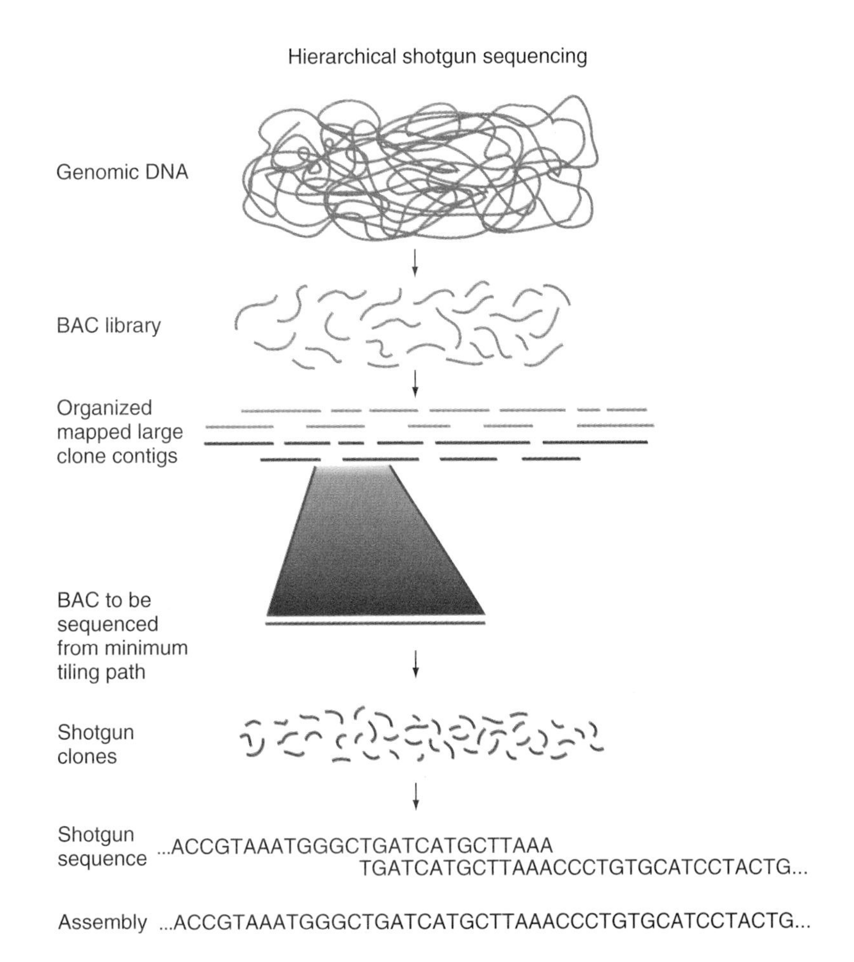

Hierarchical shotgun sequencing

Genomic DNA

BAC library

Organized
mapped large
clone contigs

BAC to be
sequenced
from minimum
tiling path

Shotgun
clones

Shotgun   ...ACCGTAAATGGGCTGATCATGCTTAAA
sequence   TGATCATGCTTAAACCCTGTGCATCCTACTG...

Assembly   ...ACCGTAAATGGGCTGATCATGCTTAAACCCTGTGCATCCTACTG...

Idealized representation of the hierarchical shotgun sequencing strategy
Figure 10.12

Hypothetical whole-genome shotgun sequencing strategy
Figure 10.13

Chromosomes

1 2 ⋯⋯ 21 22 X Y | Whole genome

| 2 kb plasmid library | 10 kb plasmid library | 200 kb BAC library | Libraries |

| Sequence ends to a 6-fold genome coverage | Sequence ends to a 3-fold genome coverage | Sequence ends to a 1-fold genome coverage | End sequences |

$(15 \times 10^6 \text{ plasmids})$ $(7.5 \times 10^6 \text{ plasmids})$ $(2.5 \times 10^6 \text{ BACs})$

Assemble sequences into chromosomal strings.

(a)

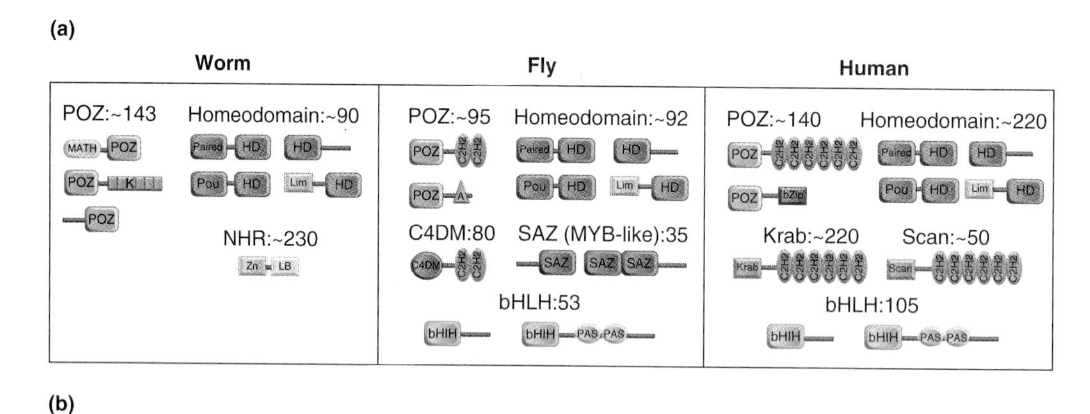

(b)

Unique and shared domain organizations in animals

How the domains and architectures of transcription factors have expanded in specific lineages
Figure 10.14

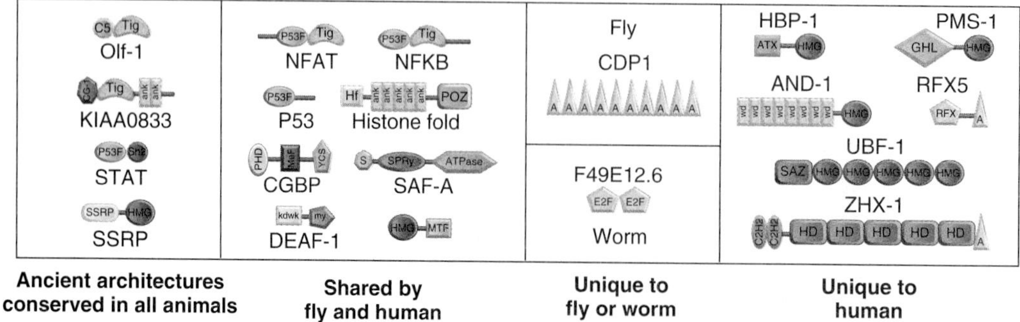

Ancient architectures conserved in all animals | Shared by fly and human | Unique to fly or worm | Unique to human

Conserved segments or syntenic blocks in the human and mouse genomes
Figure 10.15

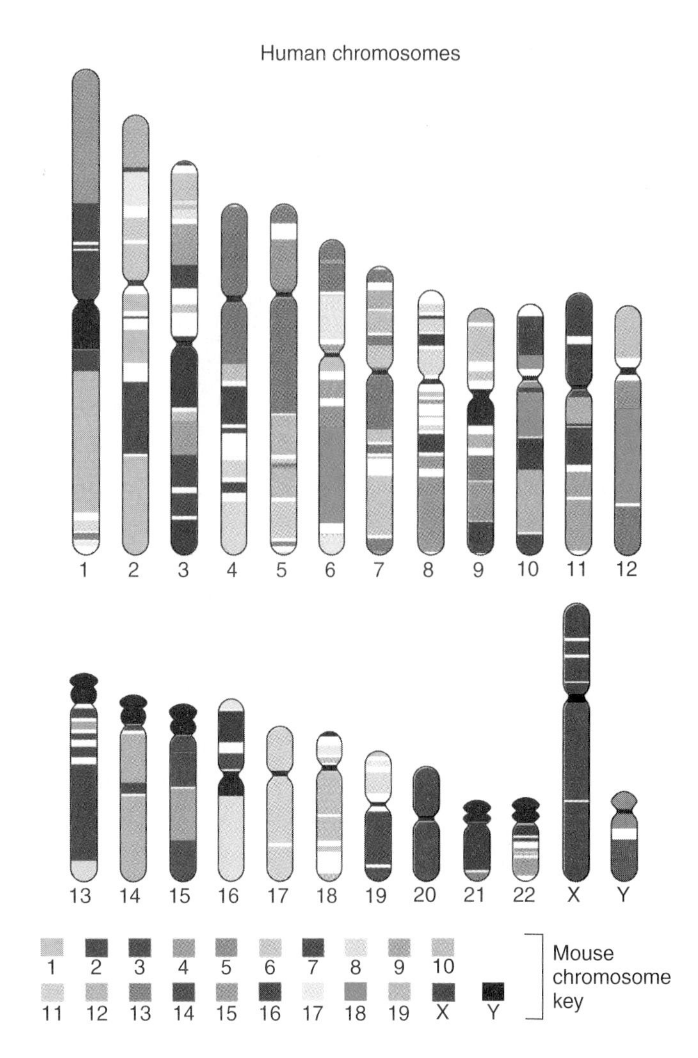

Human chromosomes

Mouse chromosome key

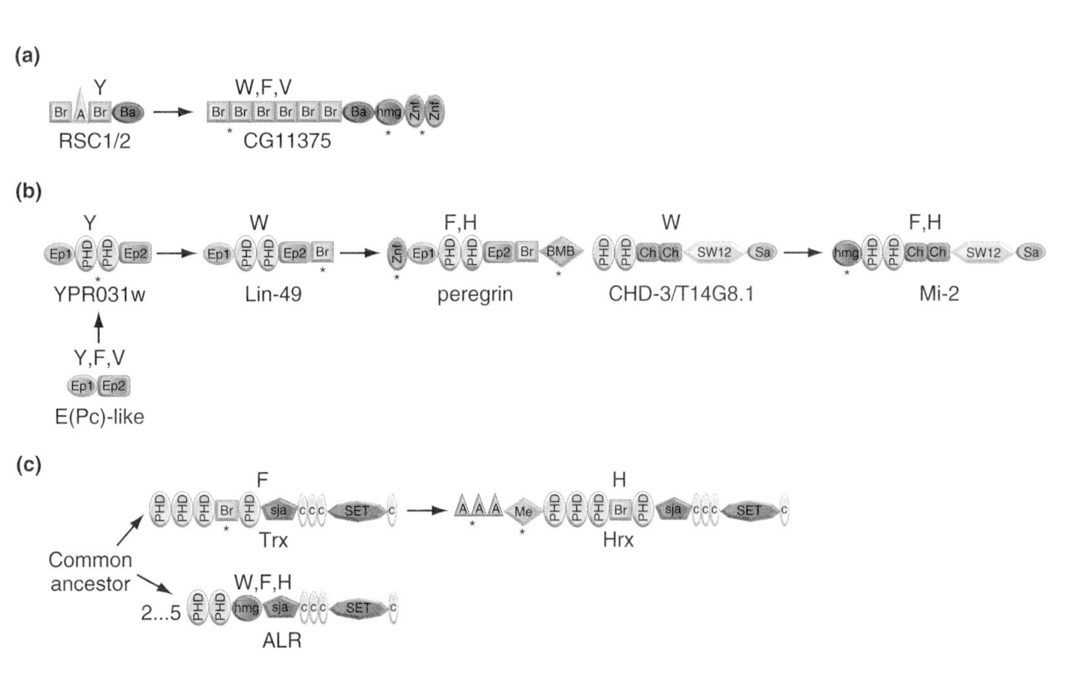

Examples of domain accretion in chromatin proteins
Figure 10.16

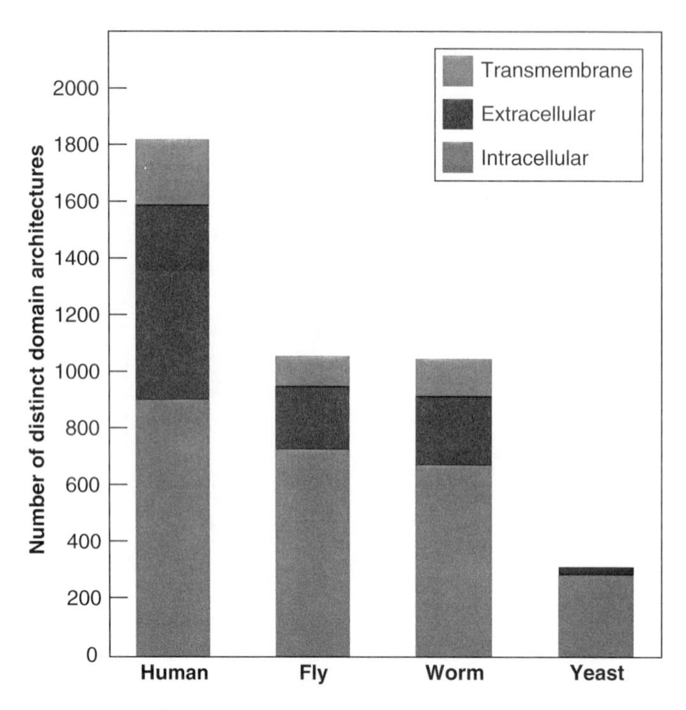

Number of distinct domain architectures in the four eukaryotic genomes
Figure 10.17

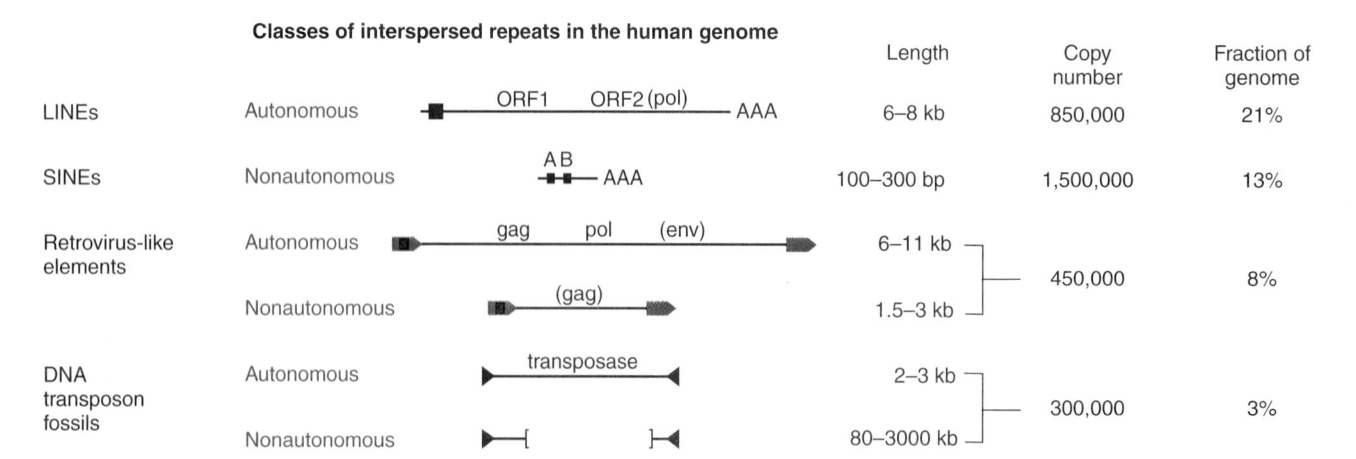

Classes of interspersed repeats in the human genome

			Length	Copy number	Fraction of genome
LINEs	Autonomous	ORF1 ORF2 (pol) AAA	6–8 kb	850,000	21%
SINEs	Nonautonomous	A B AAA	100–300 bp	1,500,000	13%
Retrovirus-like elements	Autonomous	gag pol (env)	6–11 kb	450,000	8%
	Nonautonomous	(gag)	1.5–3 kb		
DNA transposon fossils	Autonomous	transposase	2–3 kb	300,000	3%
	Nonautonomous		80–3000 kb		

Almost all transposable elements in mammals fall into one of four classes of interspersed repeats (as described in the text)
Figure 10.18

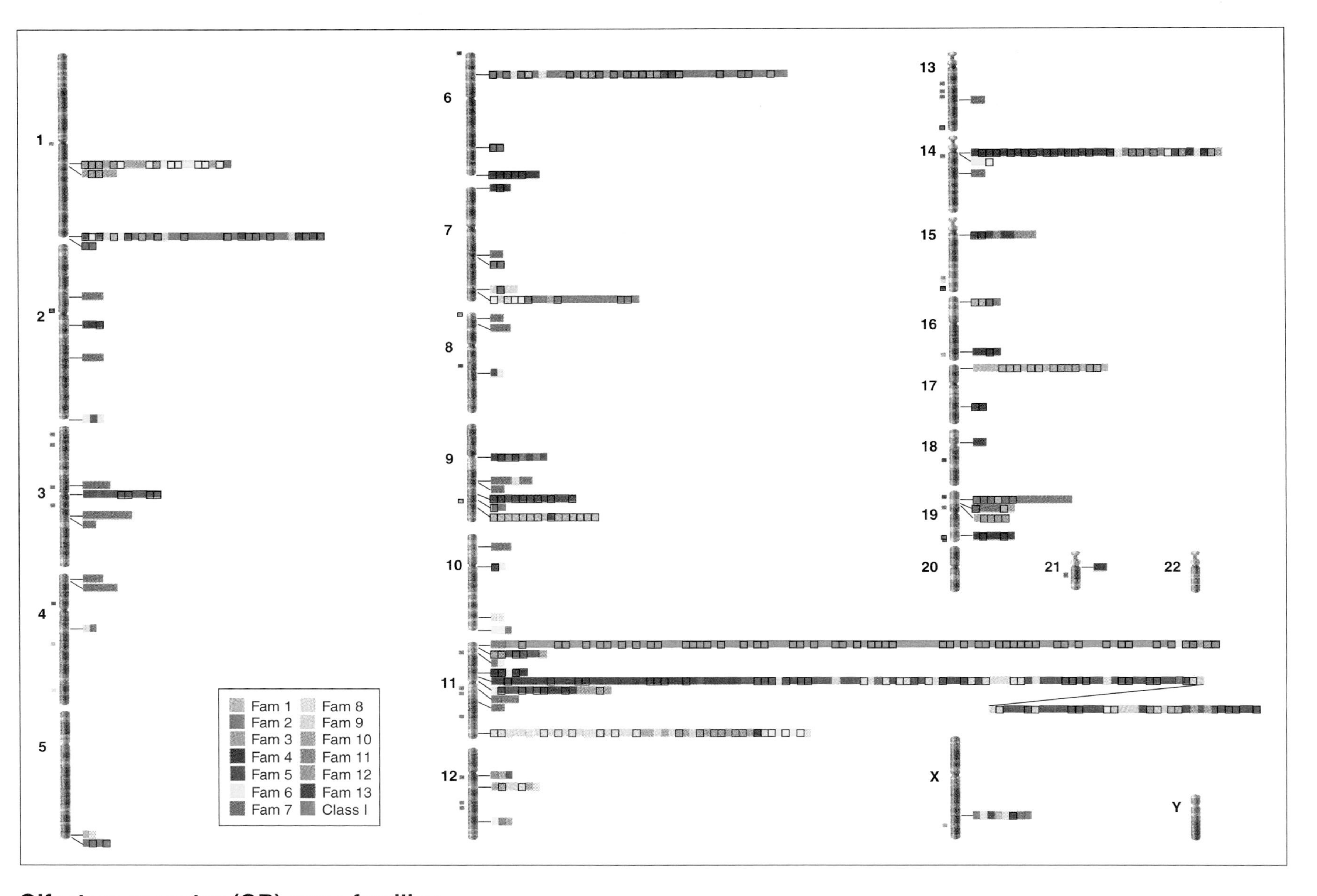

Olfactory receptor (OR) gene families
Figure 10.19

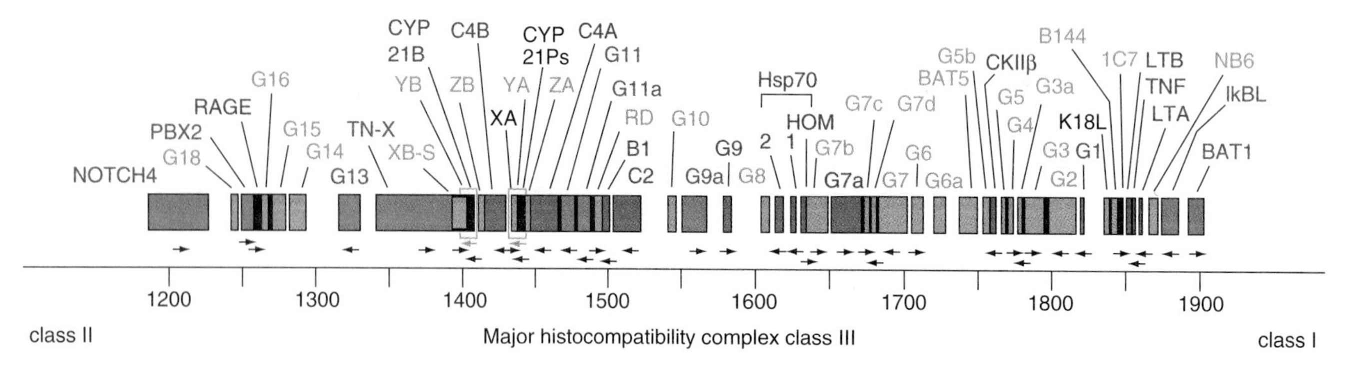

Class III region of the human major histocompatibility complex
Figure 10.20

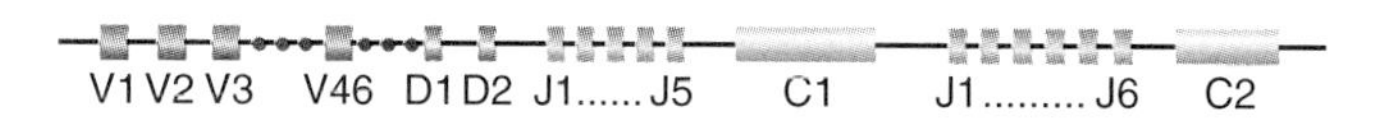

A schematic diagram of the human β T-cell receptor gene family
Figure 10.21

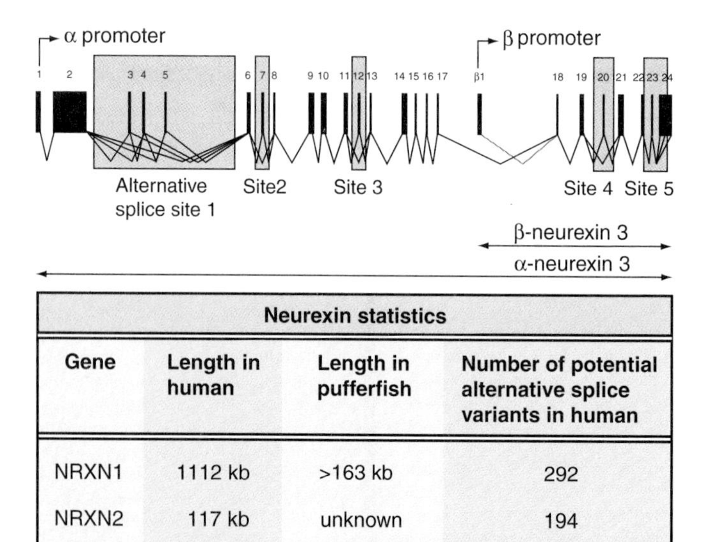

Neurexin statistics			
Gene	Length in human	Length in pufferfish	Number of potential alternative splice variants in human
NRXN1	1112 kb	>163 kb	292
NRXN2	117 kb	unknown	194
NRXN3	1692 kb	>181 kb	1764

The organization of the three neurexin genes
Figure 10.22

(a)

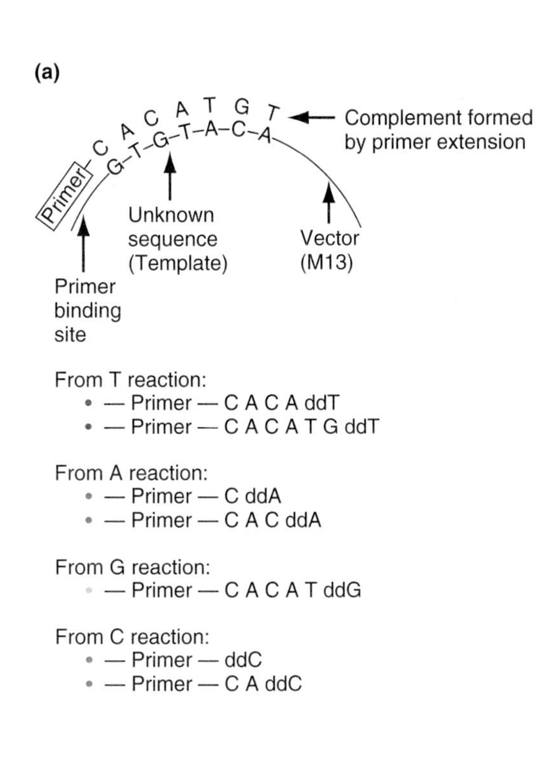

Complement formed by primer extension

Primer binding site

Unknown sequence (Template)

Vector (M13)

From T reaction:
- — Primer — C A C A ddT
- — Primer — C A C A T G ddT

From A reaction:
- — Primer — C ddA
- — Primer — C A C ddA

From G reaction:
- — Primer — C A C A T ddG

From C reaction:
- — Primer — ddC
- — Primer — C A ddC

(b)

A G C T

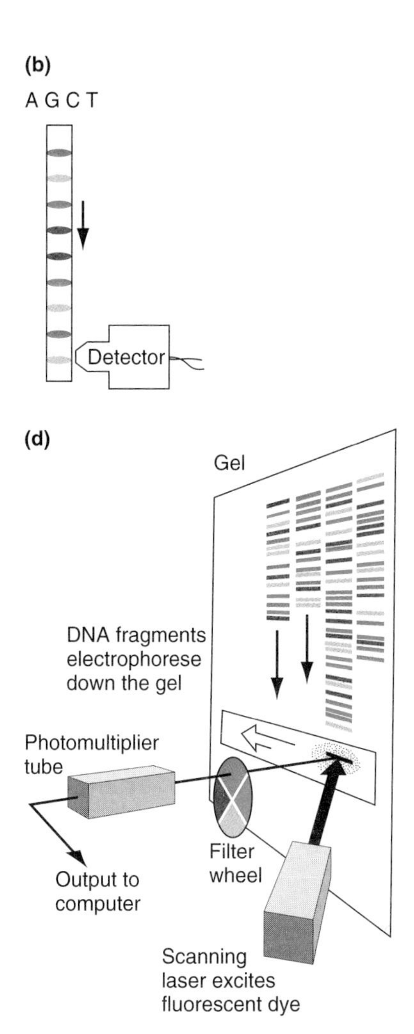

(c)

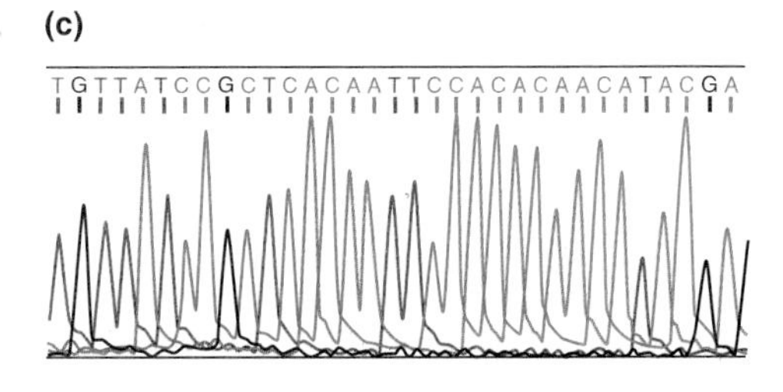

Sanger sequencing scheme
Figure 10.23

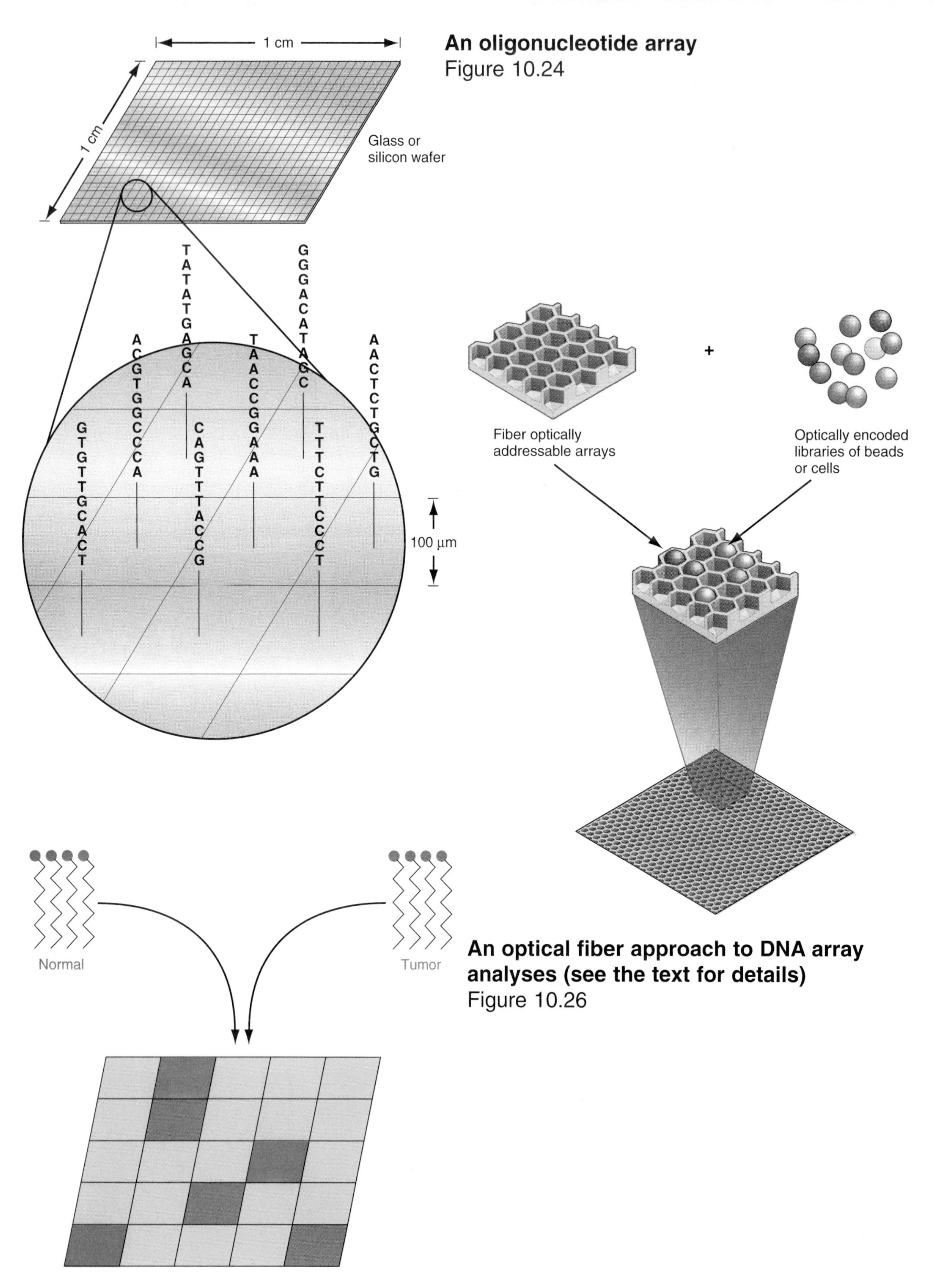

An oligonucleotide array
Figure 10.24

1 cm

1 cm

Glass or silicon wafer

T A T A T G A G C A
A C G T G G C C C A
G T G T T G C A C T

G G G A C A T A G C
T A A C C G G A A A
C A G T T A C C G

A A C T C T G C T G
T T T C T T C C C T

100 μm

Fiber optically addressable arrays

+

Optically encoded libraries of beads or cells

An optical fiber approach to DNA array analyses (see the text for details)
Figure 10.26

Normal

Tumor

Two-color DNA microarrays
Figure 10.25

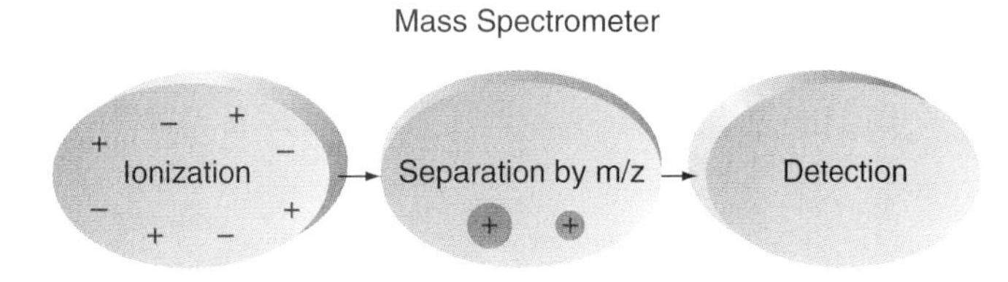

Mass spectrometer
Figure 10.27

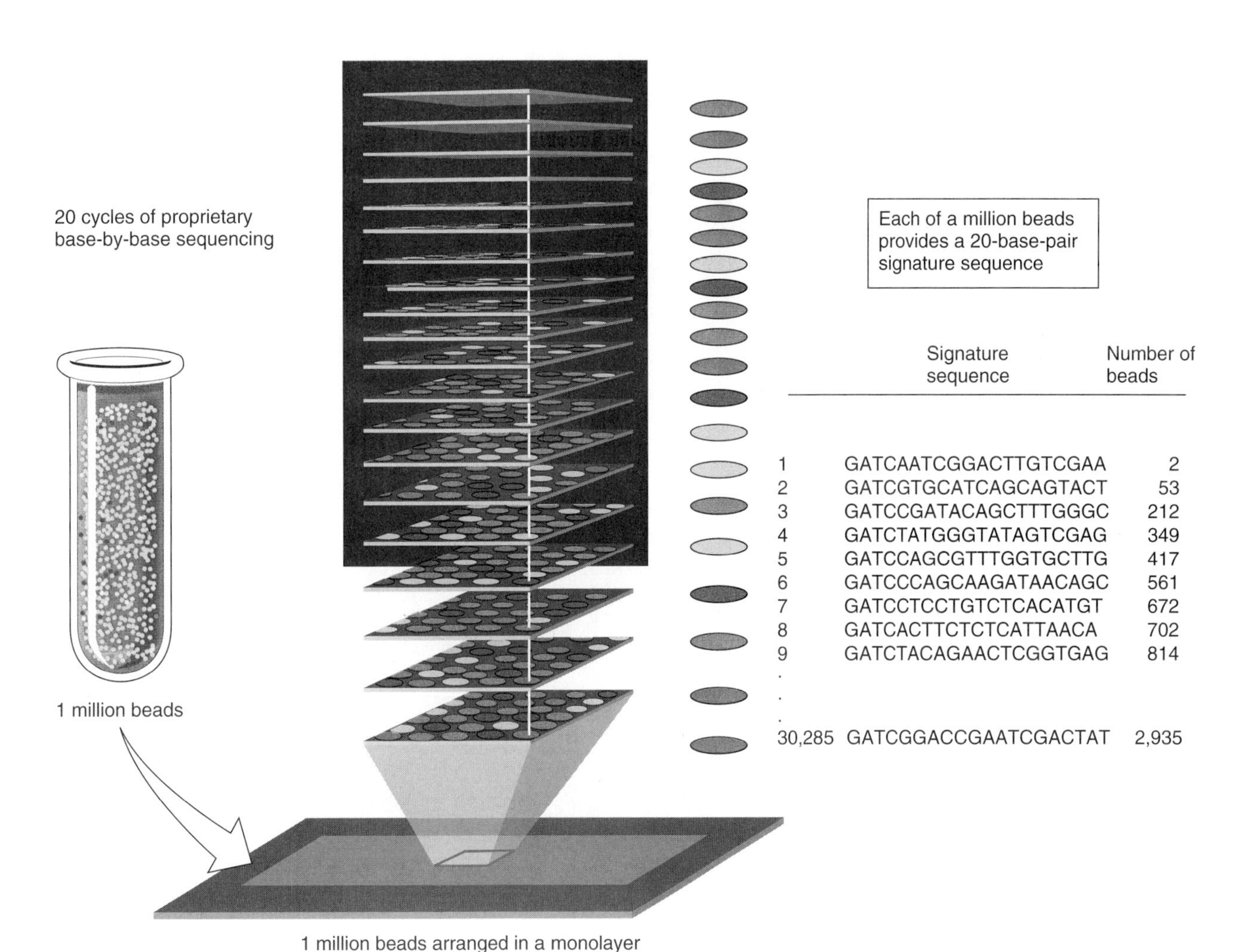

20 cycles of proprietary
base-by-base sequencing

Each of a million beads
provides a 20-base-pair
signature sequence

	Signature sequence	Number of beads
1	GATCAATCGGACTTGTCGAA	2
2	GATCGTGCATCAGCAGTACT	53
3	GATCCGATACAGCTTTGGGC	212
4	GATCTATGGGTATAGTCGAG	349
5	GATCCAGCGTTTGGTGCTTG	417
6	GATCCCAGCAAGATAACAGC	561
7	GATCCTCCTGTCTCACATGT	672
8	GATCACTTCTCTCATTAACA	702
9	GATCTACAGAACTCGGTGAG	814
.		
.		
.		
30,285	GATCGGACCGAATCGACTAT	2,935

1 million beads

1 million beads arranged in a monolayer

Lynx Therapeutics sequencing strategy of multiple parallel signature sequencing (MPSS)
Figure 10.28

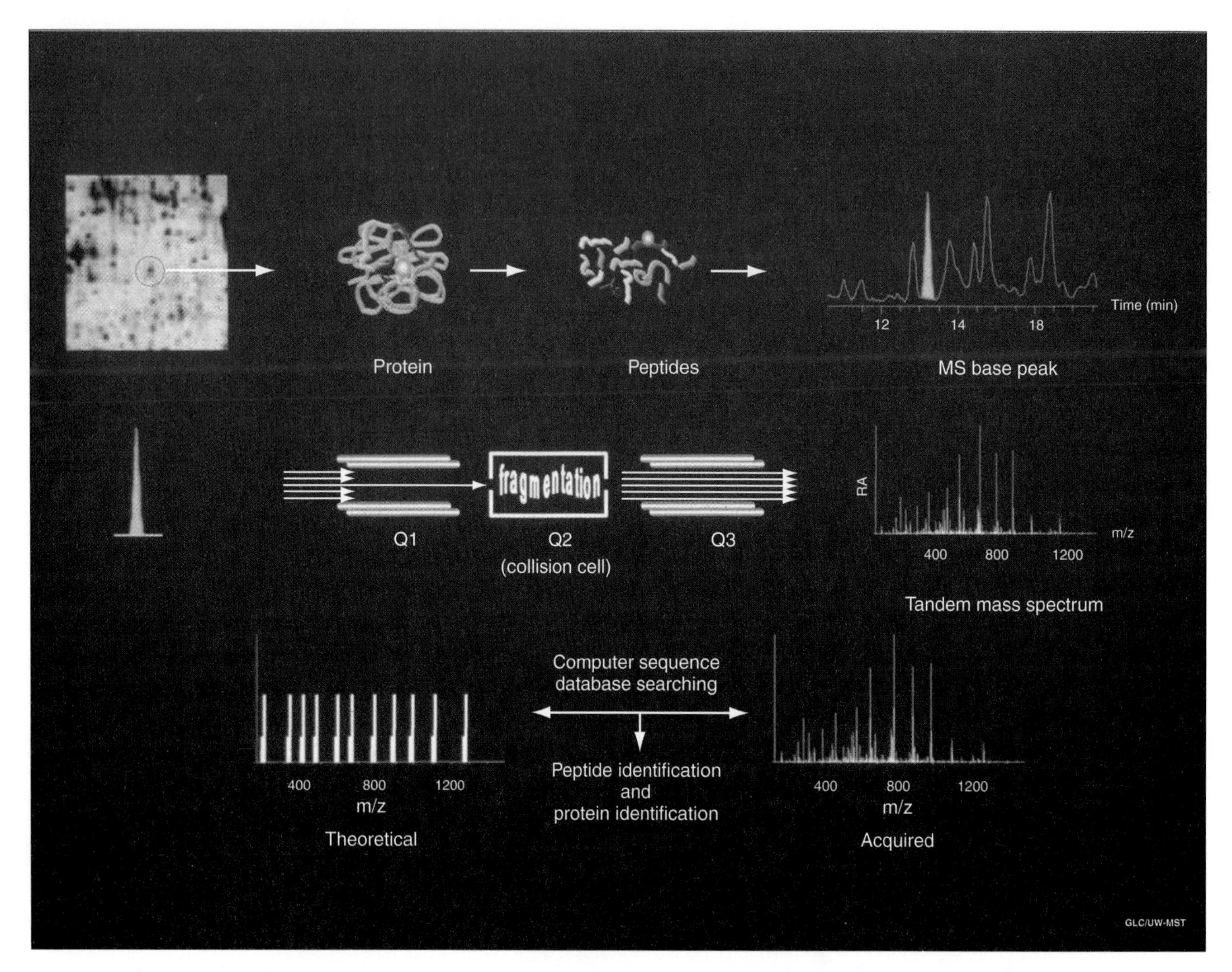

A strategy using two-dimensional gels and mass spectrometry to identify the genes encoding proteins in complex mixtures

Figure 10.29

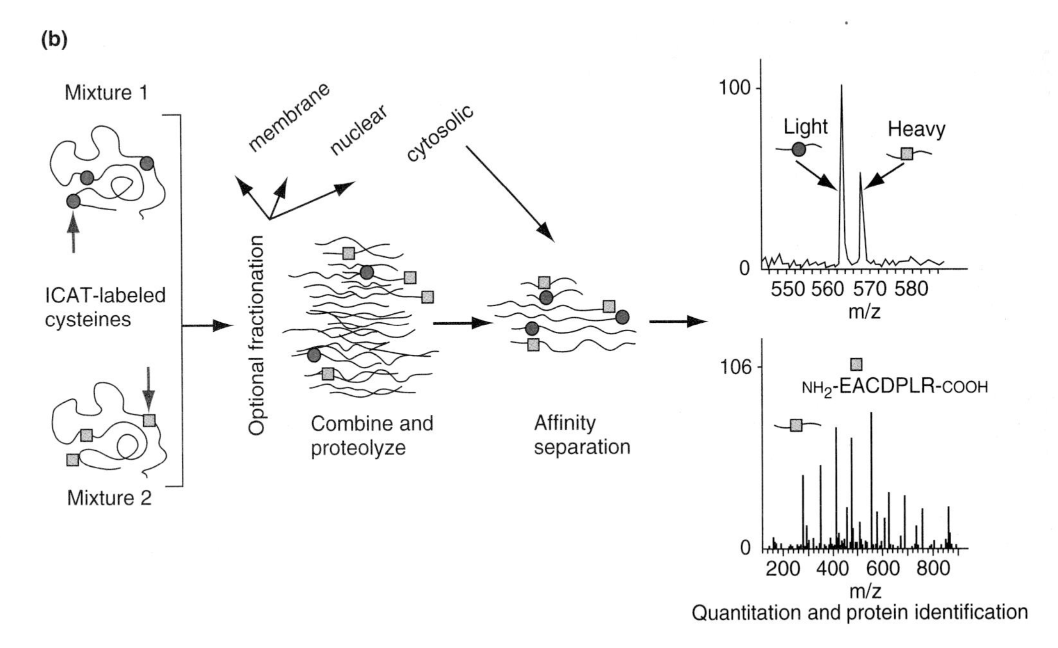

(a)

ICAT reagents: Heavy reagent: d8-ICAT (X=deuterium)
                 Light reagent: d0-ICAT (X=hydrogen)

Biotin tag   Linker
              (heavy or light)   Thiol
                                  specific
                                  reactive
                                  group

(b)

Mixture 1

membrane   nuclear   cytosolic

ICAT-labeled
cysteines

Optional fractionation

Combine and
proteolyze

Affinity
separation

Mixture 2

Light   Heavy

550 560 570 580
m/z

NH_2-EACDPLR-COOH

200 400 600 800
m/z

Quantitation and protein identification

**The isotope-coded affinity tag approach to the quantitation of
complex protein mixtures from two different states**
Figure 10.30

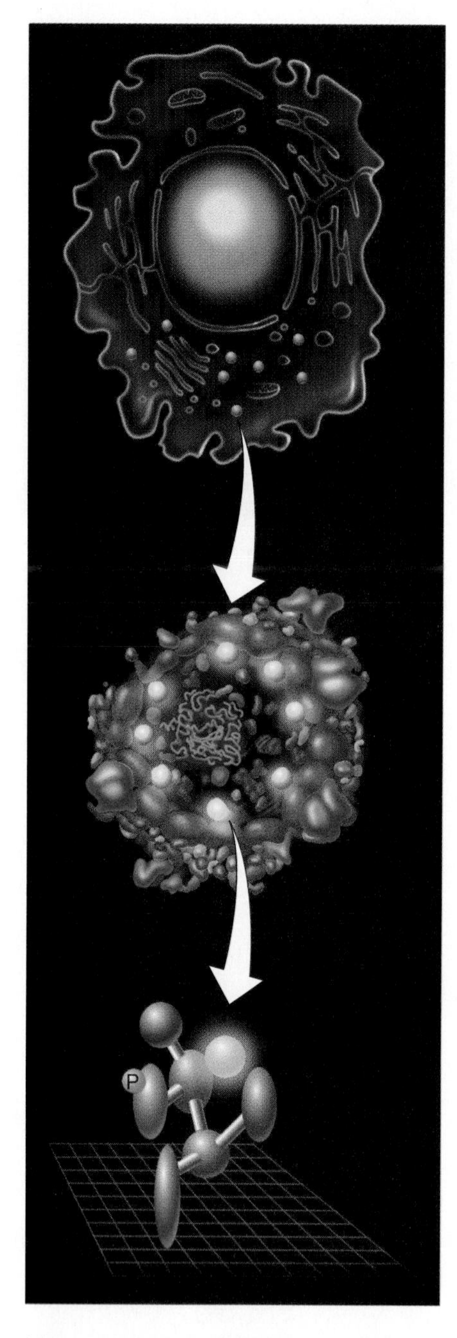

Single protein tagged in whole cell.

Cellular extract prepared.

Protein complex isolated by affinity purification.

Proteins in complex identified by mass spectrometry.

How to use affinity purification and mass spectrometry to identify protein interactions
Figure 10.31

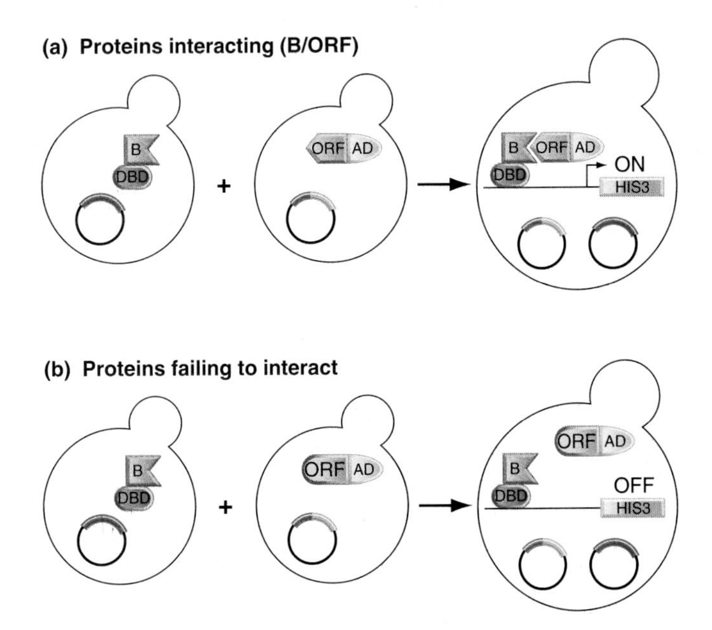

(a) Proteins interacting (B/ORF)

(b) Proteins failing to interact

The yeast two-hybrid approach to analyzing protein interactions
Figure 10.32

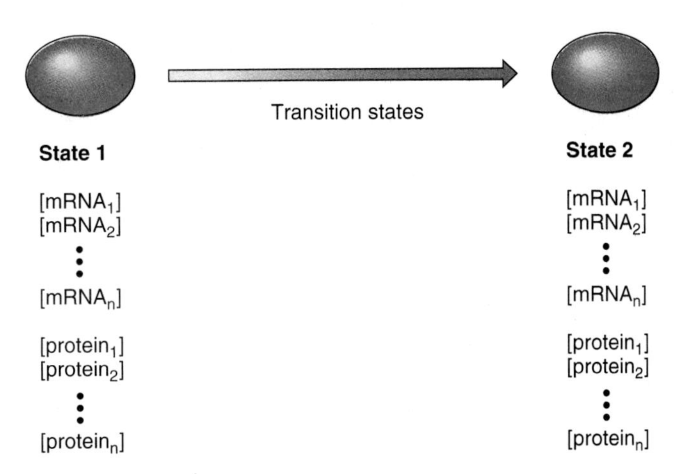

Transition states

State 1

[mRNA$_1$]
[mRNA$_2$]
⋮
[mRNA$_n$]

[protein$_1$]
[protein$_2$]
⋮
[protein$_n$]

State 2

[mRNA$_1$]
[mRNA$_2$]
⋮
[mRNA$_n$]

[protein$_1$]
[protein$_2$]
⋮
[protein$_n$]

Assumptions
- Composition and quantity of expressed genetic information defines the state.
- Comparative (subtractive) analysis of states describes biological system.

A global analysis of the mRNAs and proteins from two different states of perturbation
Figure 10.33

- Well-studied pathway involving carbon utilization in yeast
- ~ 9 genes involved in specific processing of galactose sugar: includes structural genes (e.g., enzymes) and control genes (e.g., transcription factors)

- Enzymes are transcriptionally up-regulated 1000X when cells are stimulated by galactose

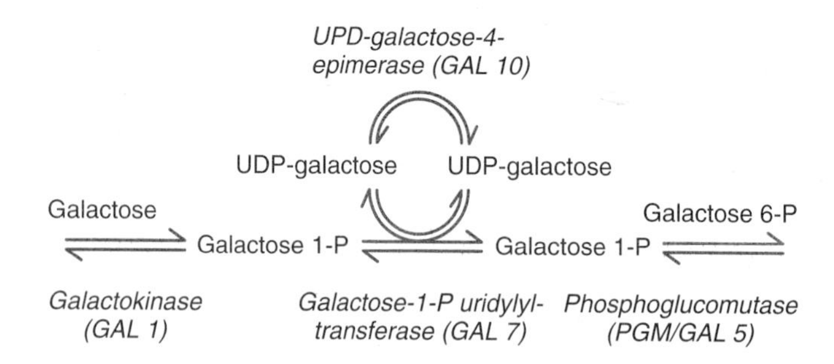

The yeast galactose-utilization system
Figure 10.34

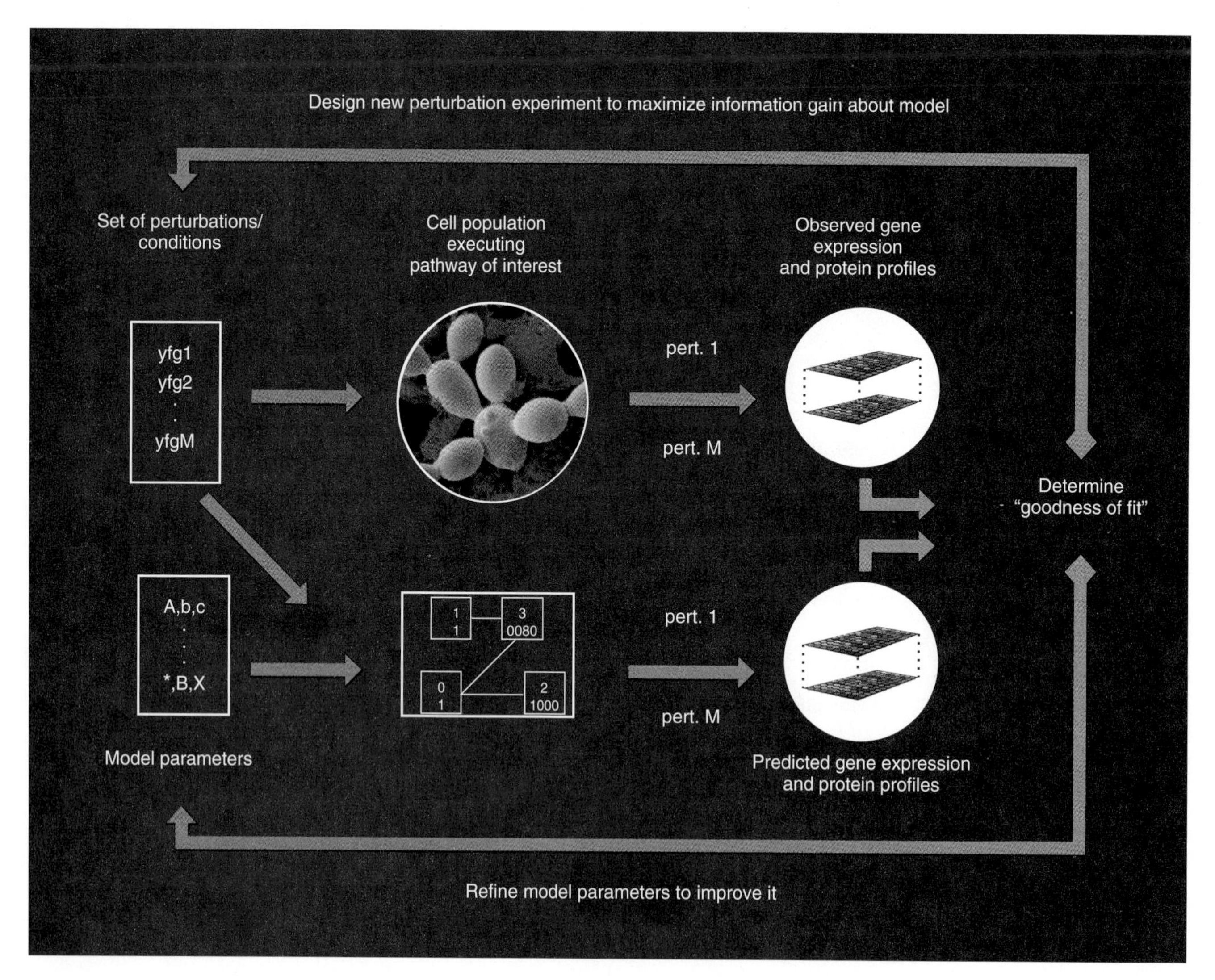

Diagram of the iterative nature of the systems approach to biology
Figure 10.35

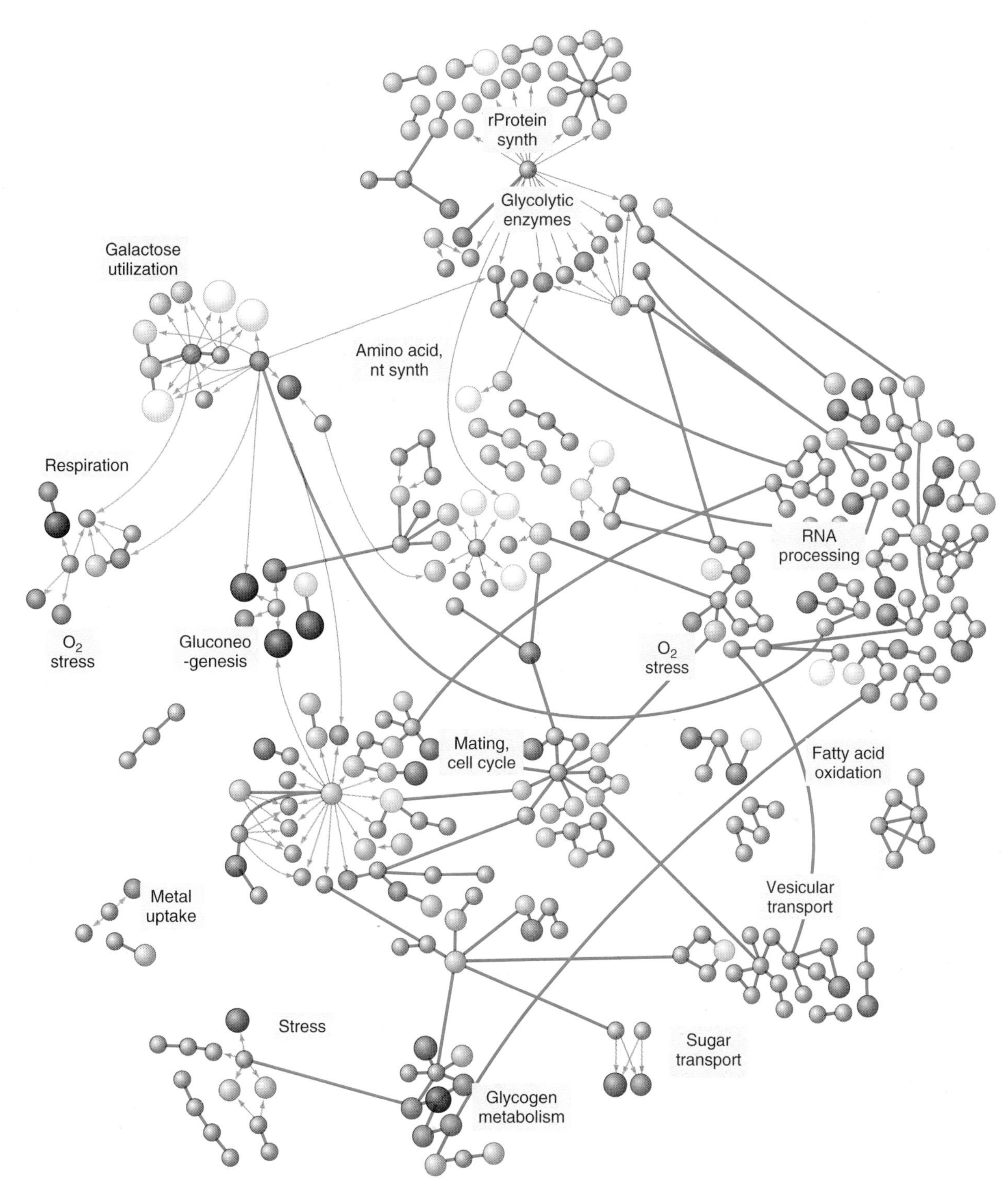

How genetic perturbation of the galactose-utilization system in yeast (the knockout of individual genes) affects the network of interactions with other metabolic and functional systems

Figure 10.36

(a)

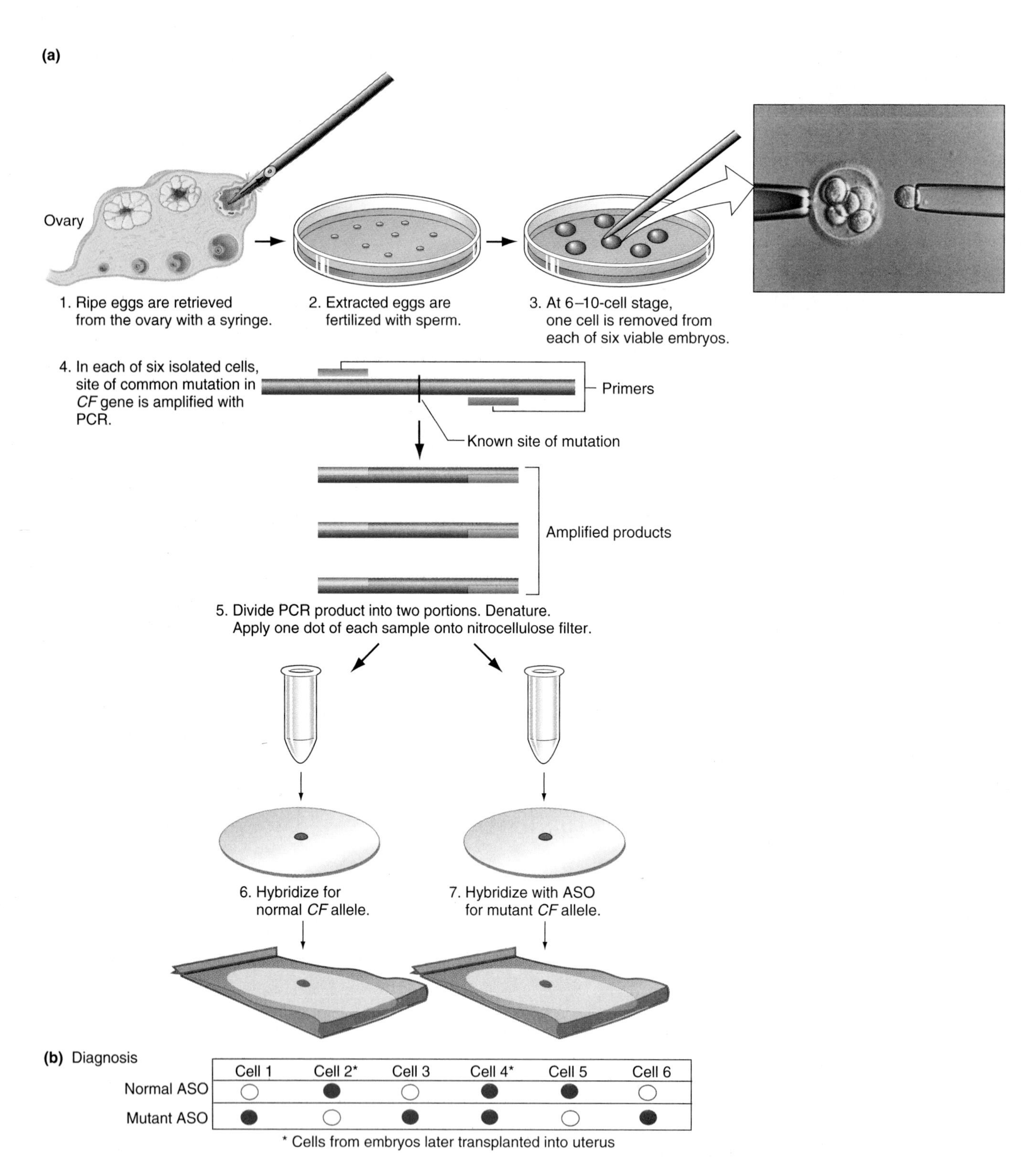

1. Ripe eggs are retrieved from the ovary with a syringe.

Ovary

2. Extracted eggs are fertilized with sperm.

3. At 6–10-cell stage, one cell is removed from each of six viable embryos.

4. In each of six isolated cells, site of common mutation in *CF* gene is amplified with PCR.

Primers

Known site of mutation

Amplified products

5. Divide PCR product into two portions. Denature. Apply one dot of each sample onto nitrocellulose filter.

6. Hybridize for normal *CF* allele.

7. Hybridize with ASO for mutant *CF* allele.

(b) Diagnosis

	Cell 1	Cell 2*	Cell 3	Cell 4*	Cell 5	Cell 6
Normal ASO	○	●	○	●	●	○
Mutant ASO	●	○	●	●	○	●

* Cells from embryos later transplanted into uterus

Preimplantation embryo diagnosis
Figure 11.1

a: Courtesy of Ronald Carson, The Reproductive Science Center of Boston/IntegraMed America, Inc.

Base-pair differences between DNA cloned from the cystic fibrosis locus of two healthy individuals
Figure 11.2

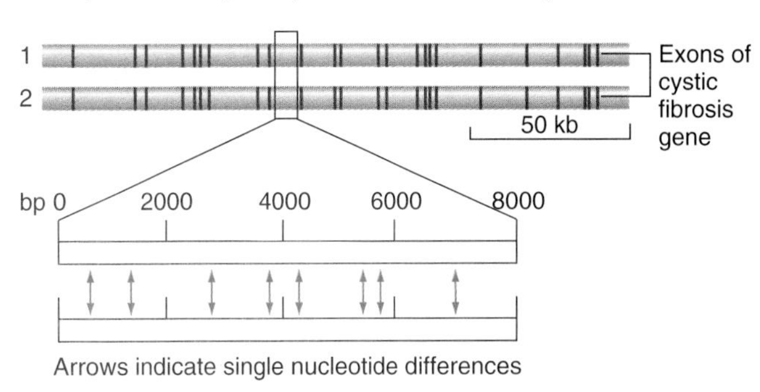

Two cystic fibrosis (*CFTR*) alleles from two healthy individuals

Exons of cystic fibrosis gene

50 kb

bp 0   2000   4000   6000   8000

Arrows indicate single nucleotide differences

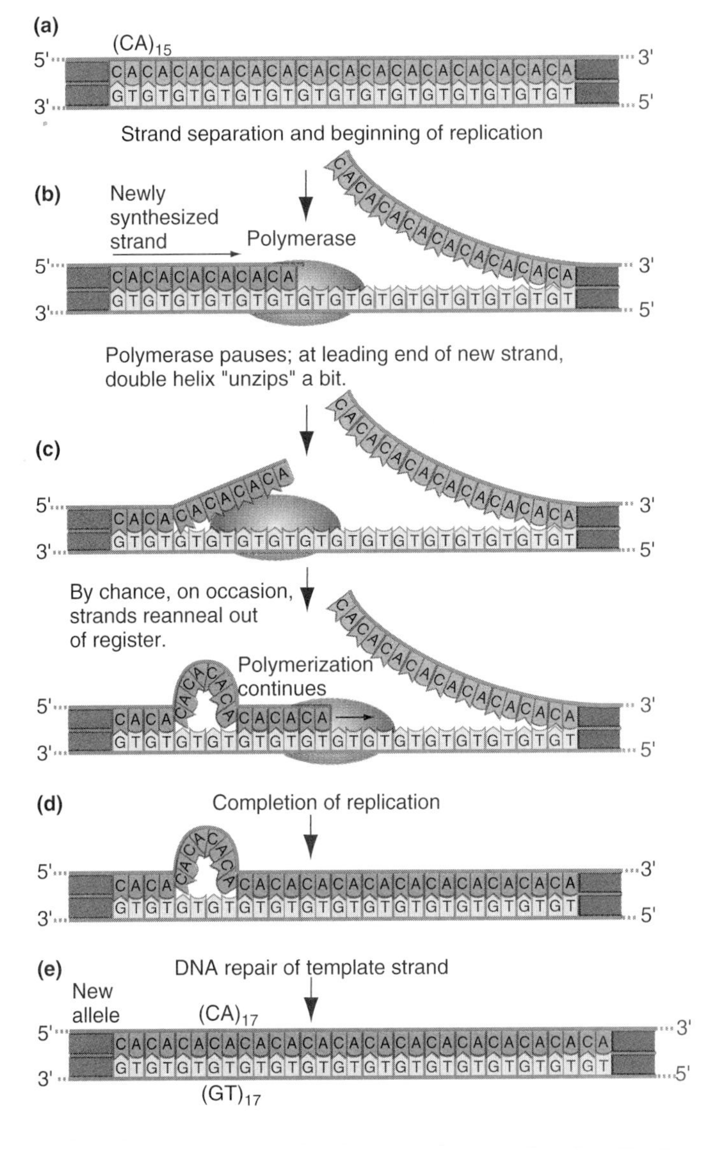

(a)

(CA)₁₅

Strand separation and beginning of replication

(b) Newly synthesized strand   Polymerase

Polymerase pauses; at leading end of new strand, double helix "unzips" a bit.

(c)

By chance, on occasion, strands reanneal out of register.

Polymerization continues

(d) Completion of replication

(e) DNA repair of template strand

New allele   (CA)₁₇

(GT)₁₇

Microsatellites are highly polymorphic because of their potential for faulty replication
Figure 11.3

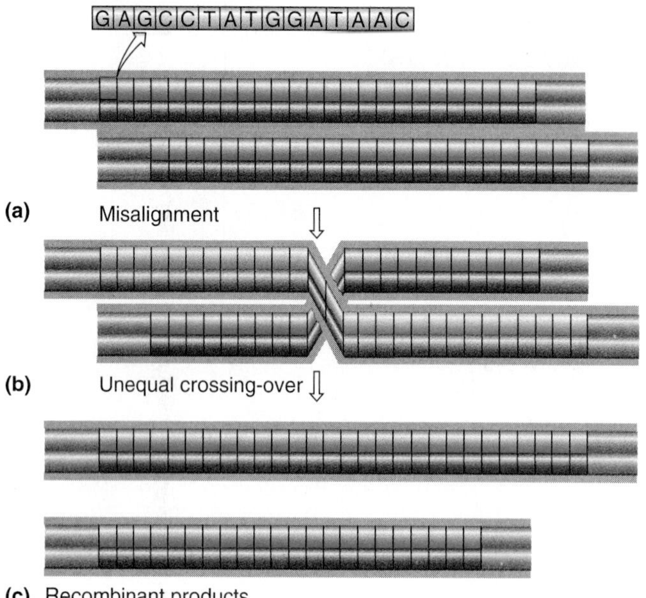

(a) Misalignment

(b) Unequal crossing-over

(c) Recombinant products

Minisatellites are highly polymorphic because of their potential for misalignment and unequal crossing-over during meiosis
Figure 11.4

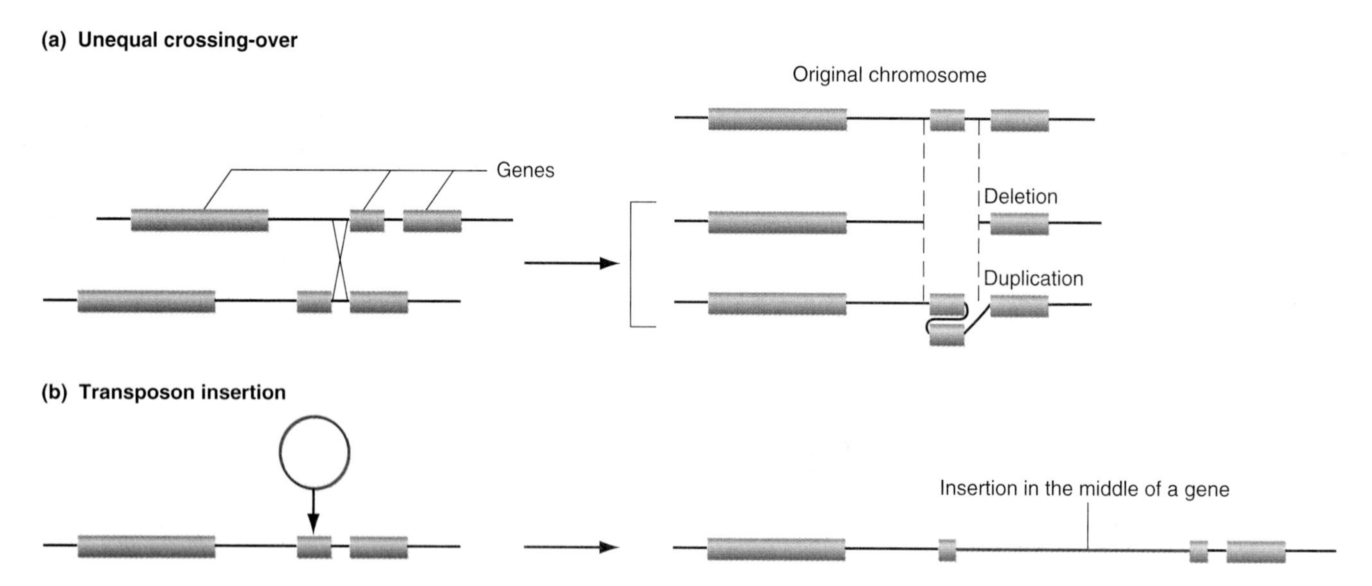

(a) Unequal crossing-over

Genes

Original chromosome

Deletion

Duplication

(b) Transposon insertion

Insertion in the middle of a gene

Deletion, duplications, and insertions cause variation in genomic content
Figure 11.5

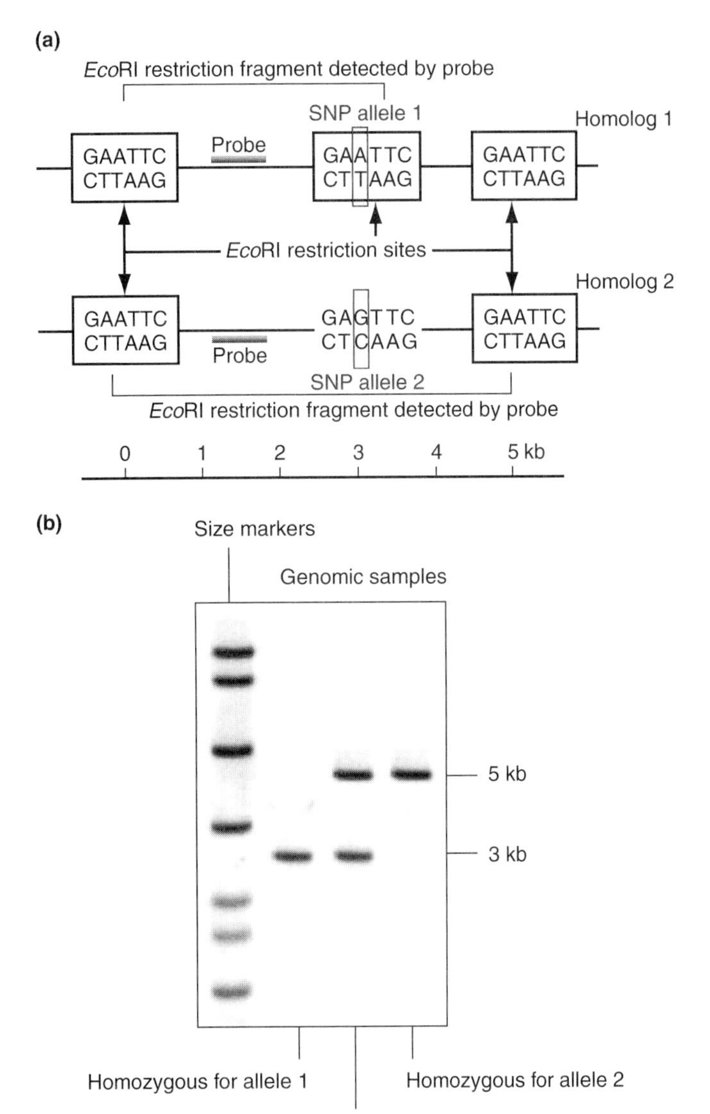

(a)

EcoRI restriction fragment detected by probe

SNP allele 1   Homolog 1

GAATTC   Probe   GA|A|TTC   GAATTC
CTTAAG   CT|T|AAG   CTTAAG

————— EcoRI restriction sites —————

Homolog 2

GAATTC   Probe   GAG TTC   GAATTC
CTTAAG   CT|C|AAG   CTTAAG

SNP allele 2
EcoRI restriction fragment detected by probe

0   1   2   3   4   5 kb

(b)

Size markers

Genomic samples

— 5 kb

— 3 kb

Homozygous for allele 1   Homozygous for allele 2

Heterozygous with alleles 1 and 2

SNP-caused restriction fragment length polymorphisms can be detected by Southern blot analysis

Figure 11.6

b: © Lee Silver, Princeton University

117

(a)

Normal allele (*A*)

Left primer

                    Pro   Glu   Glu   (translated sequence)

                    ┌─────────────┐
                    │CCTGAGG│AG   │
                    │GGACTCC│TC   │
                    └─────────────┘
                                        Right primer

                    *Mst*II restriction site

0   100   200   300   400   500 bp

Left primer

                    Pro   Val   Glu

                    ┌─────────────┐
                    │CCTG**T**GGAG │
                    │GGAC**A**CCTC │
                    └─────────────┘
                                        Right primer

Sickle-cell allele (*S*)   No restriction site

(b)

┌─────┐   ⃝
│ A S │────────────────── A S
└─────┘   │
               │
        ┌─────┐   ⃝
        │ S S │   ?
        └─────┘

                                    ── 500 bp (*S* allele)

                                    ── 300 bp ⎤
                                               ⎥ (*A* allele)
                                    ── 200 bp ⎦

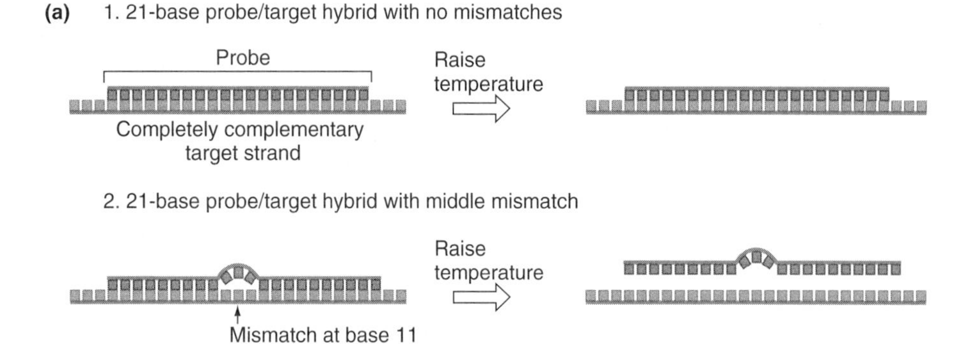

(a) 1. 21-base probe/target hybrid with no mismatches

                    Probe
                                        Raise
                                        temperature
        Completely complementary
        target strand

2. 21-base probe/target hybrid with middle mismatch

                                        Raise
                                        temperature

        Mismatch at base 11

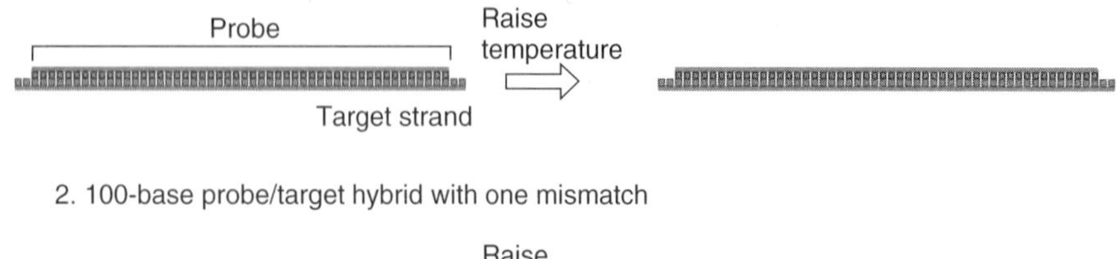

(b) 1. 50-base probe/target hybrid with no mismatches

                    Probe   Raise
                                 temperature
                    Target strand

2. 100-base probe/target hybrid with one mismatch

            Mismatch   Raise
                             temperature

Short hybridization probes can distinguish single-base mismatches
Figure 11.8

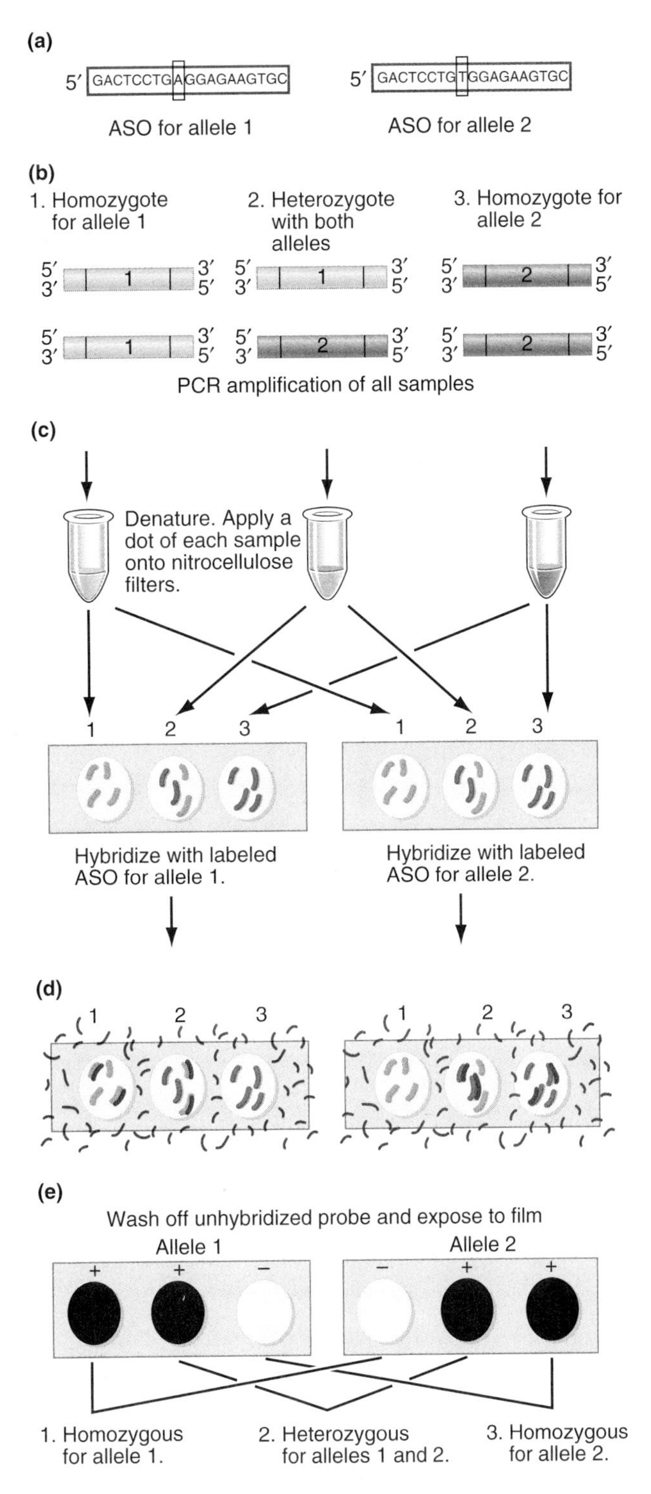

(a)

5′ GACTCCTGAGGAGAAGTGC   5′ GACTCCTGTGGAGAAGTGC

ASO for allele 1   ASO for allele 2

(b)

1. Homozygote for allele 1   2. Heterozygote with both alleles   3. Homozygote for allele 2

PCR amplification of all samples

(c)

Denature. Apply a dot of each sample onto nitrocellulose filters.

Hybridize with labeled ASO for allele 1.   Hybridize with labeled ASO for allele 2.

(d)

(e)

Wash off unhybridized probe and expose to film

Allele 1   Allele 2

1. Homozygous for allele 1.   2. Heterozygous for alleles 1 and 2.   3. Homozygous for allele 2.

ASOs can determine genotype at any SNP locus

Figure 11.9

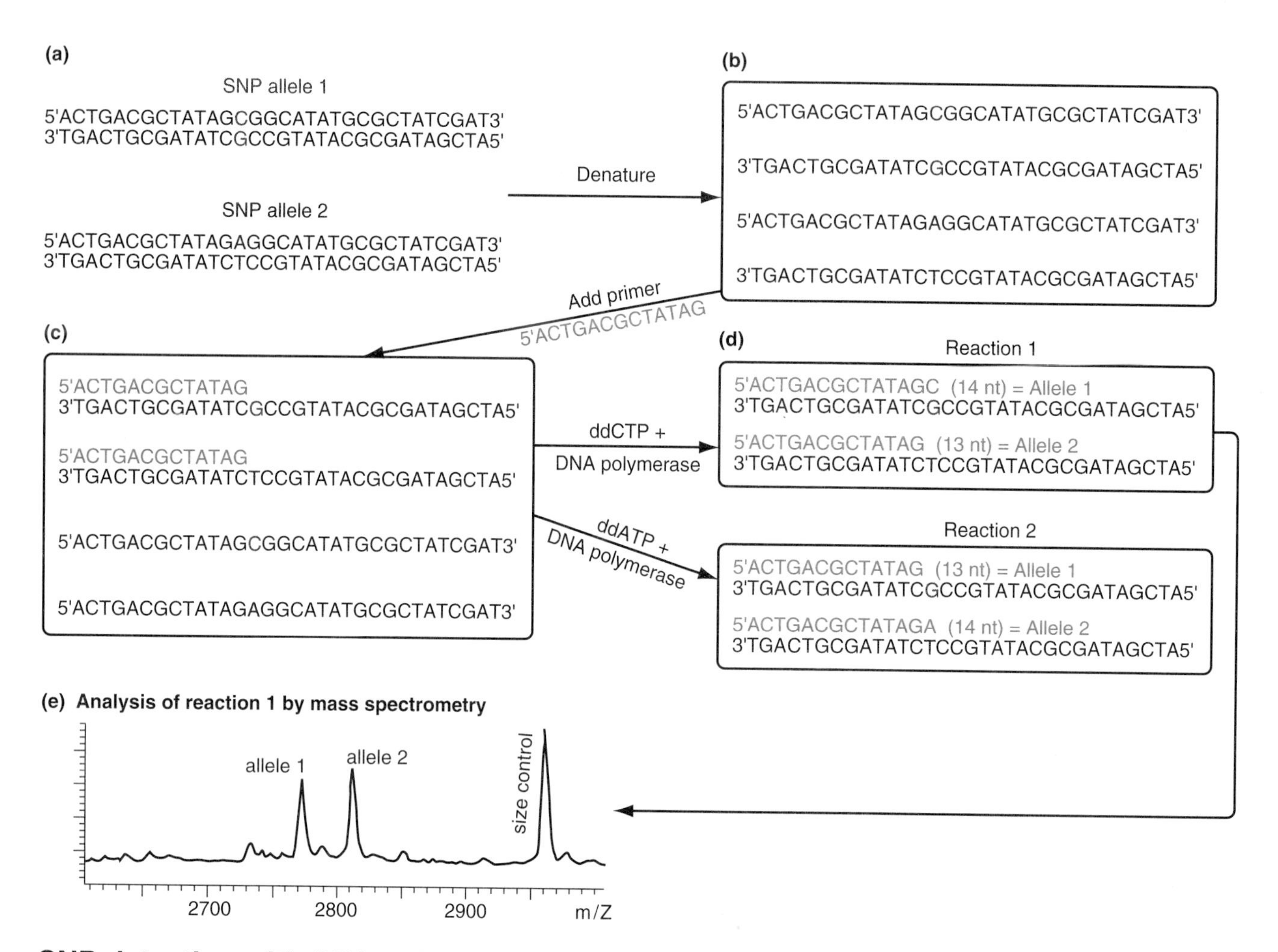

(a)

SNP allele 1

5'ACTGACGCTATAGCGGCATATGCGCTATCGAT3'
3'TGACTGCGATATCGCCGTATACGCGATAGCTA5'

SNP allele 2

5'ACTGACGCTATAGAGGCATATGCGCTATCGAT3'
3'TGACTGCGATATCTCCGTATACGCGATAGCTA5'

Denature

(b)

5'ACTGACGCTATAGCGGCATATGCGCTATCGAT3'

3'TGACTGCGATATCGCCGTATACGCGATAGCTA5'

5'ACTGACGCTATAGAGGCATATGCGCTATCGAT3'

3'TGACTGCGATATCTCCGTATACGCGATAGCTA5'

Add primer
5'ACTGACGCTATAG

(c)

5'ACTGACGCTATAG
3'TGACTGCGATATCGCCGTATACGCGATAGCTA5'

5'ACTGACGCTATAG
3'TGACTGCGATATCTCCGTATACGCGATAGCTA5'

5'ACTGACGCTATAGCGGCATATGCGCTATCGAT3'

5'ACTGACGCTATAGAGGCATATGCGCTATCGAT3'

ddCTP +
DNA polymerase

ddATP +
DNA polymerase

(d)

Reaction 1

5'ACTGACGCTATAGC (14 nt) = Allele 1
3'TGACTGCGATATCGCCGTATACGCGATAGCTA5'

5'ACTGACGCTATAG (13 nt) = Allele 2
3'TGACTGCGATATCTCCGTATACGCGATAGCTA5'

Reaction 2

5'ACTGACGCTATAG (13 nt) = Allele 1
3'TGACTGCGATATCGCCGTATACGCGATAGCTA5'

5'ACTGACGCTATAGA (14 nt) = Allele 2
3'TGACTGCGATATCTCCGTATACGCGATAGCTA5'

(e) Analysis of reaction 1 by mass spectrometry

allele 1   allele 2   size control

2700   2800   2900   m/Z

SNP detection with DNA polymerase-assisted single nucleotide primer extension
Figure 11.11

(a) Determine sequences flanking microsatellites.

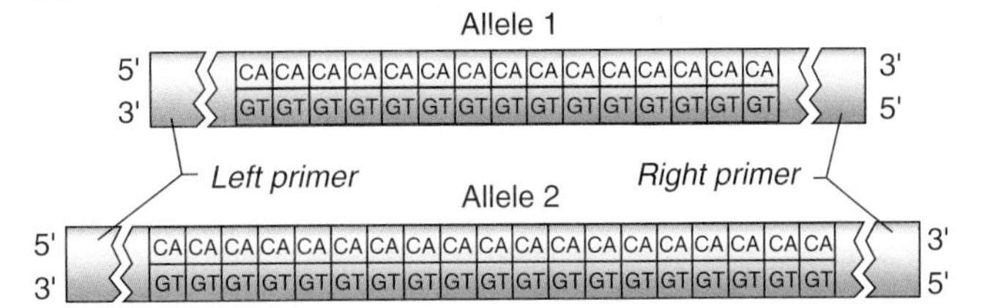

(b) Amplify alleles by PCR.

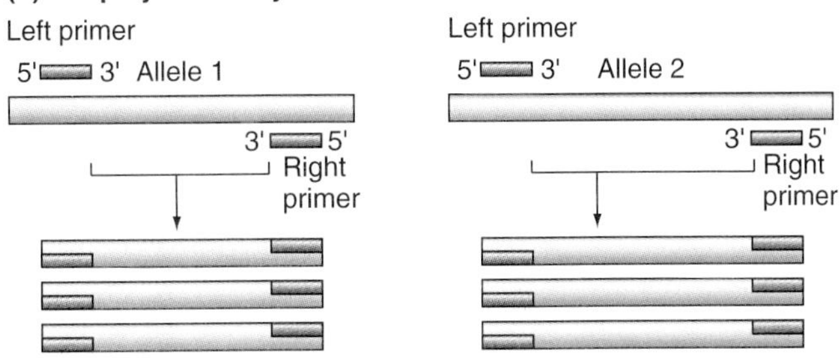

(c) Analyze PCR products by gel electrophoresis and staining.

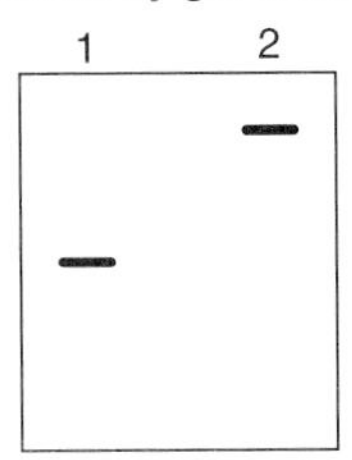

(d) Example of population with three alleles.

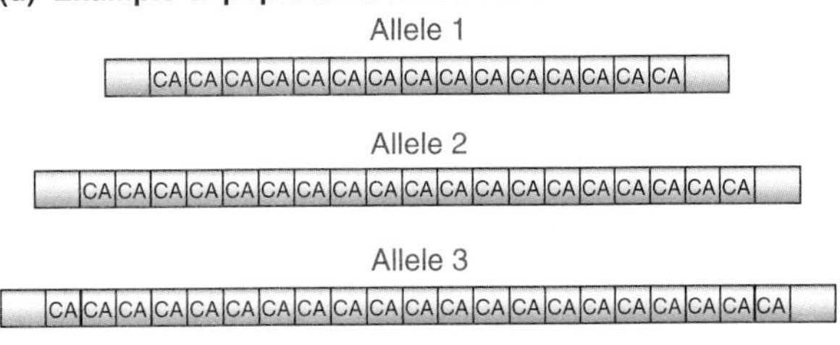

Six diploid genotypes are present in this population.

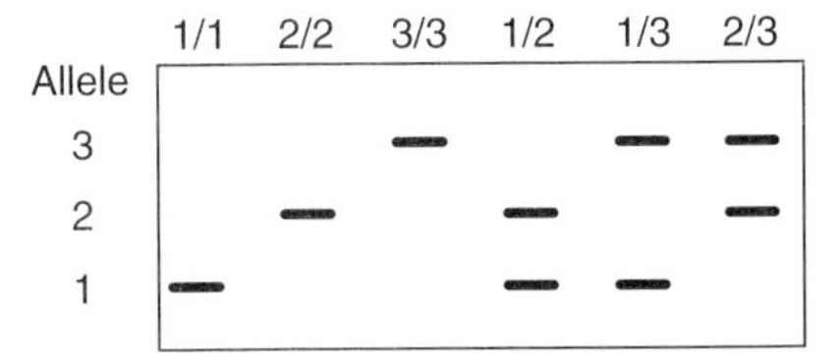

Detection of microsatellite polymorphisms by PCR and gel electrophoresis
Figure 11.12

(a) Basic structure of the *HD* gene

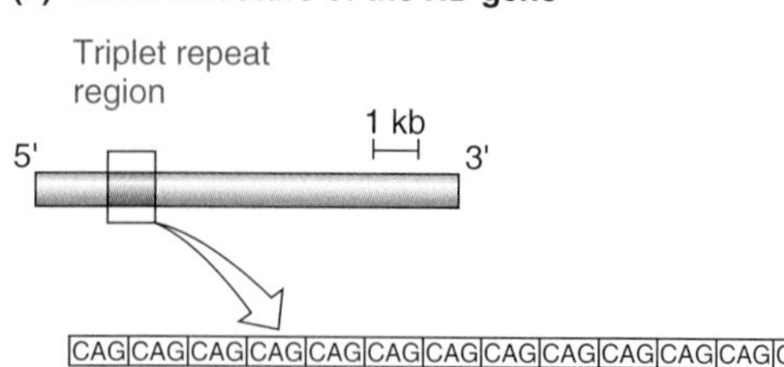

Triplet repeat region

1 kb

5'   3'

| CAG | CAG | CAG | CAG | CAG | CAG | CAG | CAG | CAG | CAG | CAG | CAG | CAG | CAG | CAG |

Each triplet encodes glutamine.

Mutations at the Huntington disease locus are caused by expansion of a triplet repeat microsatellite in a coding region
Figure 11.13

(b) Some alleles at the *HD* locus

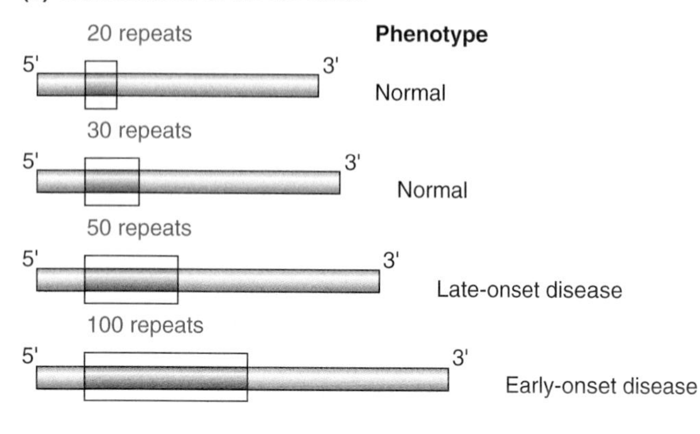

	Phenotype
20 repeats 5'—3'	Normal
30 repeats 5'—3'	Normal
50 repeats 5'—3'	Late-onset disease
100 repeats 5'—3'	Early-onset disease

(a) Digest DNA with restriction enzyme that does not cut inside microsatellite.

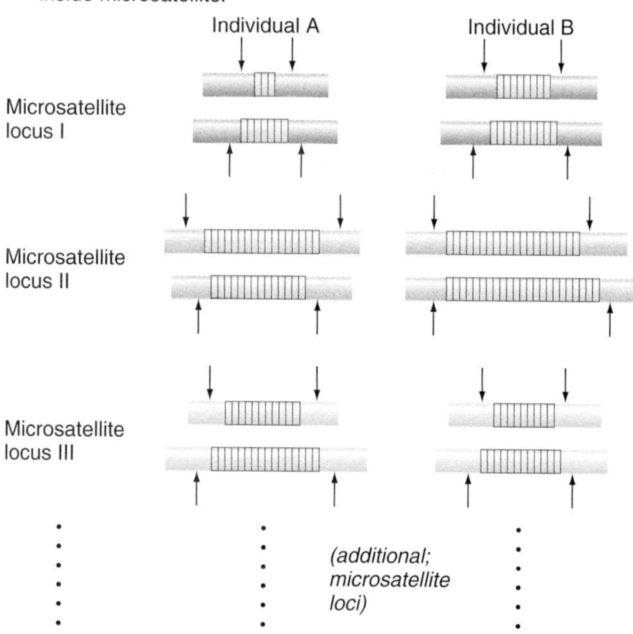

Individual A   Individual B

Microsatellite locus I

Microsatellite locus II

Microsatellite locus III

⋮   ⋮   ⋮
(additional; microsatellite loci)

(b) Run DNA samples on a gel. Perform Southern blotting. Hybridize with probe containing microsatellite sequence.

Individuals
A   B   C   D

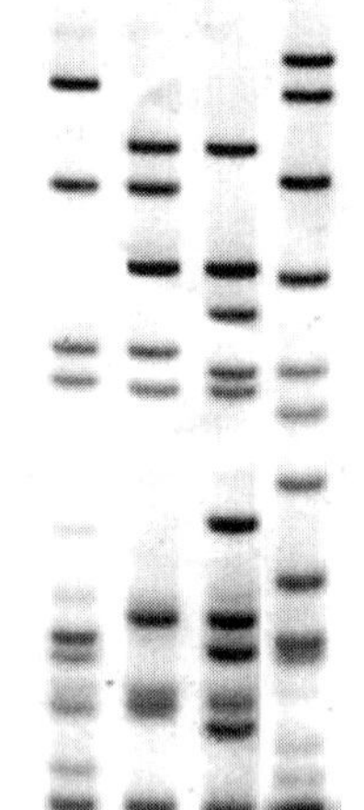

Minisatellite analysis provides a broad comparison of whole genomes
Figure 11.14

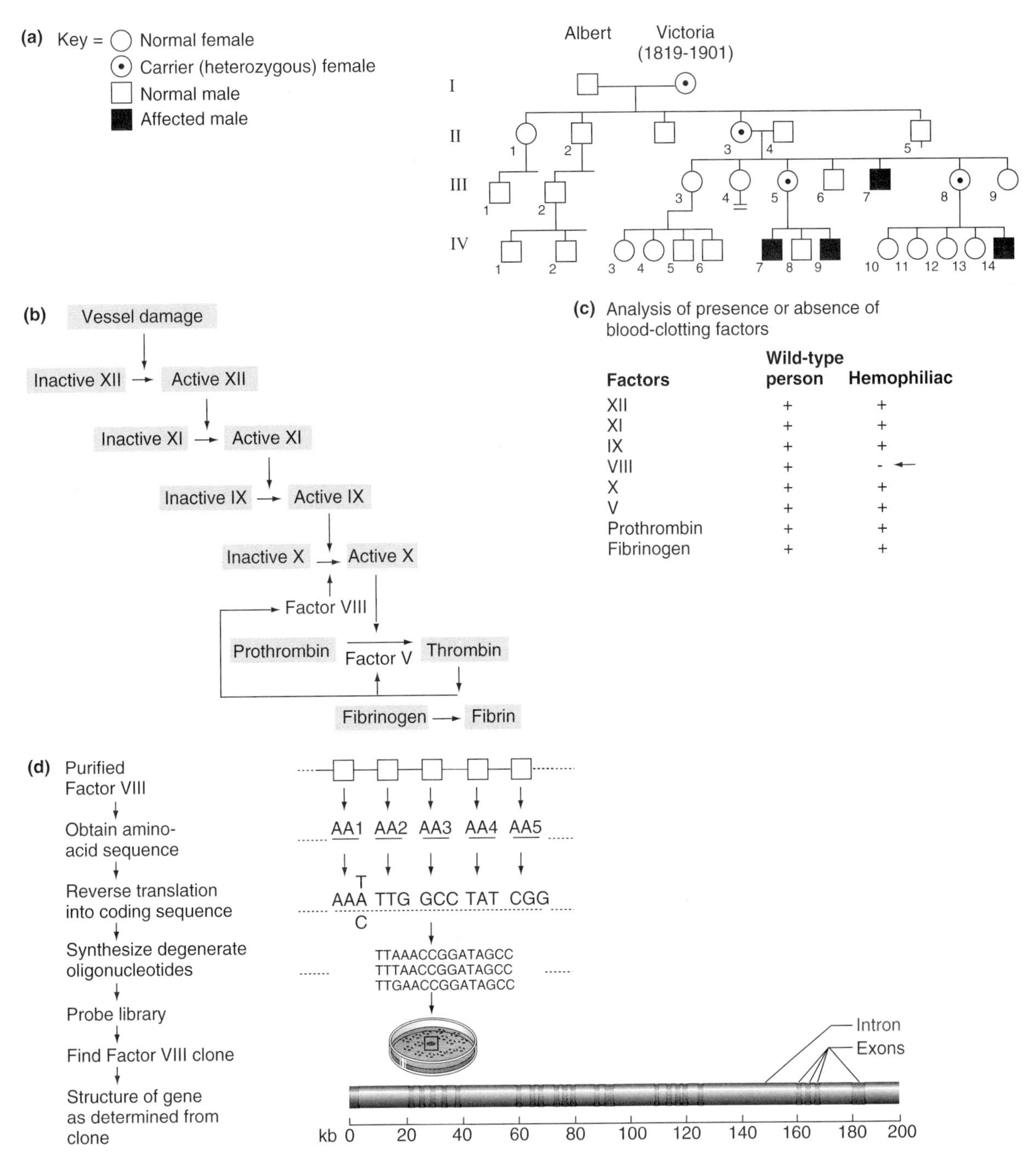

(a) Key =
- ○ Normal female
- ⊙ Carrier (heterozygous) female
- □ Normal male
- ■ Affected male

(b)
Vessel damage
↓
Inactive XII → Active XII
↓
Inactive XI → Active XI
↓
Inactive IX → Active IX
↓
Inactive X → Active X
Factor VIII ↑
Prothrombin → Factor V → Thrombin
Fibrinogen → Fibrin

(c) Analysis of presence or absence of blood-clotting factors

Factors	Wild-type person	Hemophiliac
XII	+	+
XI	+	+
IX	+	+
VIII	+	− ←
X	+	+
V	+	+
Prothrombin	+	+
Fibrinogen	+	+

(d) Purified Factor VIII
↓
Obtain amino-acid sequence
↓
Reverse translation into coding sequence
↓
Synthesize degenerate oligonucleotides
↓
Probe library
↓
Find Factor VIII clone
↓
Structure of gene as determined from clone

AA1 AA2 AA3 AA4 AA5

T
AAA TTG GCC TAT CGG
C

TTAAACCGGATAGCC
TTTAACCGGATAGCC
TTGAACCGGATAGCC

Intron
Exons

kb 0 20 40 60 80 100 120 140 160 180 200

How geneticists identified and cloned the hemophilia A gene
Figure 11.16

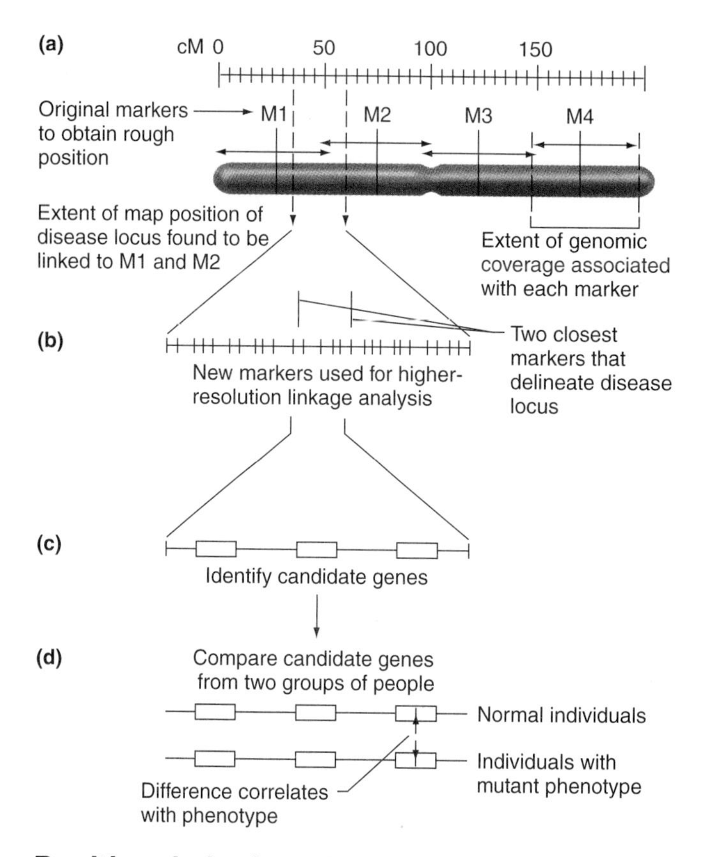

(a)

cM 0   50   100   150

Original markers
to obtain rough
position → M1 | M2 M3 M4

Extent of map position of
disease locus found to be
linked to M1 and M2

Extent of genomic
coverage associated
with each marker

(b)

Two closest
markers that
delineate disease
locus

New markers used for higher-
resolution linkage analysis

(c)

Identify candidate genes

(d)

Compare candidate genes
from two groups of people

Normal individuals

Individuals with
mutant phenotype

Difference correlates
with phenotype

**Positional cloning: From phenotype to
chromosomal location to guilty gene**
Figure 11.17

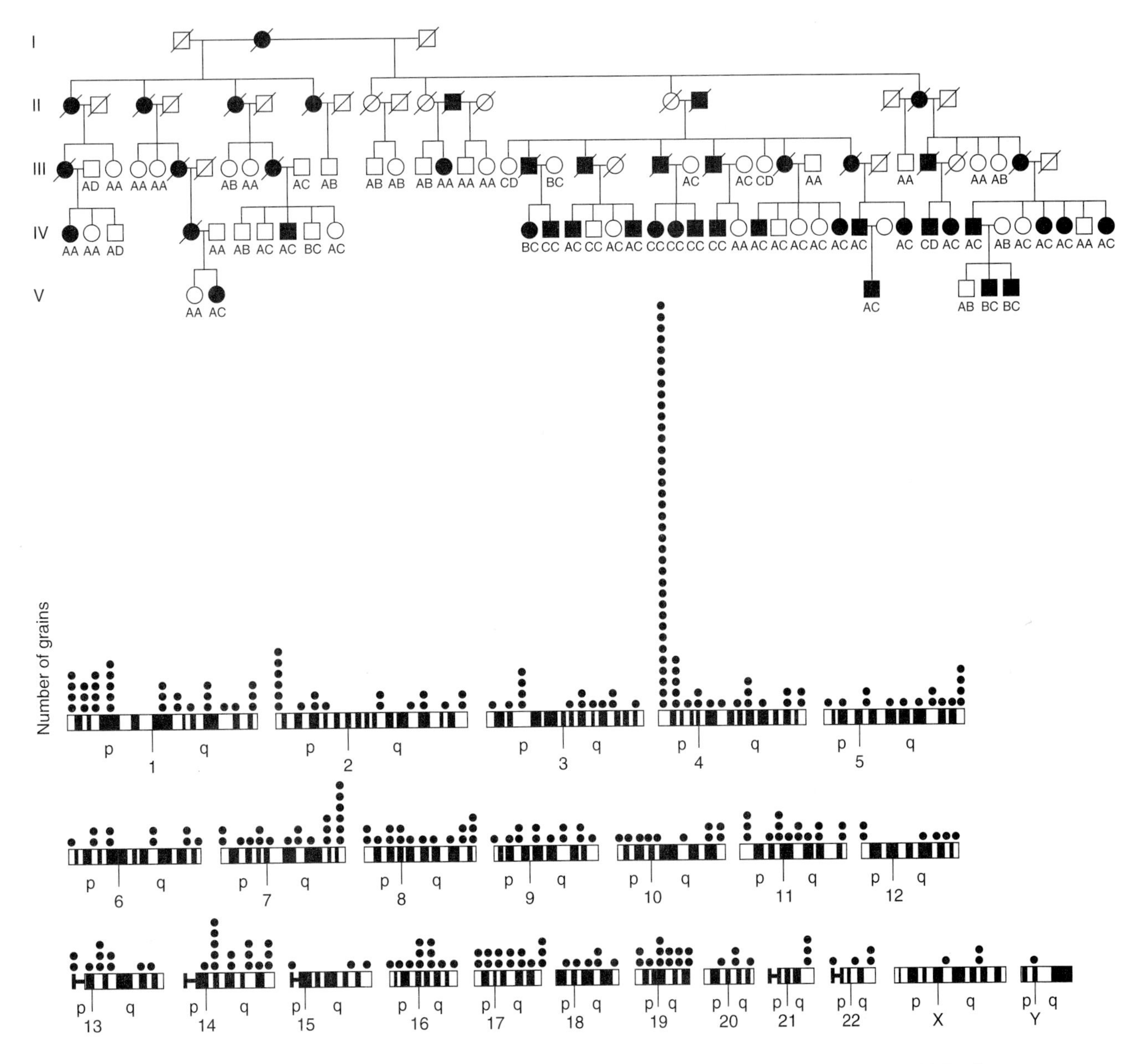

Detection of linkage between the DNA marker G8 and a locus responsible for Huntington disease (HD)

Figure 11.18

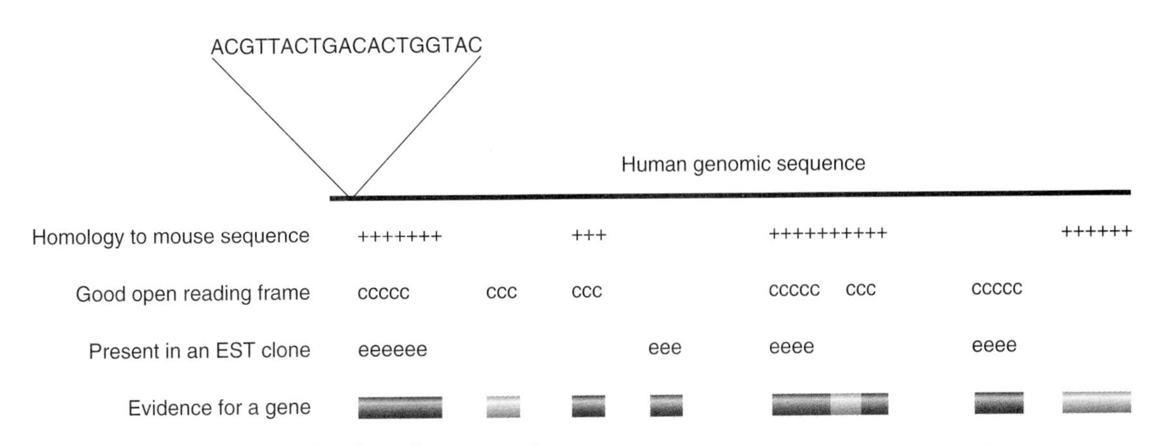

ACGTTACTGACACTGGTAC

Human genomic sequence

Homology to mouse sequence	+++++++		+++		++++++++++		++++++
Good open reading frame	ccccc	ccc	ccc		ccccc	ccc	ccccc	
Present in an EST clone	eeeeee			eee	eeee		eeee	
Evidence for a gene								

Computational analysis of genomic sequence can uncover genes

Figure 11.19

(a)

1.

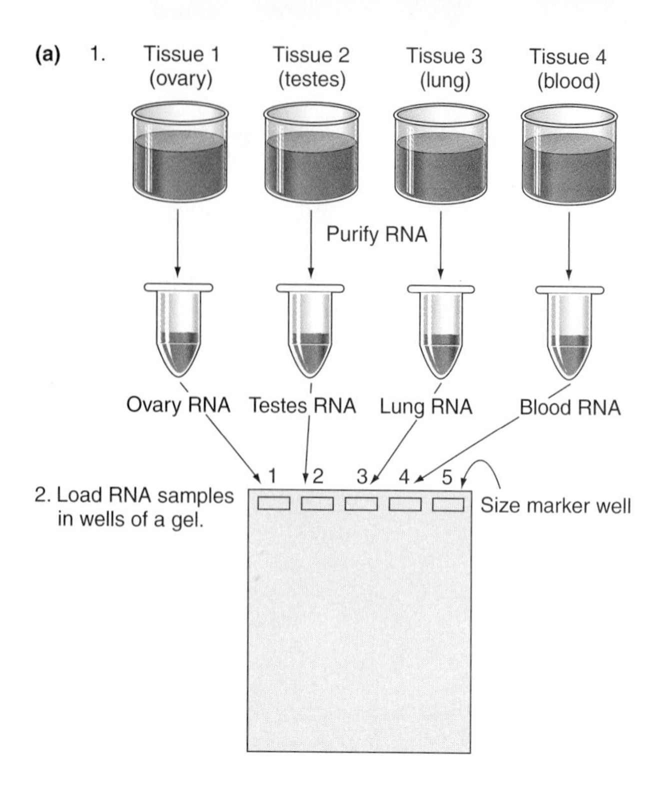

2. Load RNA samples in wells of a gel.

3. Separate RNA samples by gel electrophoresis. Blot onto filter. Expose filter to labeled hybridization probe.

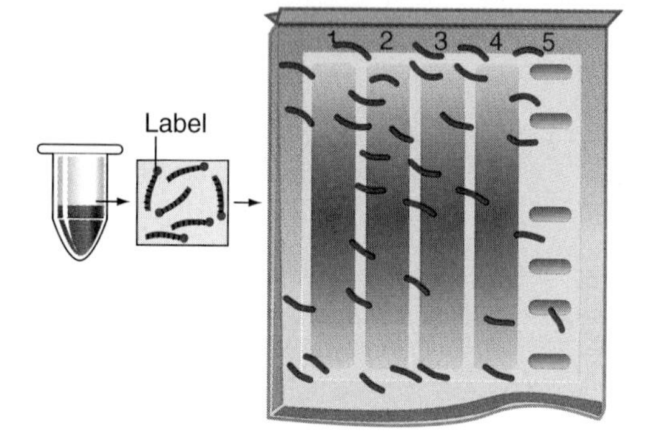

4. Wash away unhybridized probe. Make autoradiograph.

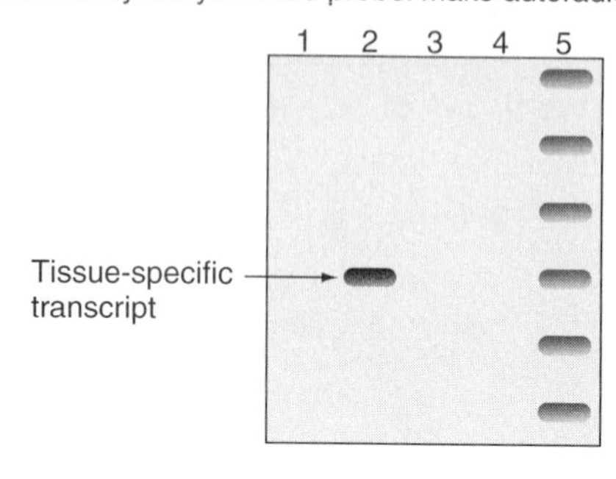

Tissue-specific transcript

(b)

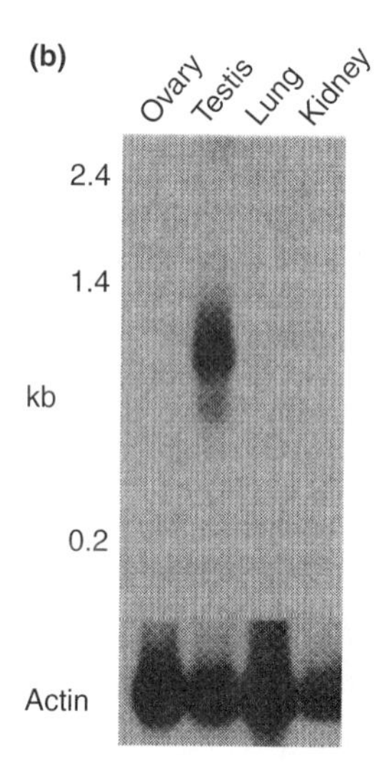

Northern blots: Snapshots of gene expression
Figure 11.20

b: Reprinted with permission from *Nature* 1990 July 19;346(6281):216-7, Sinclair et al. © 1990 Macmillian Magazines Limited

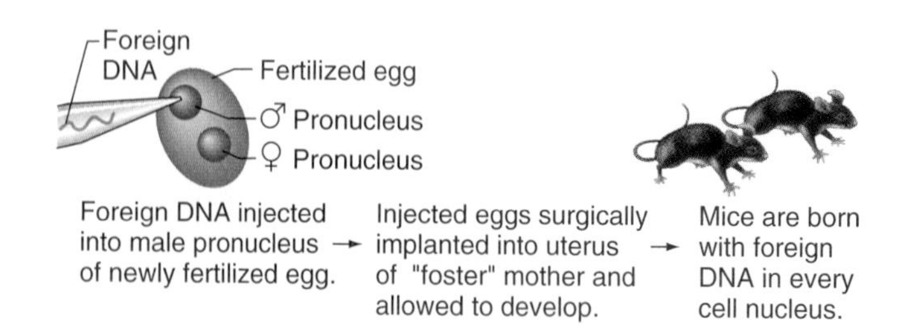

Foreign DNA injected into male pronucleus of newly fertilized egg. → Injected eggs surgically implanted into uterus of "foster" mother and allowed to develop. → Mice are born with foreign DNA in every cell nucleus.

Transgenic analysis can prove the equivalence of a candidate gene and a trait locus
Figure 11.21

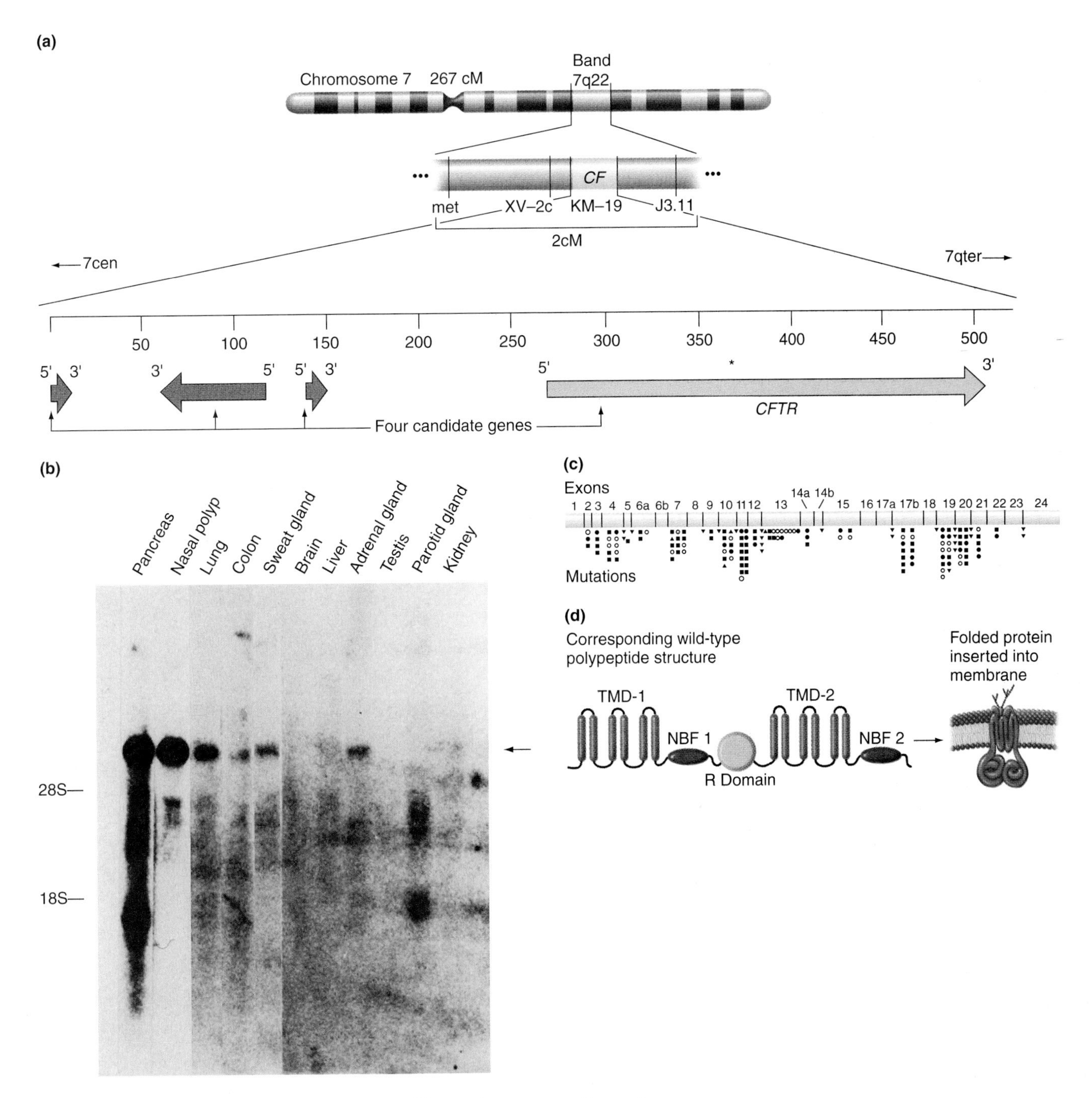

(a)

Chromosome 7   267 cM   Band 7q22

met   XV–2c   KM–19   J3.11

CF

2cM

←—7cen   7qter—→

50 100 150 200 250 300 350 400 450 500

5' 3'   3'   5' 5' 3'   5'   *   3'

— Four candidate genes —   *CFTR*

(b)

Pancreas Nasal polyp Lung Colon Sweat gland Brain Liver Adrenal gland Testis Parotid gland Kidney

28S—

18S—

(c)

Exons
1 2 3 4 5 6a 6b 7 8 9 10 11 12 13   14a 14b   15   16 17a 17b 18 19 20 21 22 23   24

Mutations

(d)

Corresponding wild-type polypeptide structure

Folded protein inserted into membrane

TMD-1   TMD-2

NBF 1   NBF 2

R Domain

Positional cloning of the cystic fibrosis gene: A review
Figure 11.22

b: © Johanna Rommens/Hospital for Sick Children, Toronto Reprinted from *Science* 245:1066, 1989. © 2001 American Association for the Advancement of Science

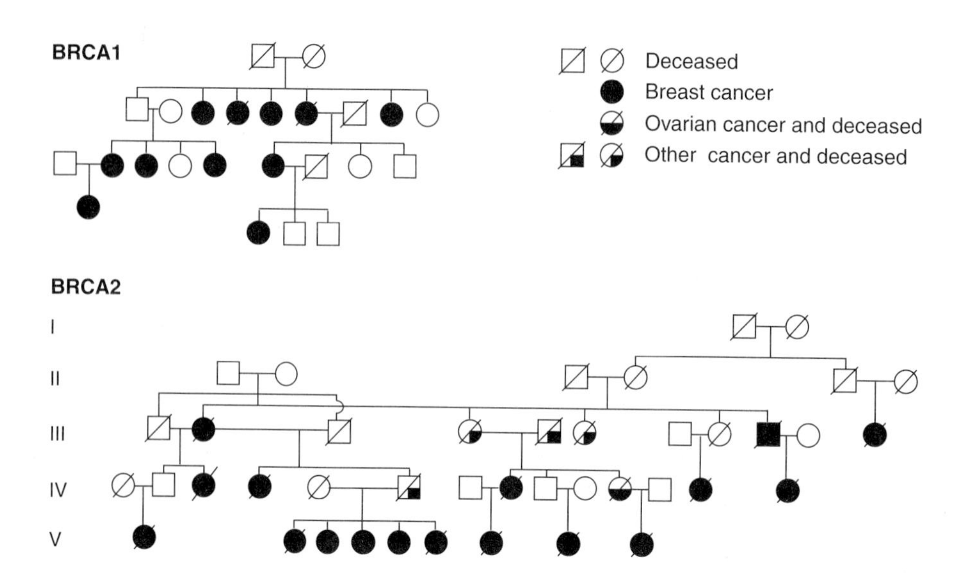

Genetic heterogeneity: Mutations at different loci can give rise to the same disease phenotype
Figure 11.23

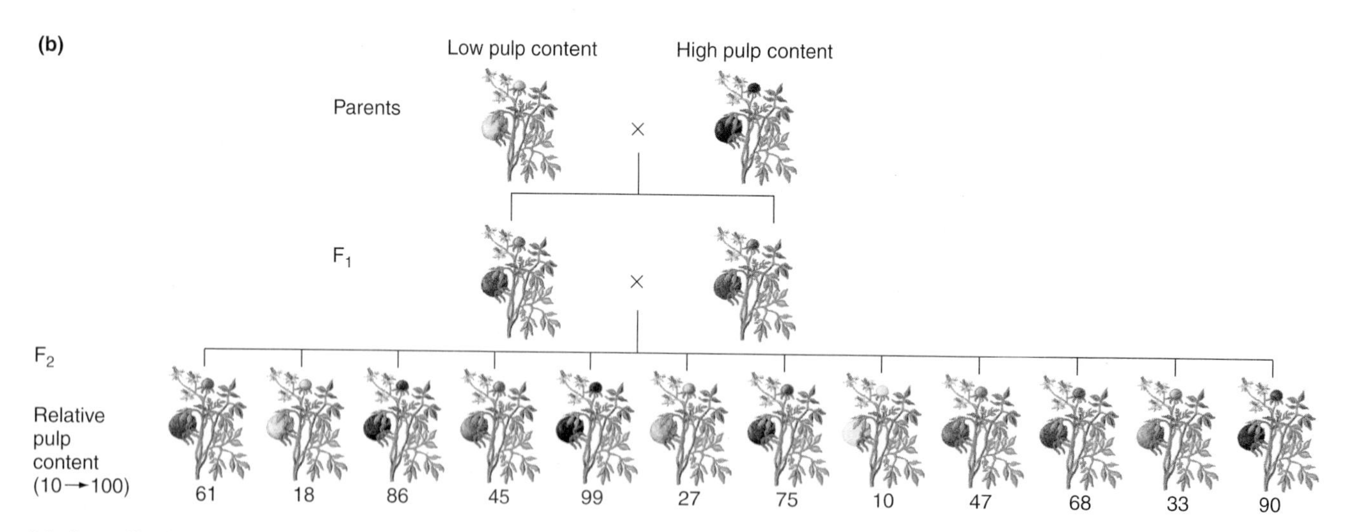

Using linkage analysis to identify loci that contribute to the expression of complex traits, such as pulp content in tomatoes
Figure 11.24

(a) Formation of haplotypes over evolutionary time

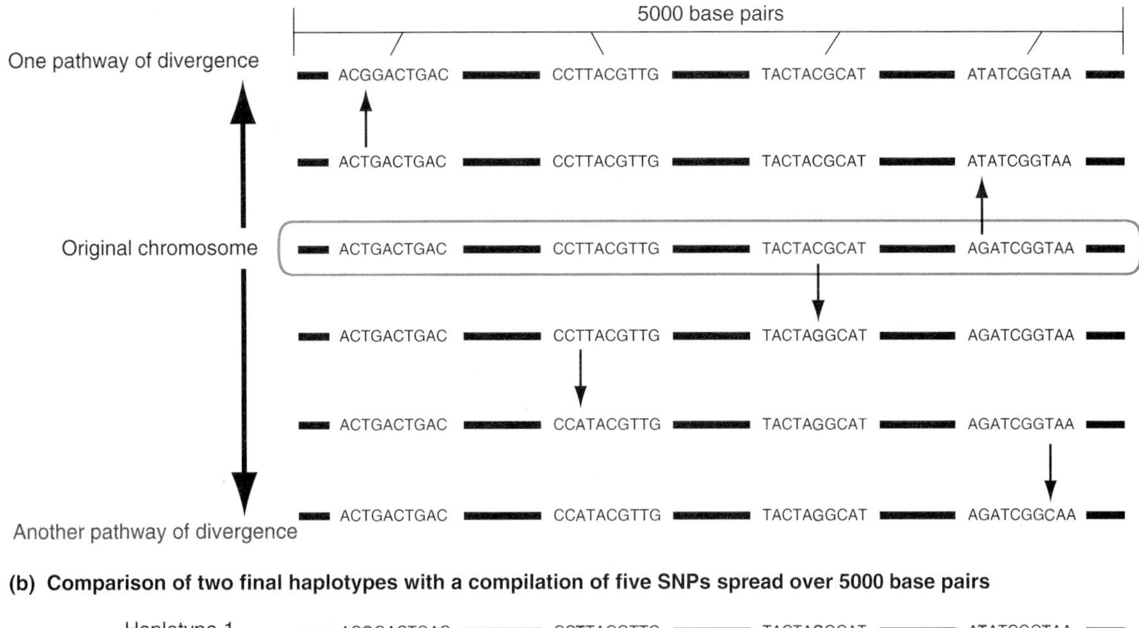

(b) Comparison of two final haplotypes with a compilation of five SNPs spread over 5000 base pairs

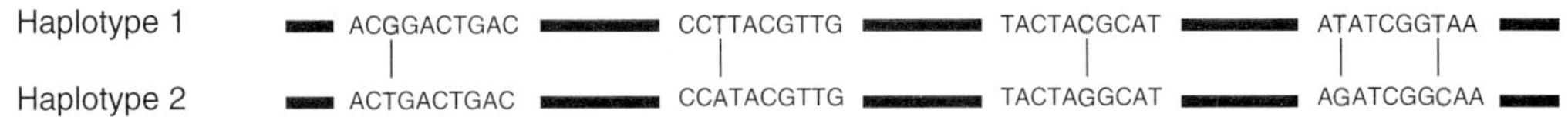

Haplotypes are formed by sequential mutations in a small genomic region
Figure 11.25

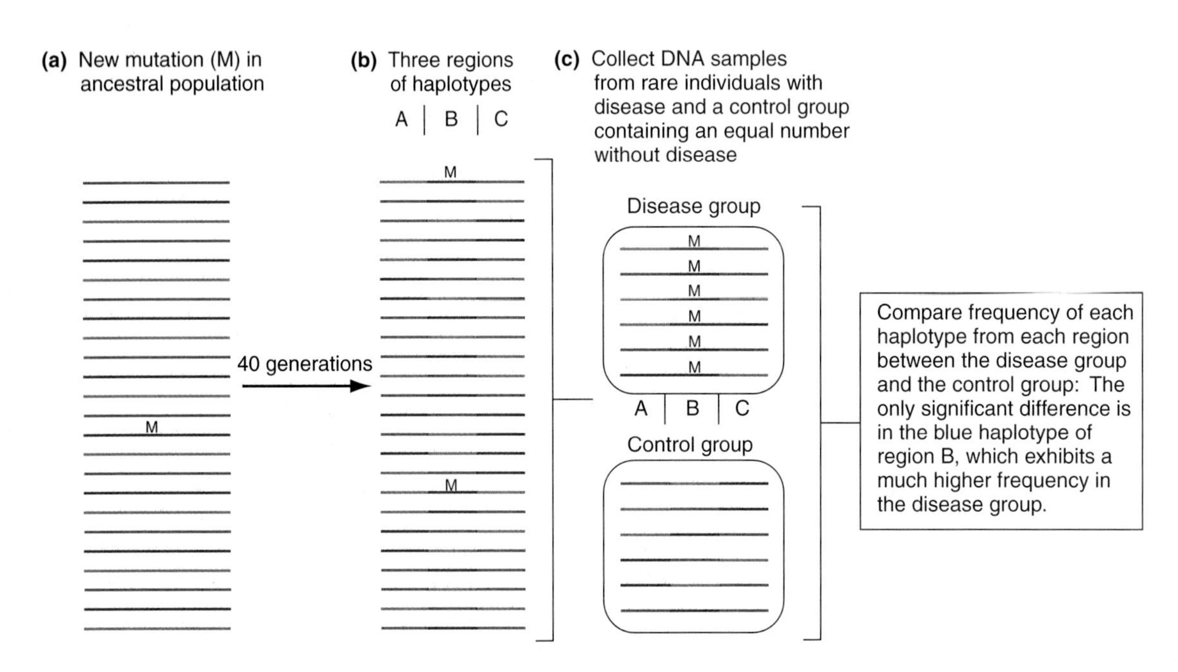

(a) New mutation (M) in ancestral population

(b) Three regions of haplotypes

A | B | C

(c) Collect DNA samples from rare individuals with disease and a control group containing an equal number without disease

40 generations

Disease group

M
M
M
M
M
M

A | B | C

Control group

Compare frequency of each haplotype from each region between the disease group and the control group: The only significant difference is in the blue haplotype of region B, which exhibits a much higher frequency in the disease group.

Haplotype association allows high-resolution gene mapping
Figure 11.26

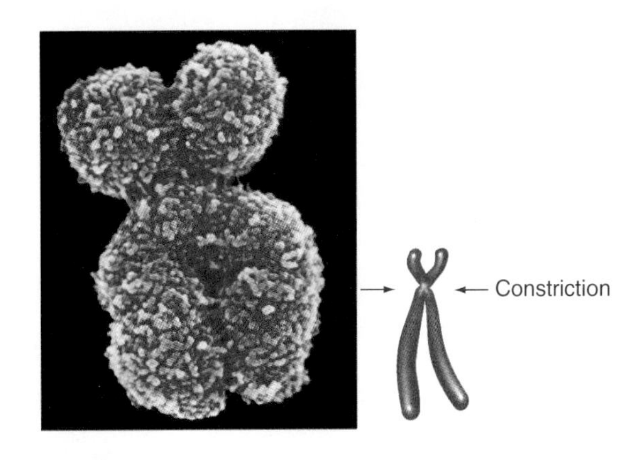

A chromosome's centromere looks like a constriction under the microscope
Figure 12.1

© Biophoto Associates/Photo Researchers, Inc.

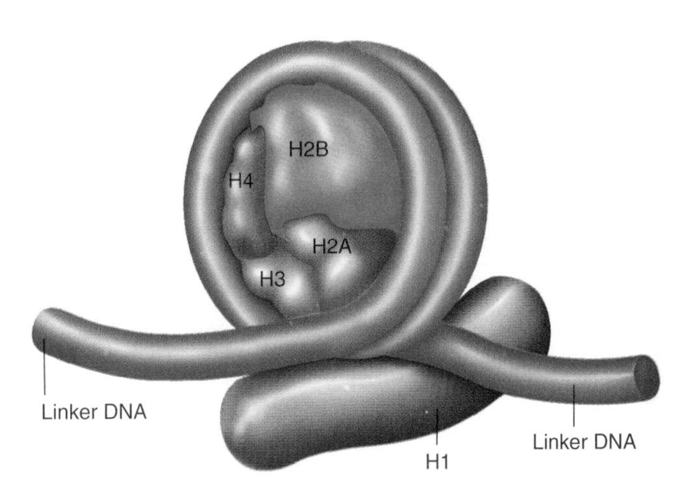

Nucleosomes: The basic building blocks of chromatin
Figure 12.3

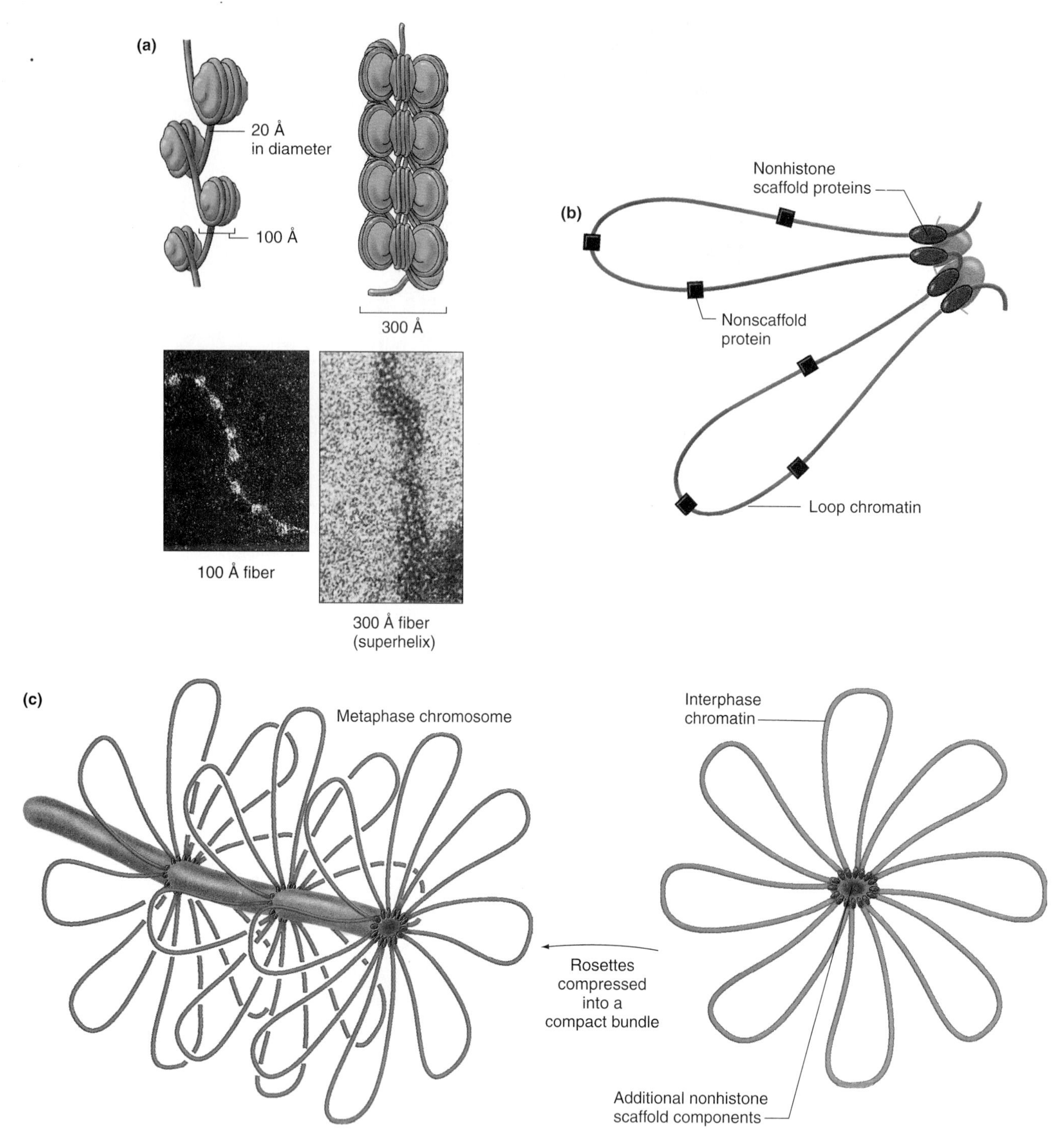

(a)

20 Å in diameter

100 Å

300 Å

100 Å fiber

300 Å fiber (superhelix)

(b)

Nonhistone scaffold proteins

Nonscaffold protein

Loop chromatin

(c)

Metaphase chromosome

Interphase chromatin

Rosettes compressed into a compact bundle

Additional nonhistone scaffold components

Models of higher-level packaging
Figure 12.4

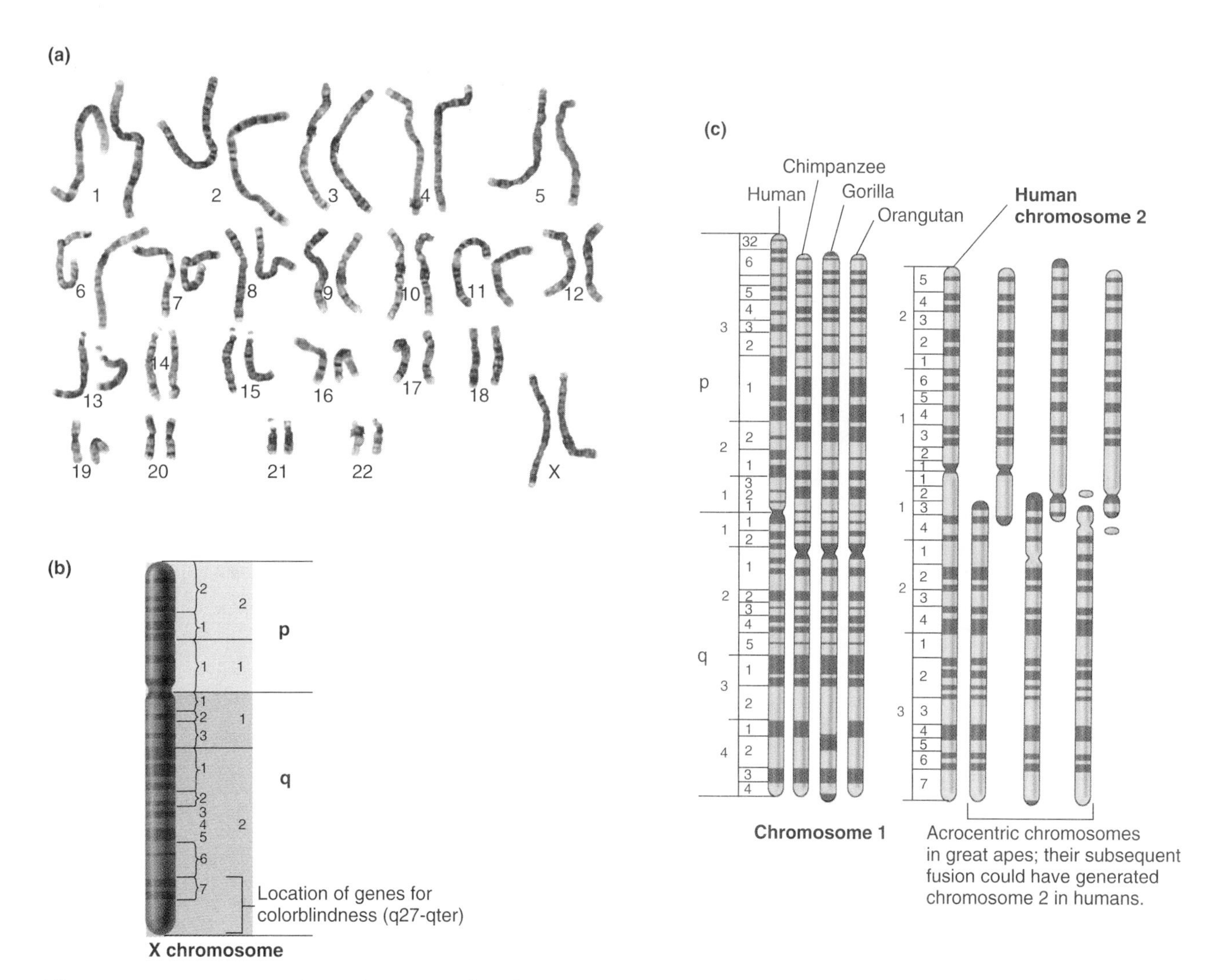

(a)

1 2 3 4 5

6 7 8 9 10 11 12

13 14 15 16 17 18

19 20 21 22 X

(b)

2
1 2
p

1 1

1
2 1
3

1
q
1
2
3 2
4
5
6
7

Location of genes for
colorblindness (q27-qter)

X chromosome

(c)

Human
Chimpanzee
Gorilla
Orangutan

**Human
chromosome 2**

Chromosome 1

Acrocentric chromosomes
in great apes; their subsequent
fusion could have generated
chromosome 2 in humans.

Chromosomes can be characterized by their banding patterns
Figure 12.6

Consensus region

5' •••CAAATTTCGTCAAAAATGCTAAGAAATAGGTTATTACTTTTATTTAAGTATTGTTTGTGCCTTTTGAAAAGCAAGCATAAAAGATCTAAACATAAAATCTGTAAAATAAC•••3'
3' •••GTTTAAAGCAGTTTTTACGATTCTTTATCCAATAATGAAAATAAATTCATAACAAACACGGAAAACTTTTCGTTCGTATTTTCTAGATTTGTATTTTAGACATTTTATTG•••5'

Eukaryotic chromosomes have multiple origins of replication
Figure 12.7

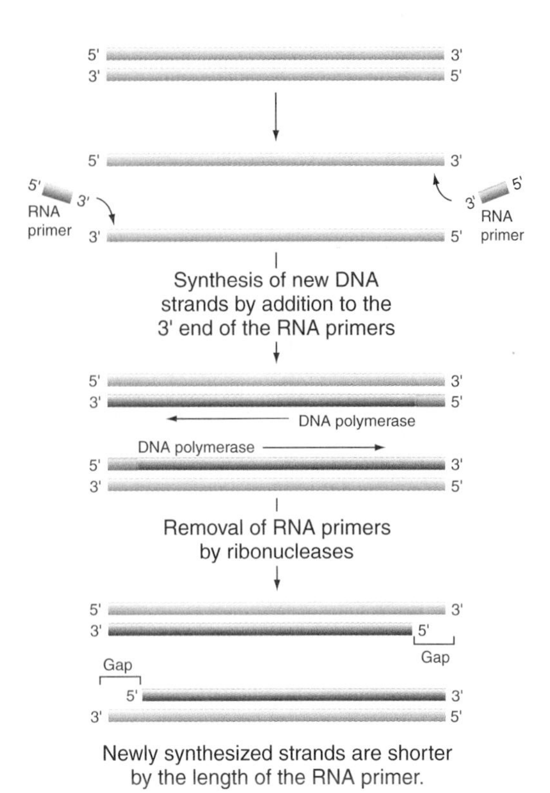

DNA polymerase cannot reconstruct the 5′ end of a DNA strand
Figure 12.9

Synthesis of new DNA strands by addition to the 3′ end of the RNA primers

DNA polymerase

DNA polymerase

Removal of RNA primers by ribonucleases

Gap

Gap

Newly synthesized strands are shorter by the length of the RNA primer.

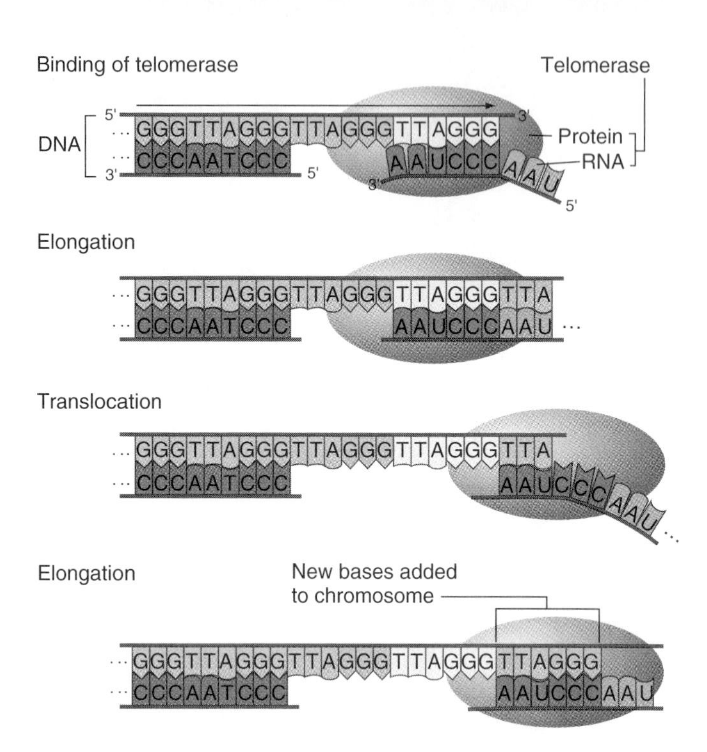

How telomerase extends telomeres
Figure 12.10

Binding of telomerase

Elongation

Translocation

Elongation — New bases added to chromosome

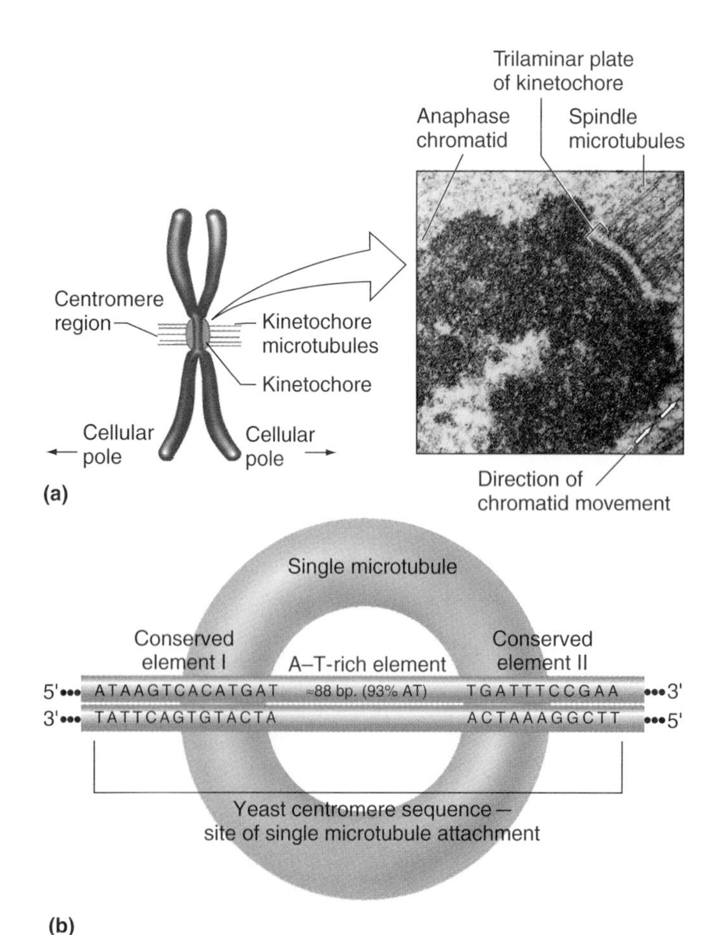

Centromere structure and function
Figure 12.11

a-2: © Dr. Jeremy David Pickett Heaps, School of Botany, University of Melbourne; b: Courtesy of Alberts et al, *Molecular Biology of the Cell, 3rd edition.* New York: Garland Publishing, 1994. Courtesy of Alberts et al, *Essential Cell Biology.* New York: Garland Publishing, 1998.

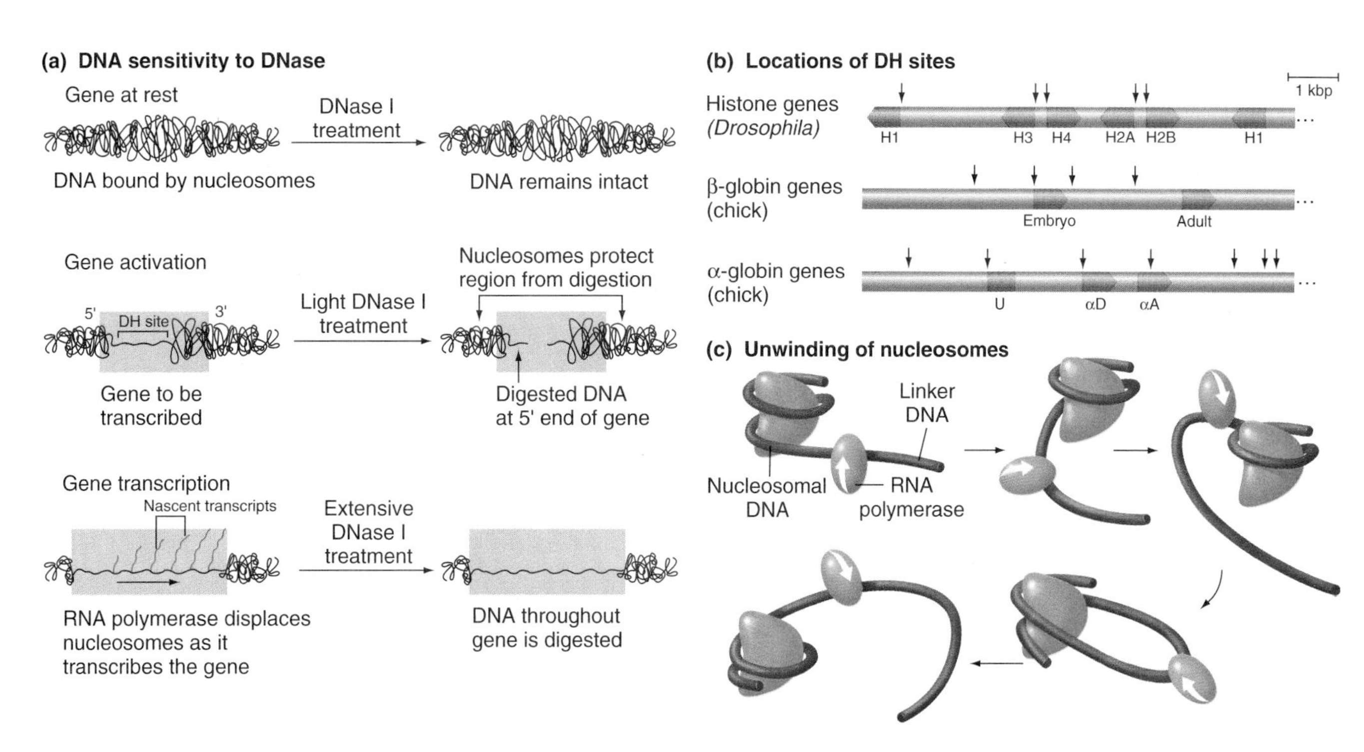

(a) DNA sensitivity to DNase

Gene at rest

DNA bound by nucleosomes

→ DNase I treatment →

DNA remains intact

Gene activation

5' DH site 3'

Gene to be transcribed

→ Light DNase I treatment →

Nucleosomes protect region from digestion

Digested DNA at 5' end of gene

Gene transcription

Nascent transcripts

RNA polymerase displaces nucleosomes as it transcribes the gene

→ Extensive DNase I treatment →

DNA throughout gene is digested

(b) Locations of DH sites

Histone genes (*Drosophila*)

H1   H3 H4   H2A H2B   H1

1 kbp

β-globin genes (chick)

Embryo   Adult

α-globin genes (chick)

U   αD   αA

(c) Unwinding of nucleosomes

Linker DNA

Nucleosomal DNA   RNA polymerase

DNA compaction and transcription
Figure 12.12

(a)

Wild-type X chromosome

Euchromatin   Heterochromatin

w^+ rst^+

Centromere

Inversion

rst^+ w^+

Inverted X chromosome

Position-effect variegation in *Drosophila* is a phenotypic effect of facultative heterochromatin
Figure 12.14

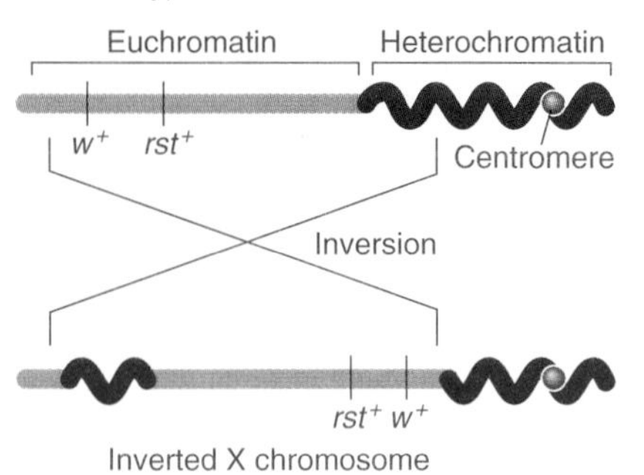

(b)

| **Appearance** | active w^+ = red
inactive w^+ = white
active rst^+ = smooth
inactive rst^+ = rough | **Interpretation** |

Red smooth sectors

rst^+ w^+

Rearrangement brings w^+ and rst^+ close to heterochromatin near centromere. Heterochromatin does not invade either gene.

White smooth sectors

rst^+ w^+

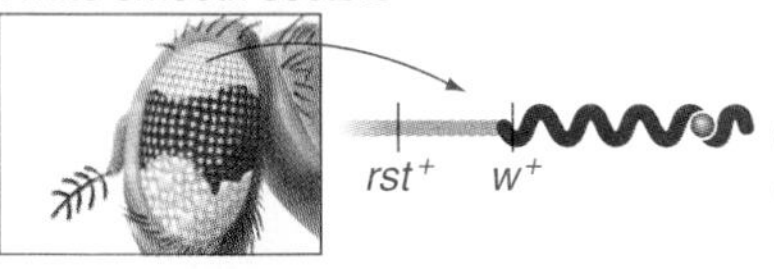

w^+ gene inactivated by spread of heterochromatin. rst^+ gene is active.

White rough sectors

rst^+ w^+

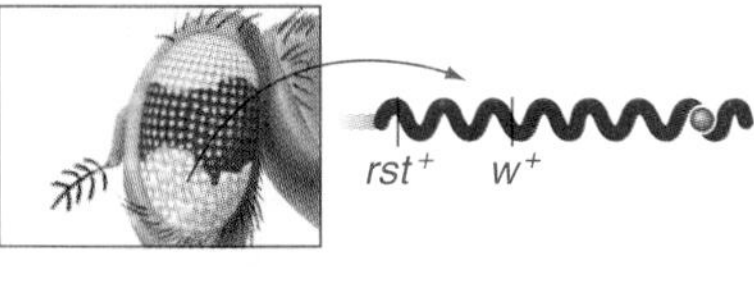

Both w^+ and rst^+ genes inactivated by spread of heterochromatin.

Red rough sectors

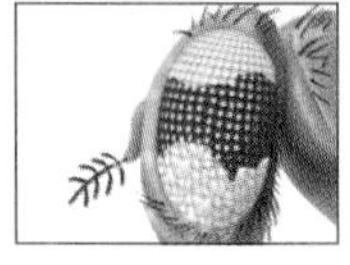

Never observed

rst^+ w^+

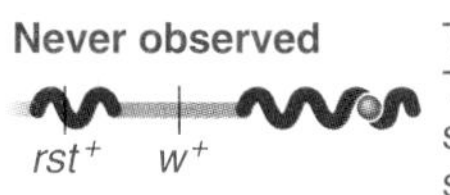

This is never observed. Therefore, heterochromatin spreads linearly without skipping genes.

— Interband

— Band

1 µm

Right arm of chromosome 3

X chromosome

Chromocenter (heterochromatin)

Chromosome 4

Right arm of chromosome 2

Left arm of chromosome 2

Left arm of chromosome 3

20 µm

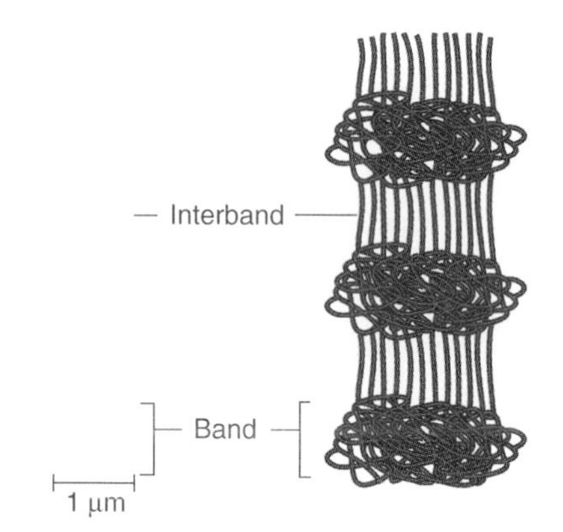

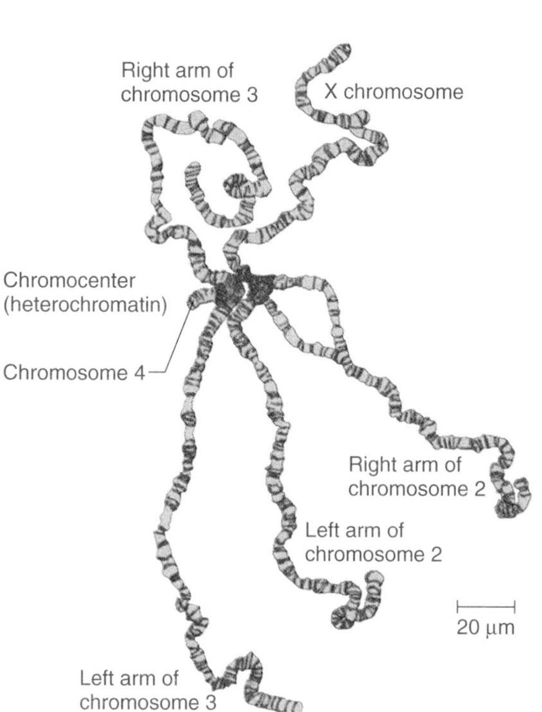

Polytene chromosomes in the salivary glands of *Drosophila* larvae
Figure 12.15

Direction of transcription ⟶

18S 5.8S 28S   18S 5.8S 28S   18S 5.8S 28S

Nontranscribed spacers

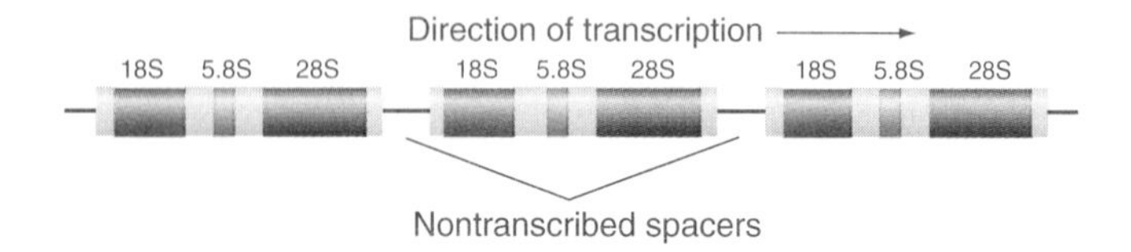

The nucleolus
Figure 12.16

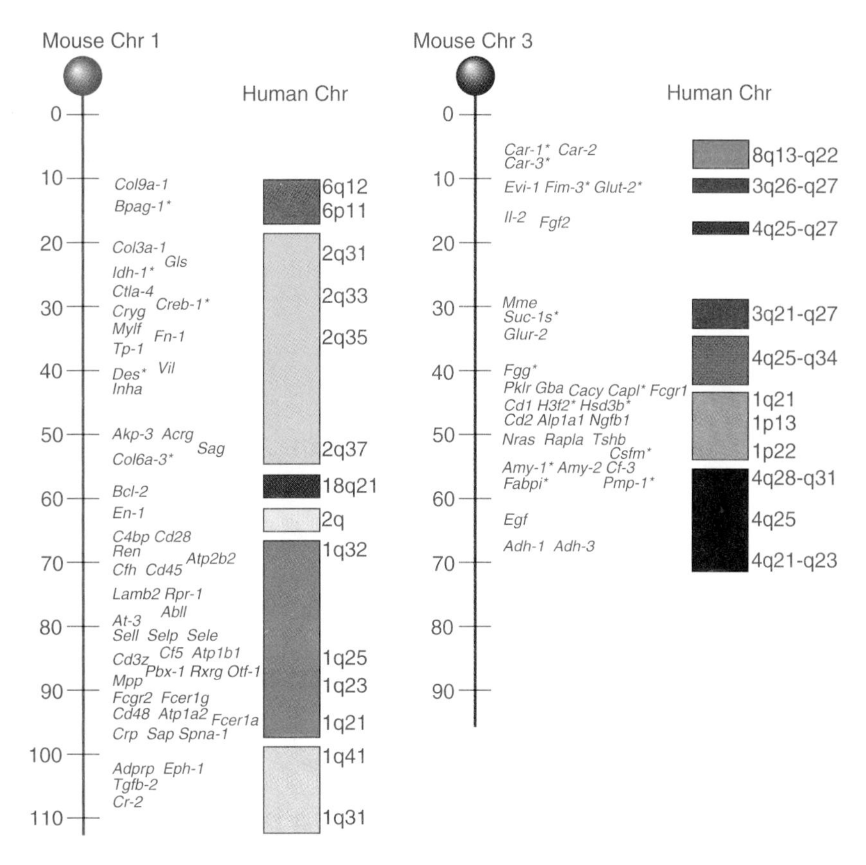

Comparing the mouse and human genomes
Figure 13.1

(a) DNA breakage may cause deletions

X rays break
both strands
of DNA
$A\ B\ C\ D\ E\ F\ G$ → $A\ B\ F\ G$

$C\ D\ E$

Deletion of region *CDE*

(b) Detecting deletions using PCR

Wild-type PCR product

No deletion

PCR primer 1
5′
3′ Wild type
PCR primer 2

Deleted DNA

PCR primer 1
5′
3′ Deletion *(Del)*
PCR primer 2

Deletion PCR product

Deletions: Origin and detection
Figure 13.2

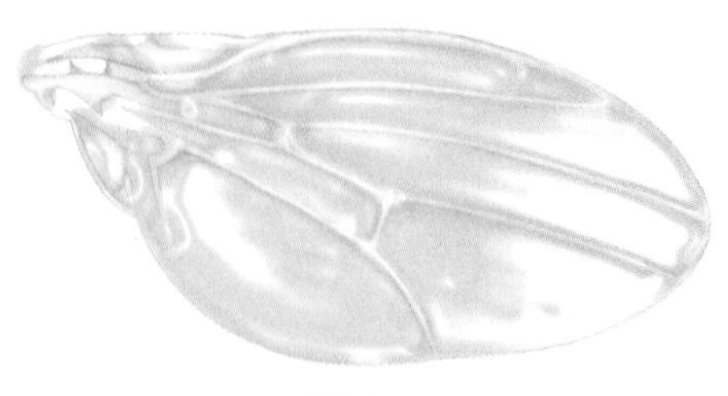

Wild type
(two copies of *Notch*⁺)

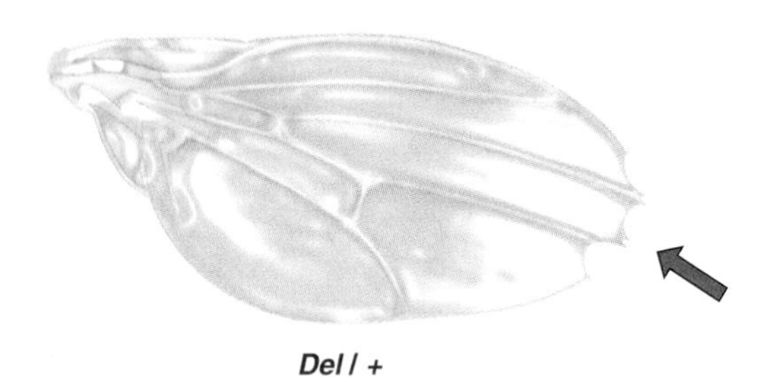

Del / +
(one copy of *Notch*⁺)

Heterozygosity for deletions may have phenotypic consequences
Figure 13.3

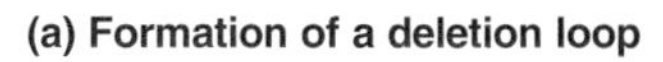

(a) Formation of a deletion loop

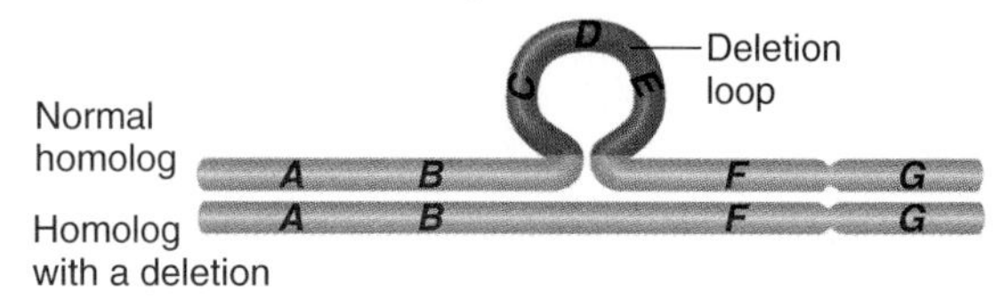

Deletion loops form in the chromosomes of deletion heterozygotes
Figure 13.4

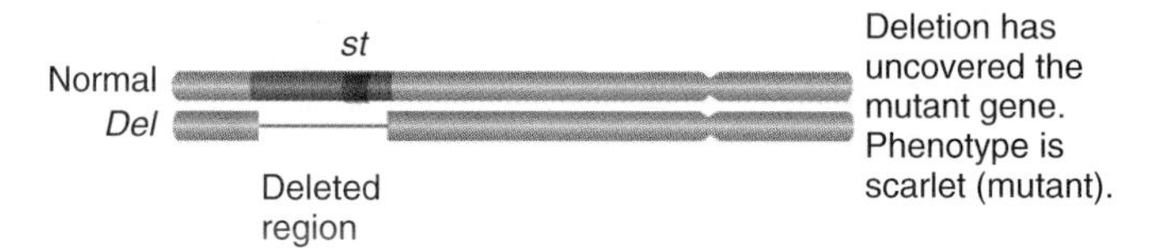

In deletion heterozygotes, pseudodominance shows that a deletion has removed a particular gene
Figure 13.5

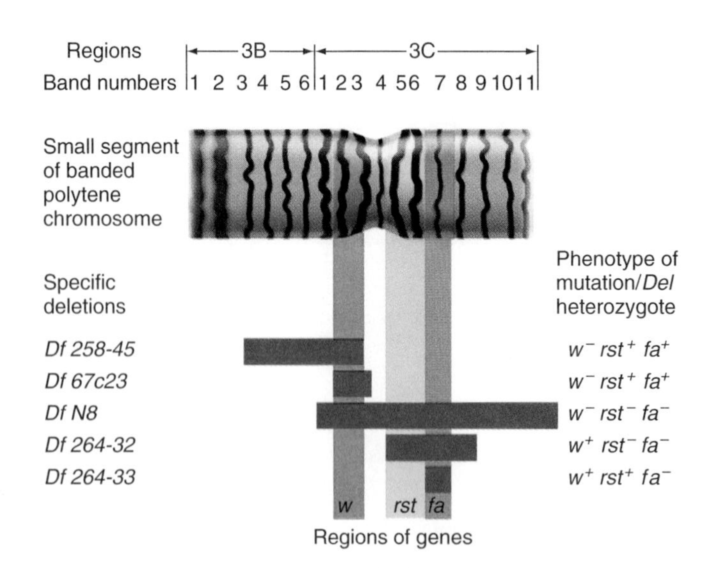

Using deletions to assign genes to bands on *Drosophila* polytene chromosomes
Figure 13.6

(a) Characterizing deletions with *in situ* hybridization.

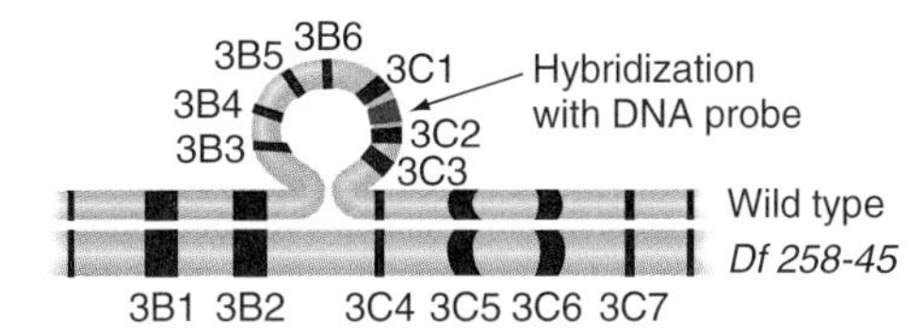

(b) Probing the location of a deletion breakpoint.

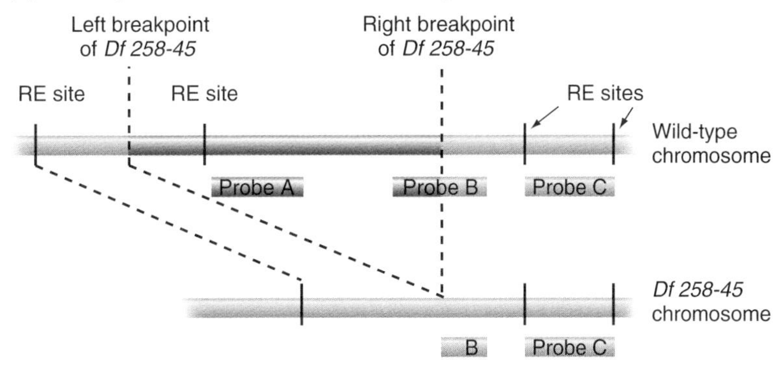

(c) Southern blot results

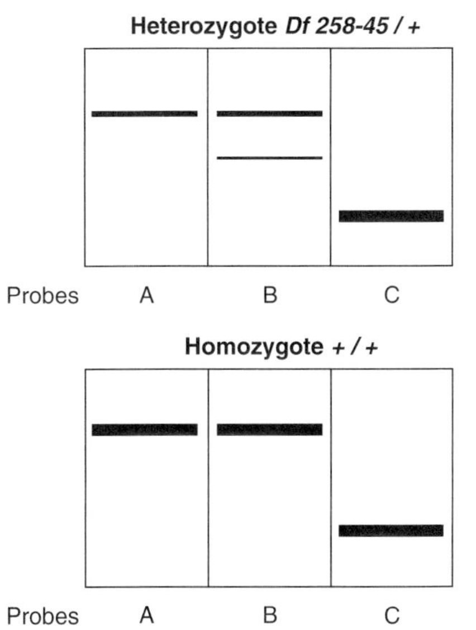

Using deletions to locate genes at the molecular level

Figure 13.7

(a) Types of duplications

Tandem duplications

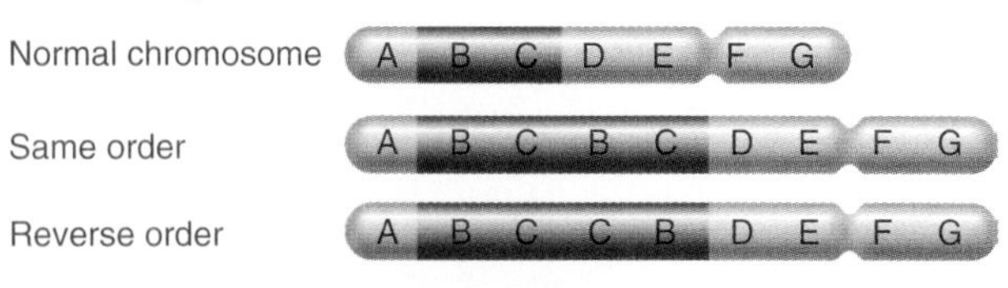

Normal chromosome

Same order

Reverse order

Nontandem (dispersed) duplications

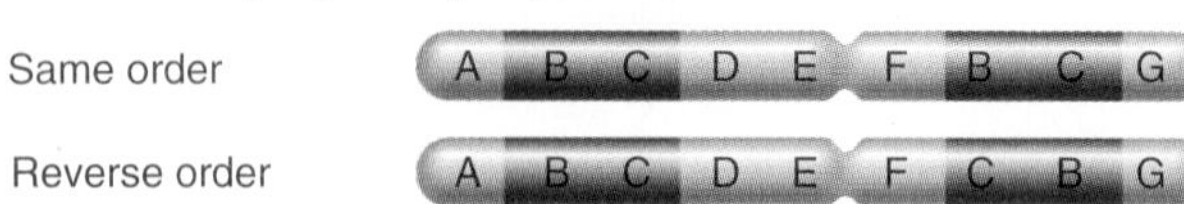

Same order

Reverse order

(c) Different kinds of duplication loops

Duplicated chromosome

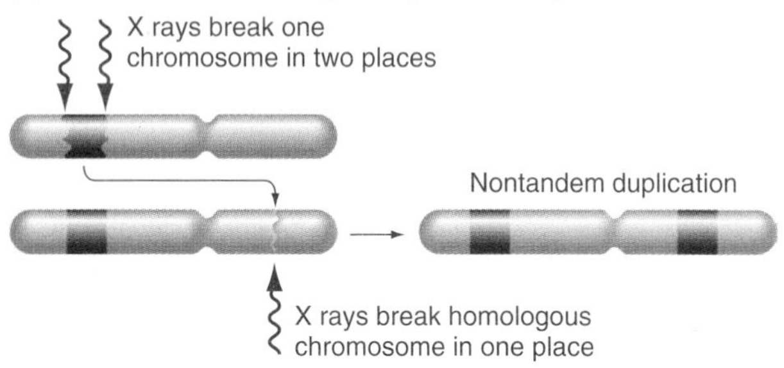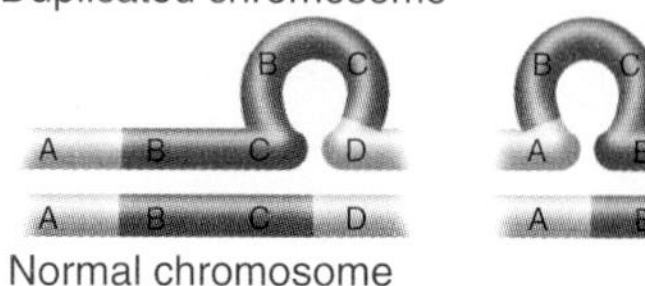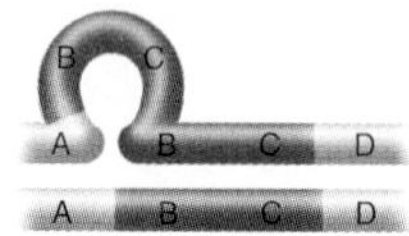

Normal chromosome

(b) Chromosome breakage can produce duplications

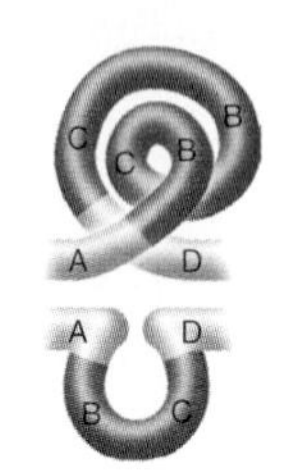

X rays break one
chromosome in two places

Nontandem duplication

X rays break homologous
chromosome in one place

Duplications: Structure, origin, and detection
Figure 13.8

(a) Duplication heterozygosity can cause visible phenotypes.

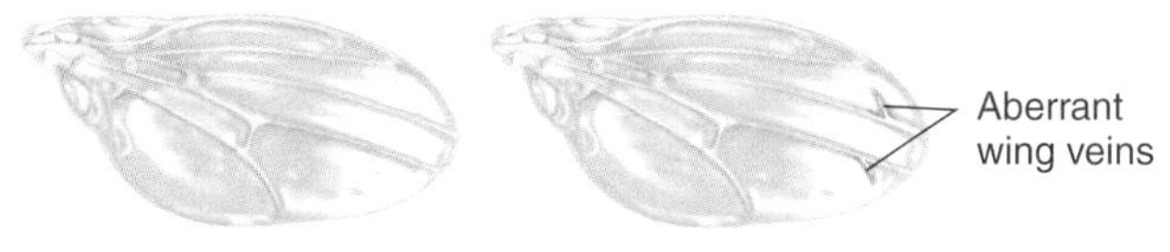

Aberrant
wing veins

Wild-type wing:
two copies of *Notch⁺* gene

Three copies of *Notch⁺ gene*

(b) For rare genes, survival requires exactly two copies.

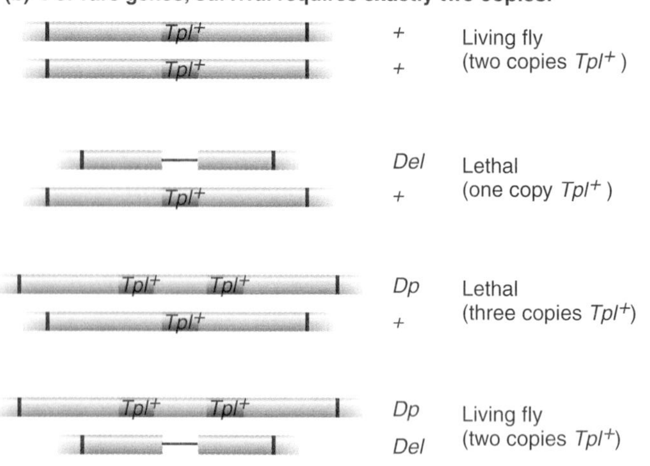

The phenotypic consequences of duplications
Figure 13.9

Genotype of X chromosomes

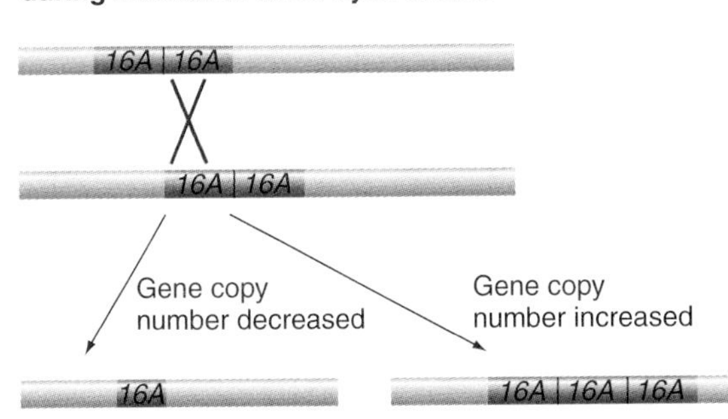

16A

Wild type

16A | 16A

Bar

16A | 16A | 16A

Double-Bar

Out-of-register pairing during meiosis in a Bar-eyed female

16A | 16A

16A | 16A

Gene copy number decreased

Gene copy number increased

16A

16A | 16A | 16A

Unequal crossing-over can increase or decrease copy number
Figure 13.10

(a) Chromosome breakage can produce inversions.

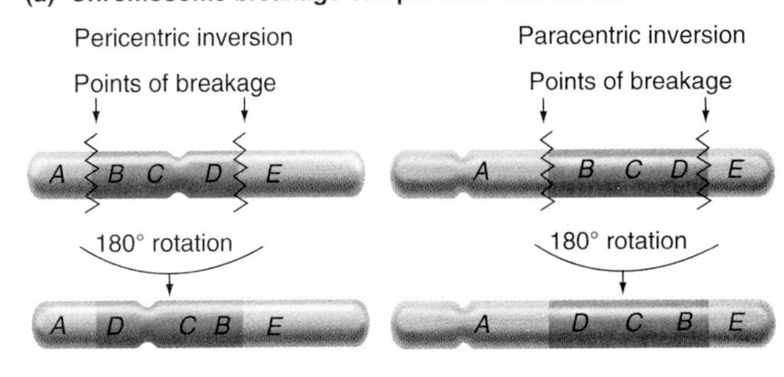

Pericentric inversion

Points of breakage

A B C D E

180° rotation

A D C B E

Paracentric inversion

Points of breakage

A B C D E

180° rotation

A D C B E

(b) Intrachromosomal recombination can also cause inversions.

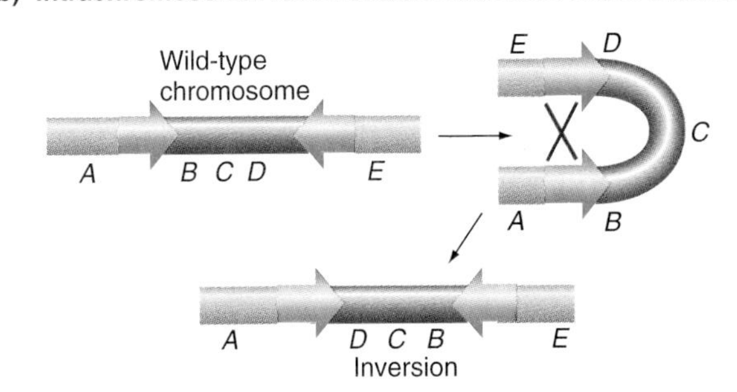

Wild-type chromosome

A   B C D   E

E   D

C

A   B

A   D C B   E

Inversion

(c) Inversions can disrupt gene function.

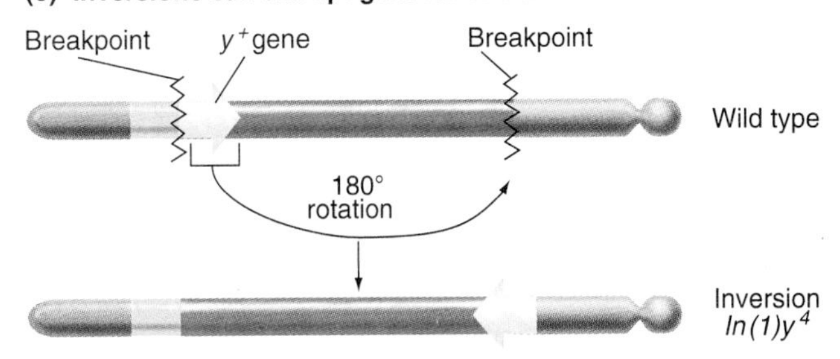

Breakpoint y^+ gene   Breakpoint

Wild type

180° rotation

Inversion
$In(1)y^4$

Inversions: Origins, types, and phenotypic effects
Figure 13.11

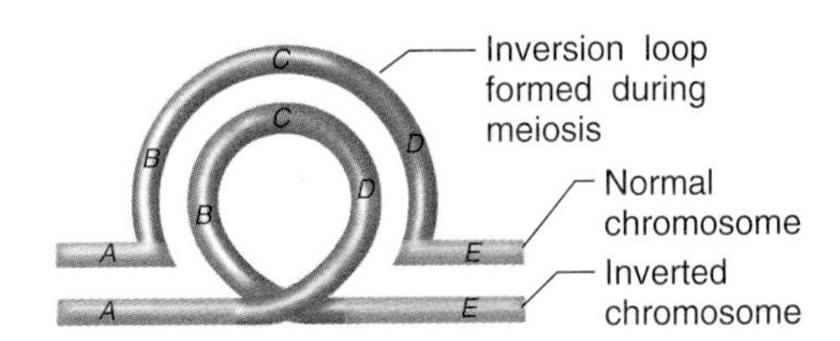

Inversion loop formed during meiosis

Normal chromosome

Inverted chromosome

C

C

B   D

B   D

A   E

A   E

Inversion loops form in inversion heterozygotes
Figure 13.12

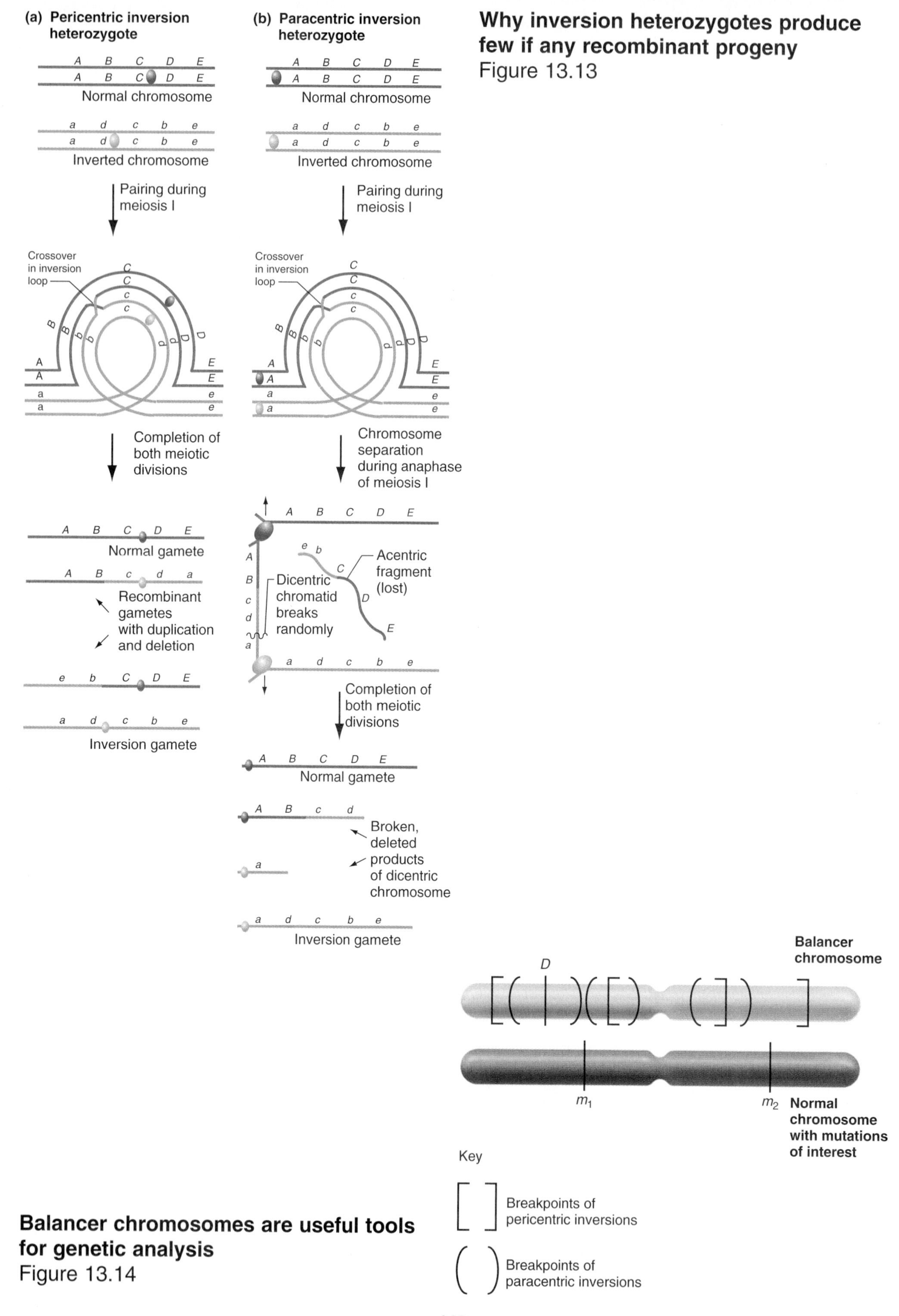

(a) Pericentric inversion heterozygote

A B C D E
A B C D E
Normal chromosome

a d c b e
a d c b e
Inverted chromosome

↓ Pairing during meiosis I

Crossover in inversion loop

C
C
c
c
B
B
b
b
A
A
a
a
E
E
e
e
D
D

↓ Completion of both meiotic divisions

A B C D E
Normal gamete

A B c d a
Recombinant gametes with duplication and deletion

e b C D E

a d c b e
Inversion gamete

(b) Paracentric inversion heterozygote

A B C D E
A B C D E
Normal chromosome

a d c b e
a d c b e
Inverted chromosome

↓ Pairing during meiosis I

Crossover in inversion loop

C
C
c
c
B
B
b
b
A
A
a
a
E
E
e
e
D
D

↓ Chromosome separation during anaphase of meiosis I

A B C D E

A
B
c
d
a
e b
C
D
E

Acentric fragment (lost)

Dicentric chromatid breaks randomly

a d c b e

↓ Completion of both meiotic divisions

A B C D E
Normal gamete

A B c d
a
Broken, deleted products of dicentric chromosome

a d c b e
Inversion gamete

Why inversion heterozygotes produce few if any recombinant progeny
Figure 13.13

D

m₁

m₂

Balancer chromosome

Normal chromosome with mutations of interest

Key

[ ] Breakpoints of pericentric inversions

( ) Breakpoints of paracentric inversions

Balancer chromosomes are useful tools for genetic analysis
Figure 13.14

142

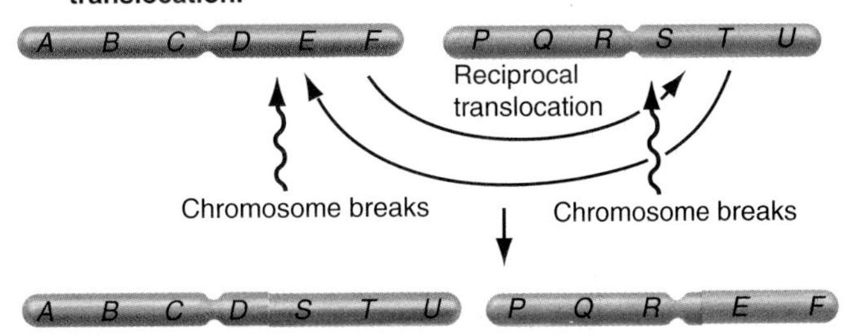

(a) Two chromosome breaks can produce a reciprocal translocation.

Reciprocal translocations are exchanges between nonhomologous chromosomes
Figure 13.15

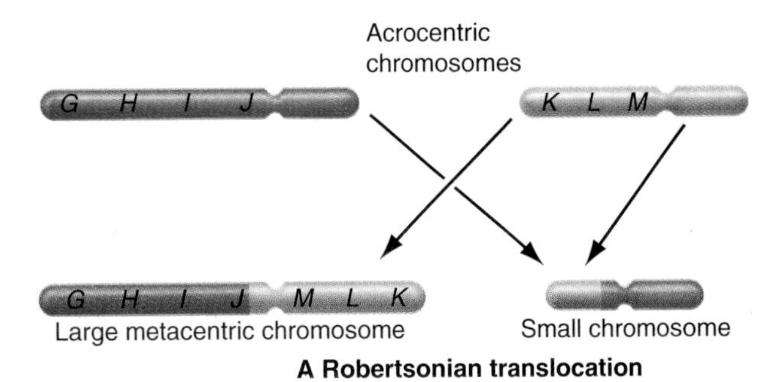

A Robertsonian translocation

Robertsonian translocations can reshape genomes
Figure 13.16

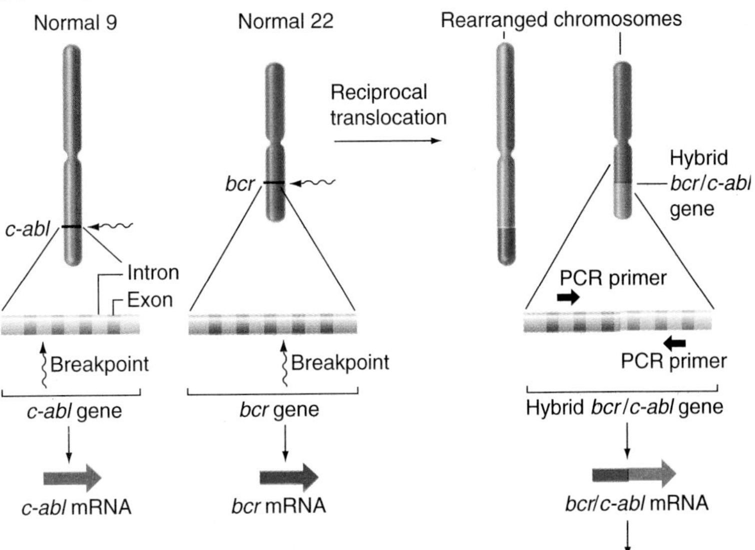

(b) The genetic basis for chronic myelogenous leukemia.

How a reciprocal translocation helps cause one kind of leukemia
Figure 13.17

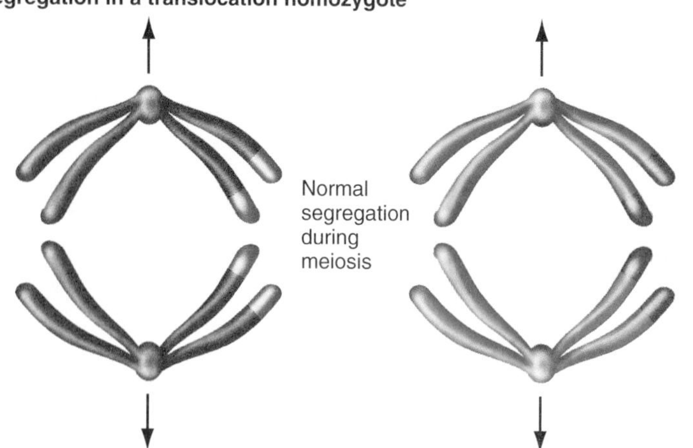

(a) Segregation in a translocation homozygote

Normal segregation during meiosis

(b) Chromosome pairing in a translocation heterozygote

(c) Segregation in a translocation heterozygote

Segregation pattern	Alternate		Adjacent - 1		Adjacent - 2 (less frequent)	
	Balanced N1 + N2	Balanced T1 + T2	Unbalanced T1 + N2	Unbalanced N1 + T2	Unbalanced N1 + T1	Unbalanced N2 + T2
Gametes	a b c d e f / p q r s t u	A B C D S T U / P Q R E F	A B C D S T U / p q r s t u	a b c d e f / P Q R E F	a b c d e f / A B C D S T U	p q r s t u / P Q R E F
Type of progeny when mated with normal abcdefpqrstu homozygote	abcdef pqrstu	ABCDEF PQRSTU	None surviving	None surviving	None surviving	None surviving

The meiotic segregation of chromosomes that have sustained reciprocal translocations

Figure 13.18

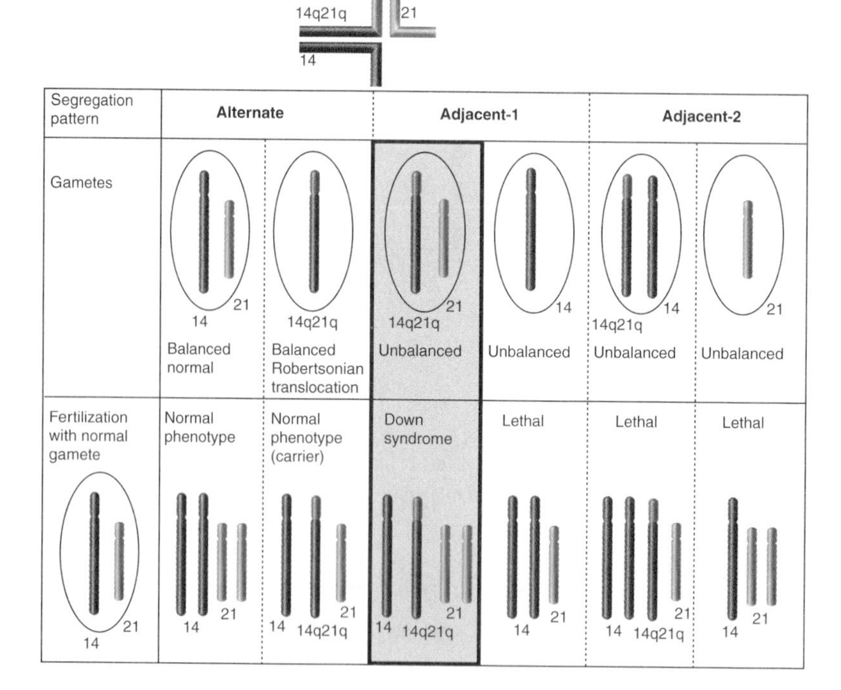

Segregation pattern	Alternate		Adjacent-1		Adjacent-2	
Gametes	14 21 Balanced normal	14q21q Balanced Robertsonian translocation	14q21q 21 Unbalanced	14 Unbalanced	14q21q 14 Unbalanced	21 Unbalanced
Fertilization with normal gamete	Normal phenotype	Normal phenotype (carrier)	Down syndrome	Lethal	Lethal	Lethal
	14 21	14 14q21q 21	14 14q21q 21	14 21	14 14q21q 21	14 21

How translocation Down syndrome arises

Figure 13.19

144

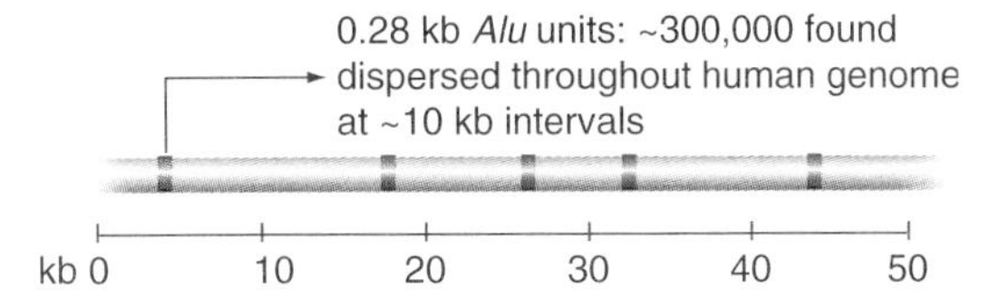

(a) *Alu* SINEs in the human genome.

0.28 kb *Alu* units: ~300,000 found
dispersed throughout human genome
at ~10 kb intervals

kb 0   10   20   30   40   50

TEs in human and corn genomes
Figure 13.22

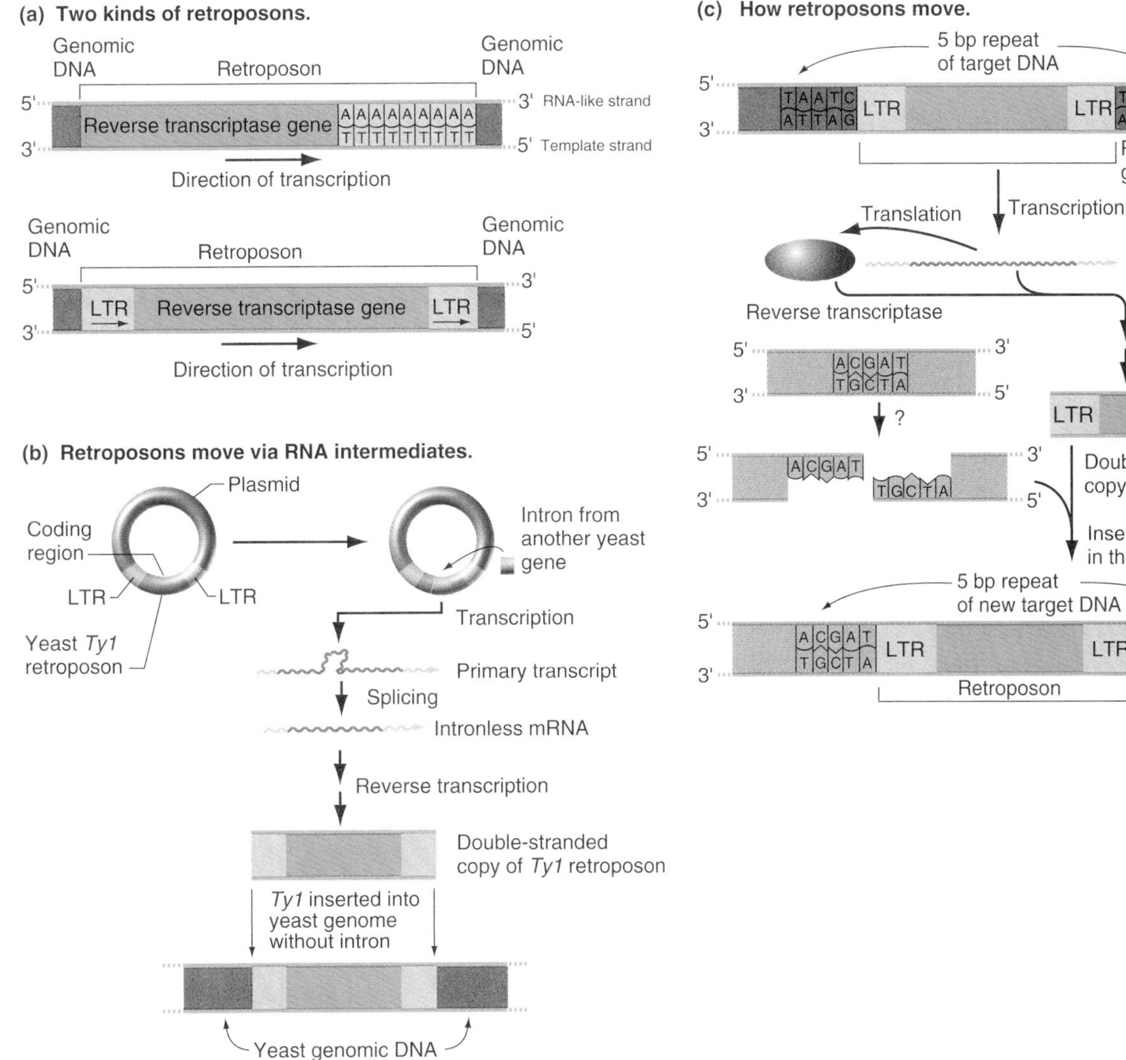

(a) Two kinds of retroposons.

Genomic DNA — Retroposon — Genomic DNA

5' | Reverse transcriptase gene | AAAAAAAAA | 3' RNA-like strand
3' | | TTTTTTTTTT | 5' Template strand

Direction of transcription

Genomic DNA — Retroposon — Genomic DNA

5' | LTR | Reverse transcriptase gene | LTR | 3'
3' | | | | 5'

Direction of transcription

(b) Retroposons move via RNA intermediates.

Plasmid
Coding region
LTR — LTR
Yeast *Ty1* retroposon

Intron from another yeast gene

Transcription

Primary transcript
Splicing
Intronless mRNA
Reverse transcription

Double-stranded copy of *Ty1* retroposon

Ty1 inserted into yeast genome without intron

Yeast genomic DNA

(c) How retroposons move.

5 bp repeat of target DNA

5' | TAATC / ATTAG | LTR | | LTR | TAATC / ATTAG | 3'
3' | | | | | | 5'

Retroposon at original genomic position

Translation ← Transcription →

RNA transcript

Reverse transcriptase

5' | ACGAT / TGCTA | 3'
3' | | 5'

? ↓

5' | ACGAT | TGCTA | 3'
3' | | 5'

Reverse transcription

LTR | | LTR

Double-stranded cDNA copy of retroposon

Insertion into new site in the genome

5 bp repeat of new target DNA

5' | ACGAT / TGCTA | LTR | | LTR | ACGAT / TGCTA | 3'
3' | | | | | | 5'

Retroposon

Retroposons: Structure and movement
Figure 13.23

(a) Transposon structure

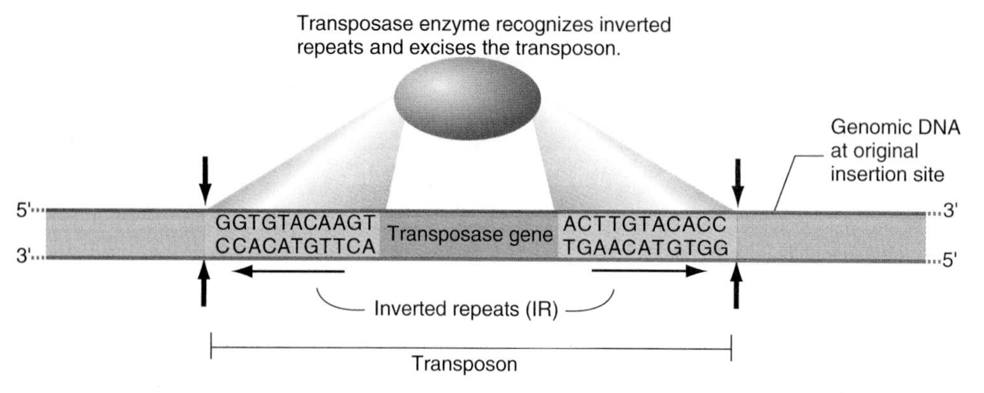

Transposase enzyme recognizes inverted repeats and excises the transposon.

Genomic DNA at original insertion site

5'

GGTGTACAAGT   Transposase gene   ACTTGTACACC
CCACATGTTCA   TGAACATGTGG

3'

Inverted repeats (IR)

Transposon

(b) How *P* element transposons move

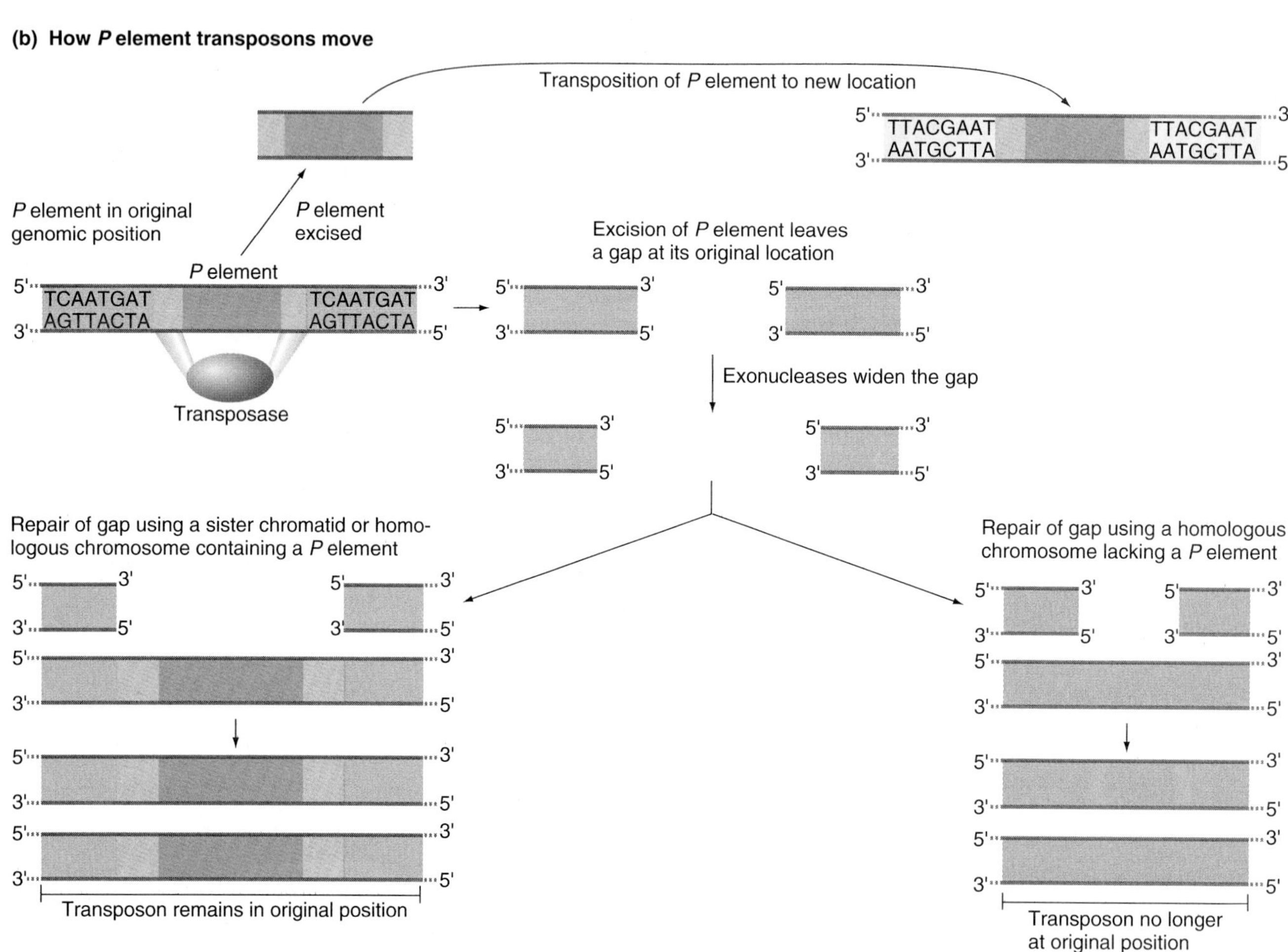

Transposition of *P* element to new location

P element in original genomic position

P element excised

5' TTACGAAT   TTACGAAT 3'
3' AATGCTTA   AATGCTTA 5'

P element

5' TCAATGAT   TCAATGAT 3'
3' AGTTACTA   AGTTACTA 5'

Transposase

Excision of *P* element leaves a gap at its original location

Exonucleases widen the gap

Repair of gap using a sister chromatid or homologous chromosome containing a *P* element

Transposon remains in original position

Repair of gap using a homologous chromosome lacking a *P* element

Transposon no longer at original position

Transposons: Structure and movement
Figure 13.24

TEs can cause mutations on insertion into a gene
Figure 13.25

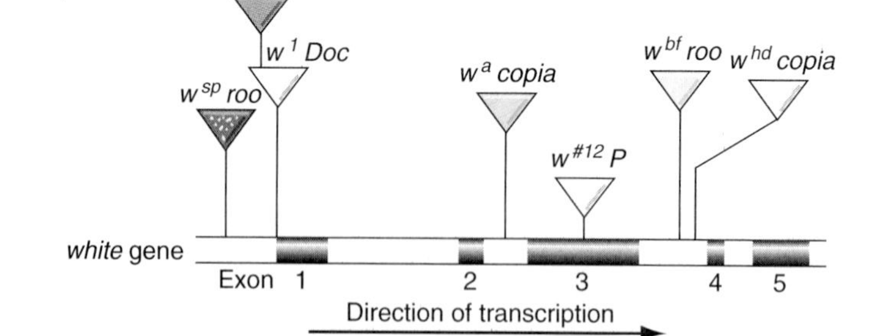

w^{ch} *pogo* (in *Doc*)

w^1 *Doc*

w^{sp} *roo*

w^a *copia*

w^{bf} *roo* w^{hd} *copia*

$w^{#12}$ *P*

white gene

Exon 1   2   3   4   5

Direction of transcription

(a) Unequal crossing-over between TEs.

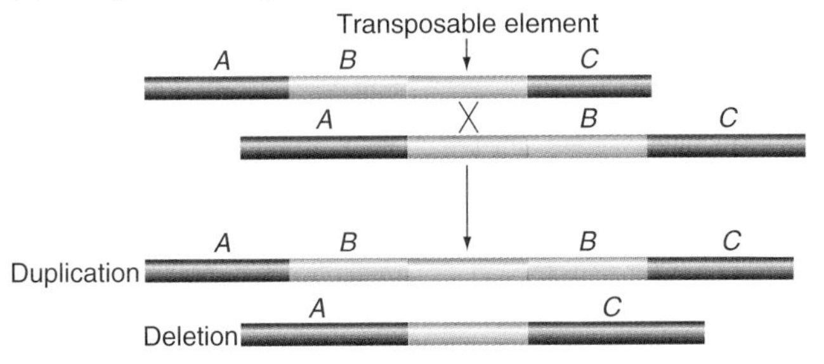

Duplication

Deletion

(b) Two transposons can form a large, composite transposon.

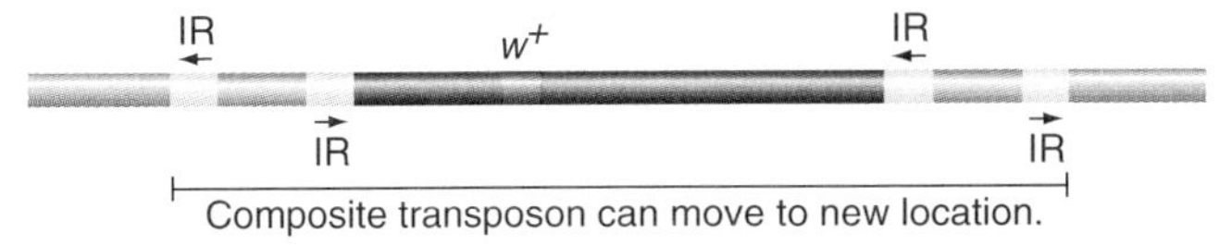

Composite transposon can move to new location.

How TEs generate chromosomal rearrangements and relocate genes
Figure 13.26

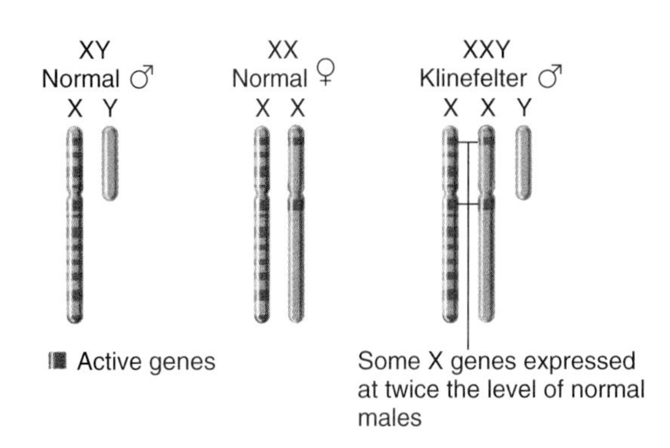

Active genes

Some X genes expressed at twice the level of normal males

Why aneuploidy for the X chromosome can have phenotypic consequences
Figure 13.27

(a) Nondisjunction can occur during either meiotic division.

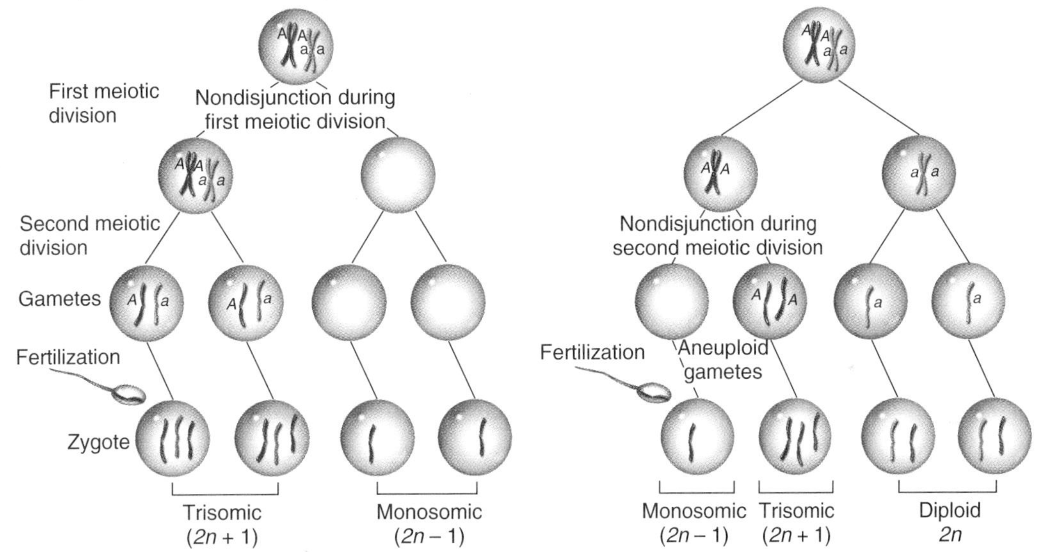

(b) Aneuploids beget aneuploid progeny.

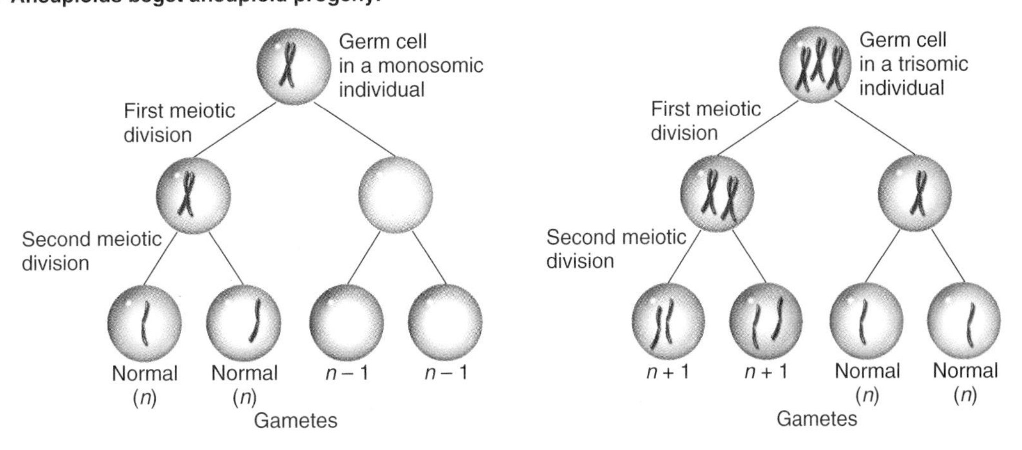

Aneuploidy is caused by problems in meiotic chromosome segregation
Figure 13.28

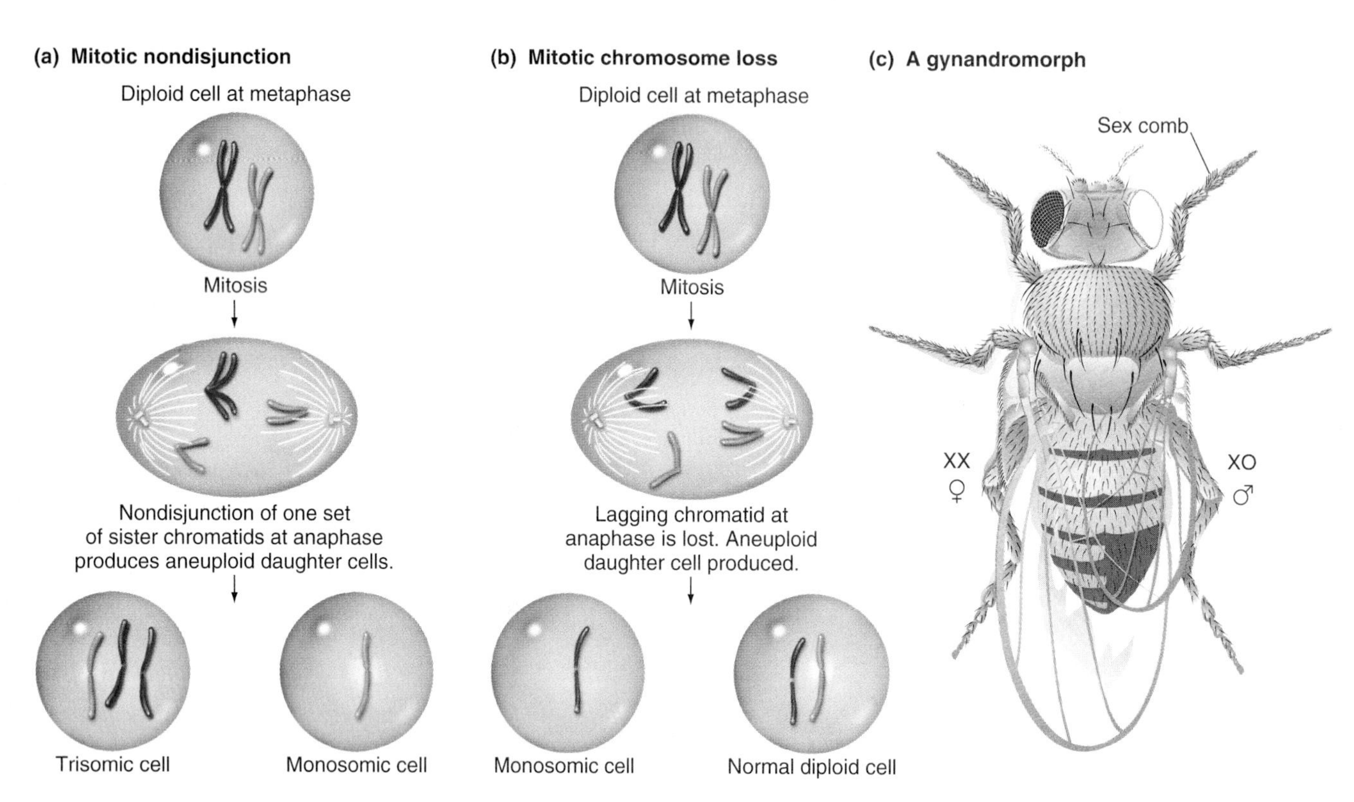

(a) Mitotic nondisjunction

Diploid cell at metaphase

Mitosis

Nondisjunction of one set
of sister chromatids at anaphase
produces aneuploid daughter cells.

Trisomic cell   Monosomic cell

(b) Mitotic chromosome loss

Diploid cell at metaphase

Mitosis

Lagging chromatid at
anaphase is lost. Aneuploid
daughter cell produced.

Monosomic cell   Normal diploid cell

(c) A gynandromorph

Sex comb

XX   XO
♀   ♂

Mistakes during mitosis can generate clones of aneuploid cells
Figure 13.29

(a) How to create a monoploid plant

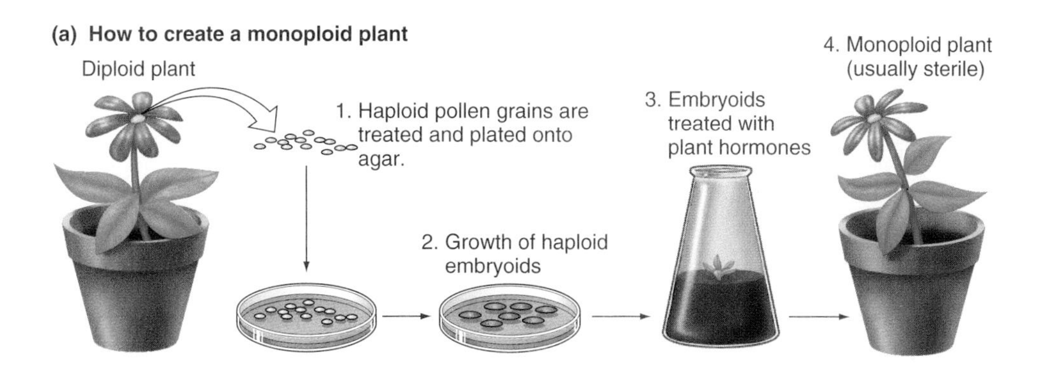

Diploid plant

1. Haploid pollen grains are treated and plated onto agar.

2. Growth of haploid embryoids

3. Embryoids treated with plant hormones

4. Monoploid plant (usually sterile)

(b) Using monoploloid plants to select for herbicide resistance

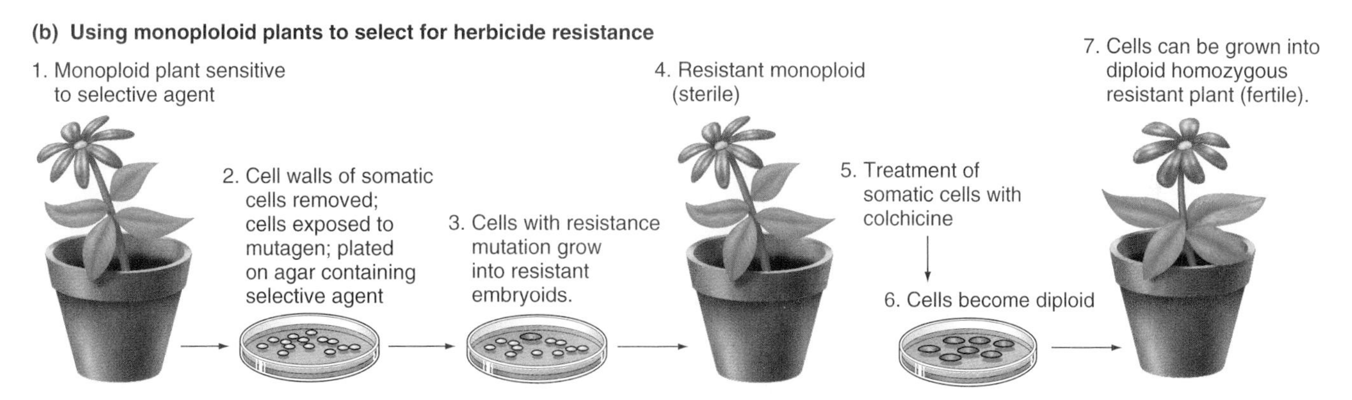

1. Monoploid plant sensitive to selective agent

2. Cell walls of somatic cells removed; cells exposed to mutagen; plated on agar containing selective agent

3. Cells with resistance mutation grow into resistant embryoids.

4. Resistant monoploid (sterile)

5. Treatment of somatic cells with colchicine

6. Cells become diploid

7. Cells can be grown into diploid homozygous resistant plant (fertile).

(c) Using colchicine to double chromosome numbers

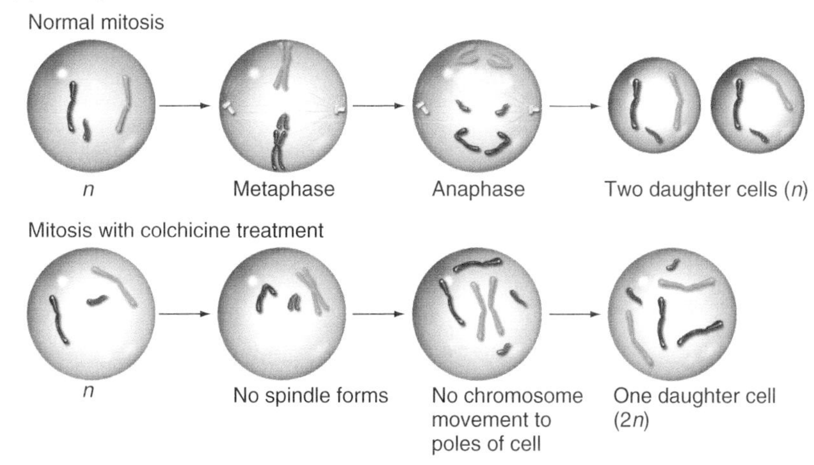

Normal mitosis

n Metaphase   Anaphase   Two daughter cells (n)

Mitosis with colchicine treatment

n No spindle forms   No chromosome movement to poles of cell   One daughter cell ($2n$)

The creation and use of monoploid plants
Figure 13.30

(a) Formation of a triploid organism

Meiosis in tetraploid (4x)
parent

Meiosis in diploid (2x)
parent

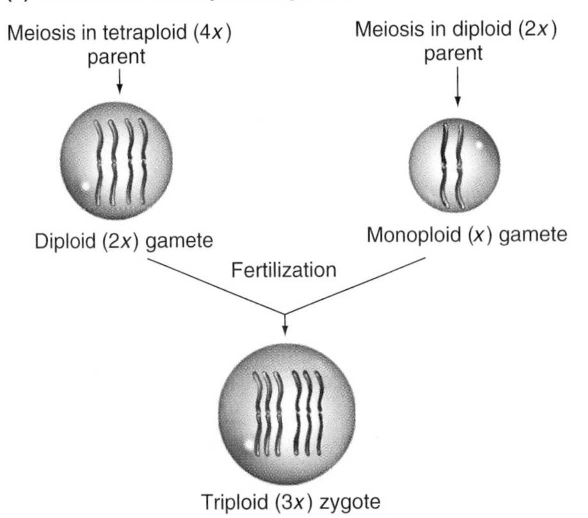

Diploid (2x) gamete

Monoploid (x) gamete

Fertilization

Triploid (3x) zygote

(b) Meiosis in a triploid organism

Triploid cell

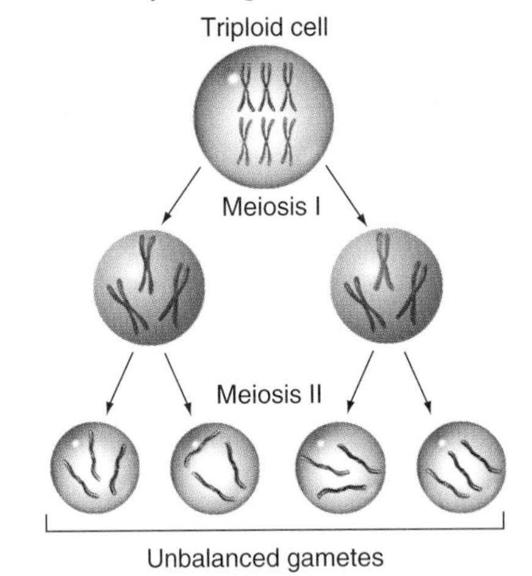

Meiosis I

Meiosis II

Unbalanced gametes

The genetics of triploidy
Figure 13.32

(a) Generation of tetraploid (4x) cells

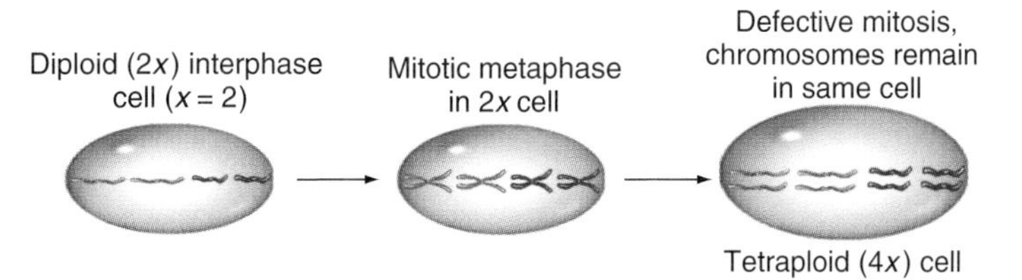

Diploid (2x) interphase cell (x = 2) → Mitotic metaphase in 2x cell → Defective mitosis, chromosomes remain in same cell → Tetraploid (4x) cell

(b) Pairing of chromosomes as bivalents

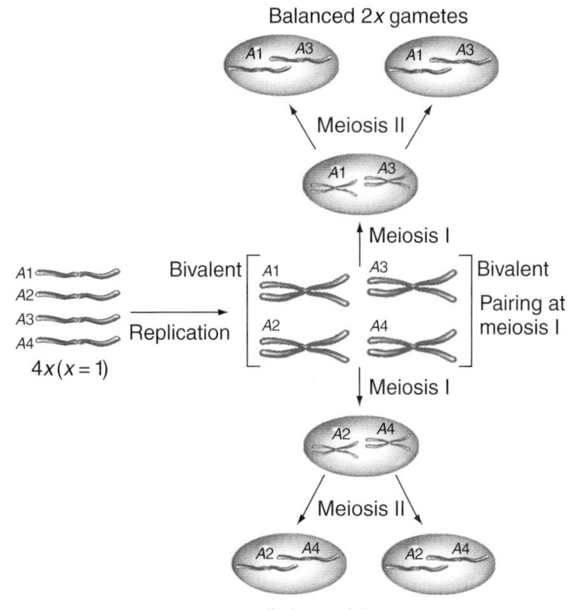

Balanced 2x gametes

A1 A3   A1 A3

Meiosis II

A1 A3

Meiosis I

A1   A3
Bivalent   Bivalent
A2   A4   Pairing at meiosis I

Replication

4x (x = 1)

A1
A2
A3
A4

Meiosis I

A2 A4

Meiosis II

A2 A4   A2 A4

Balanced 2x gametes

(c) Gametes formed by *A A a a* tetraploids

Chromosomes	Pairing			Gametes produced by random spindle attachment	
1. *A* 2. *A* 3. *a* 4. *a*	1 ↑*A* 3 ↑*a* 2 ↓*A* 4 ↓*a*	or	1 ↑*A* 4 ↑*a* 2 ↓*A* 3 ↓*a*	1 + 3 *A a* 2 + 4 *A a*	or 1 + 4 *A a* 2 + 3 *A a*
	1 ↑*A* 2 ↑*A* 3 ↓*a* 4 ↓*a*	or	1 ↑*A* 4 ↑*a* 3 ↓*a* 2 ↓*A*	1 + 2 *A A* 3 + 4 *a a*	or 1 + 4 *A a* 2 + 3 *A a*
	1 ↑*A* 2 ↑*A* 4 ↓*a* 3 ↓*a*	or	1 ↑*A* 3 ↑*a* 4 ↓*a* 2 ↓*A*	1 + 2 *A A* 3 + 4 *a a*	or 1 + 3 *A a* 2 + 4 *A a*

Total:
8*A a* : 2*A A* : 2*a a* = 4*A a* : 1*A A* : 1*a a*

The genetics of tetraploidy
Figure 13.33

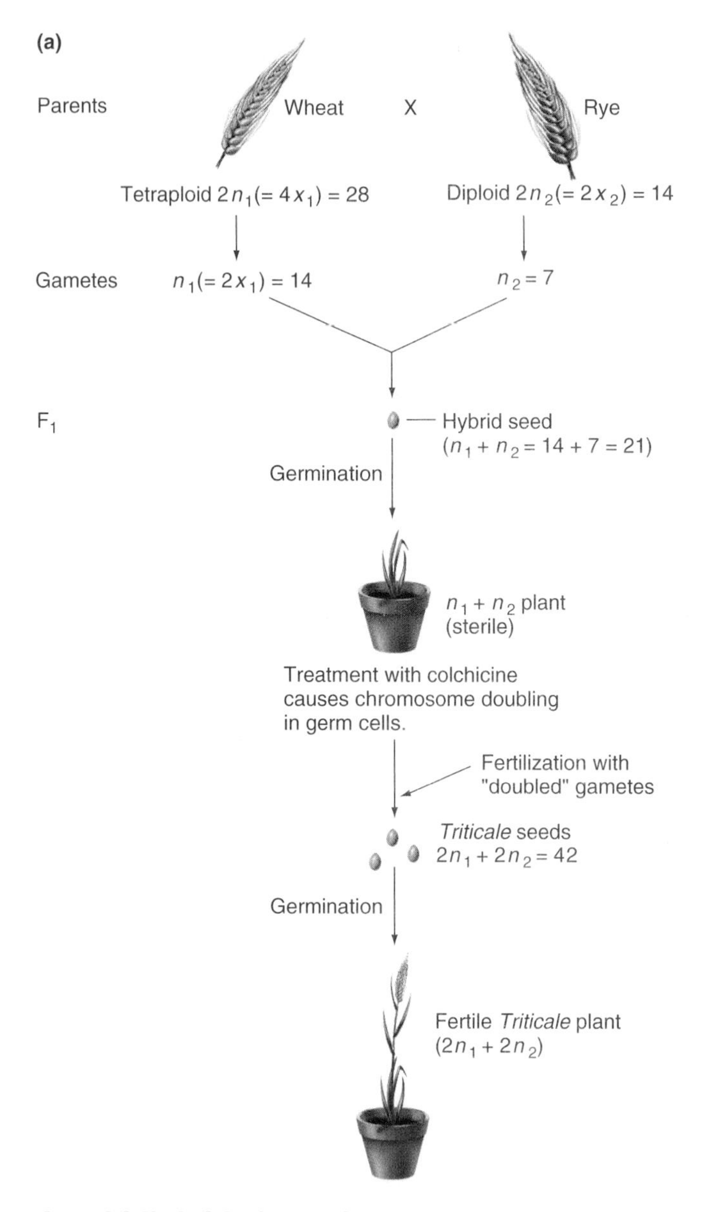

(a)

Parents   Wheat   X   Rye

Tetraploid $2n_1(= 4x_1) = 28$ Diploid $2n_2(= 2x_2) = 14$

Gametes $n_1(= 2x_1) = 14$ $n_2 = 7$

F_1 Hybrid seed
($n_1 + n_2 = 14 + 7 = 21$)

Germination

$n_1 + n_2$ plant
(sterile)

Treatment with colchicine
causes chromosome doubling
in germ cells.

Fertilization with
"doubled" gametes

Triticale seeds
$2n_1 + 2n_2 = 42$

Germination

Fertile *Triticale* plant
($2n_1 + 2n_2$)

Amphidiploids in agriculture
Figure 13.34

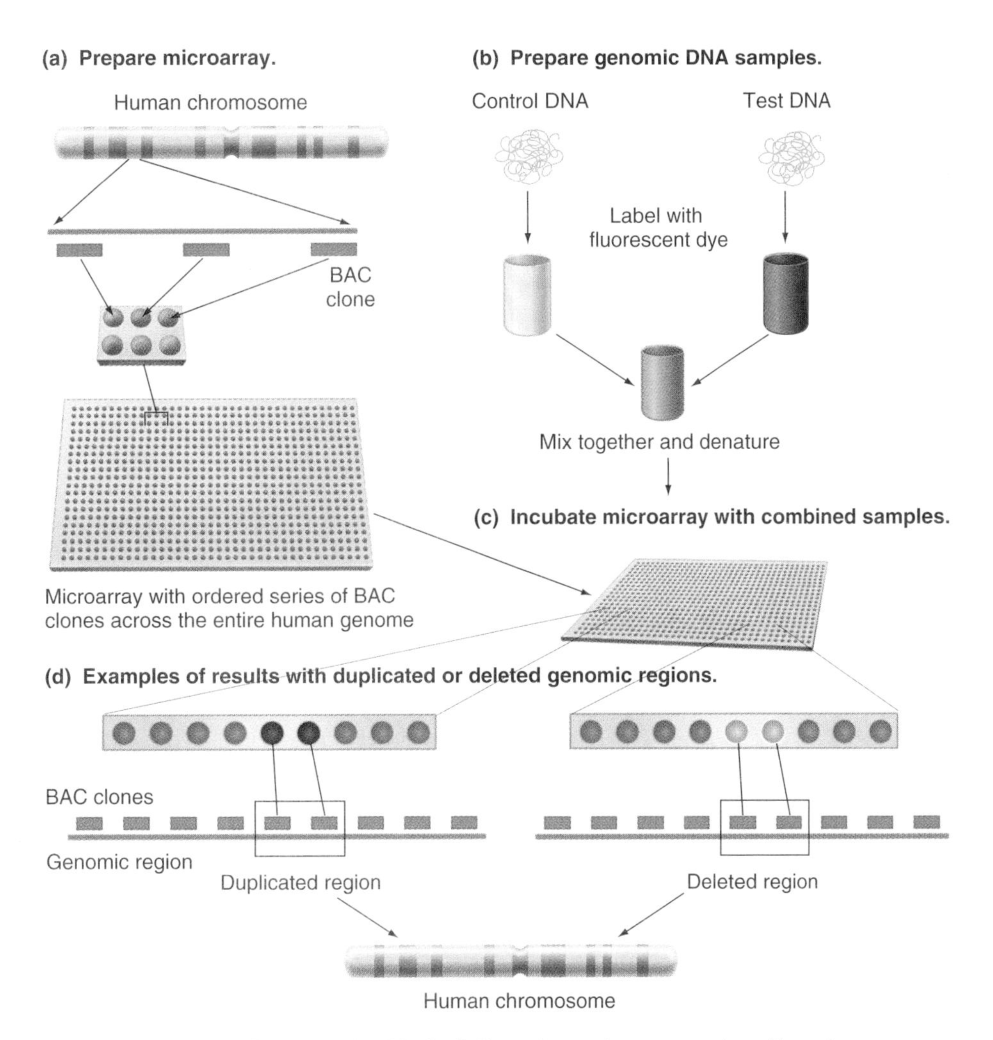

(a) Prepare microarray.

Human chromosome

BAC clone

Microarray with ordered series of BAC clones across the entire human genome

(b) Prepare genomic DNA samples.

Control DNA

Test DNA

Label with fluorescent dye

Mix together and denature

(c) Incubate microarray with combined samples.

(d) Examples of results with duplicated or deleted genomic regions.

BAC clones

Genomic region

Duplicated region

Deleted region

Human chromosome

Comparative Genomic Hybridization detects duplications, deletions, and aneuploidy
Figure 13.35

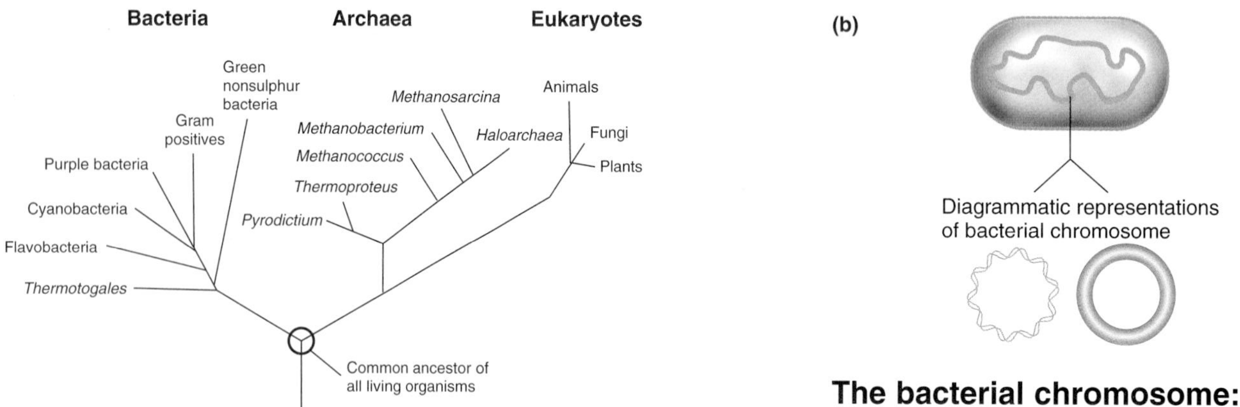

A family tree of living organisms
Figure 14.1

(b)

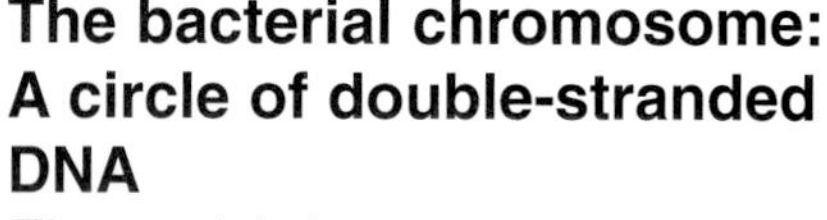

Diagrammatic representations
of bacterial chromosome

**The bacterial chromosome:
A circle of double-stranded
DNA**
Figure 14.4

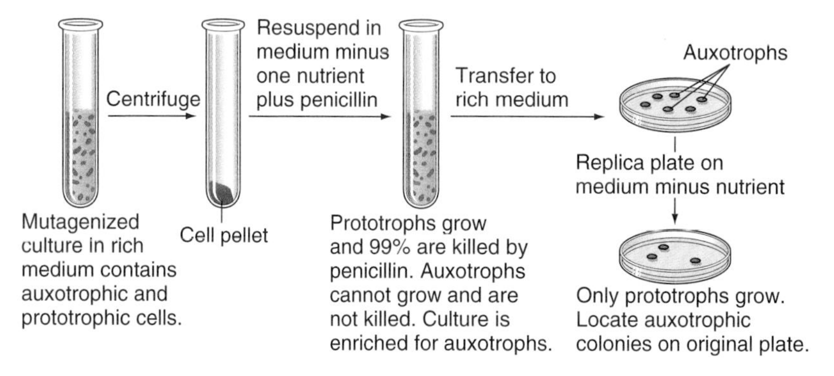

Penicillin enrichment for auxotrophic mutants
Figure 14.5

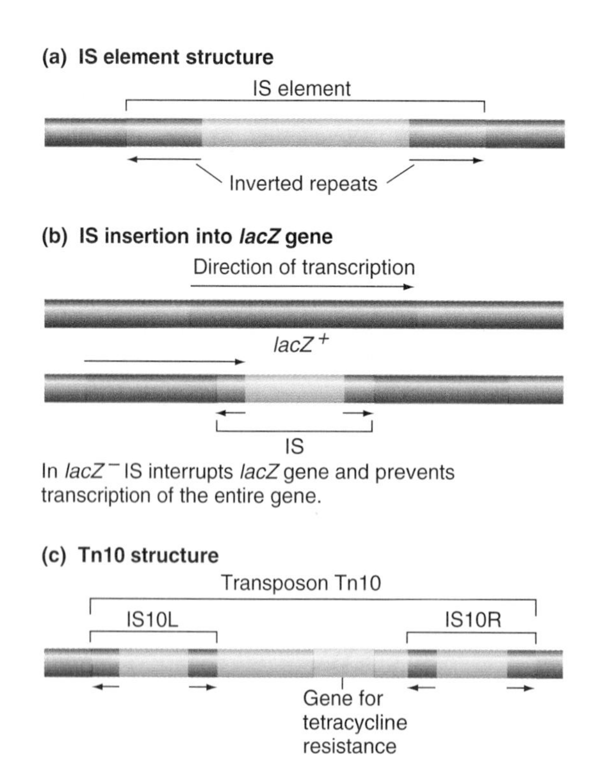

(a) IS element structure

(b) IS insertion into *lacZ* gene

In *lacZ*⁻ IS interrupts *lacZ* gene and prevents
transcription of the entire gene.

(c) Tn10 structure

Transposable elements in bacteria
Figure 14.6

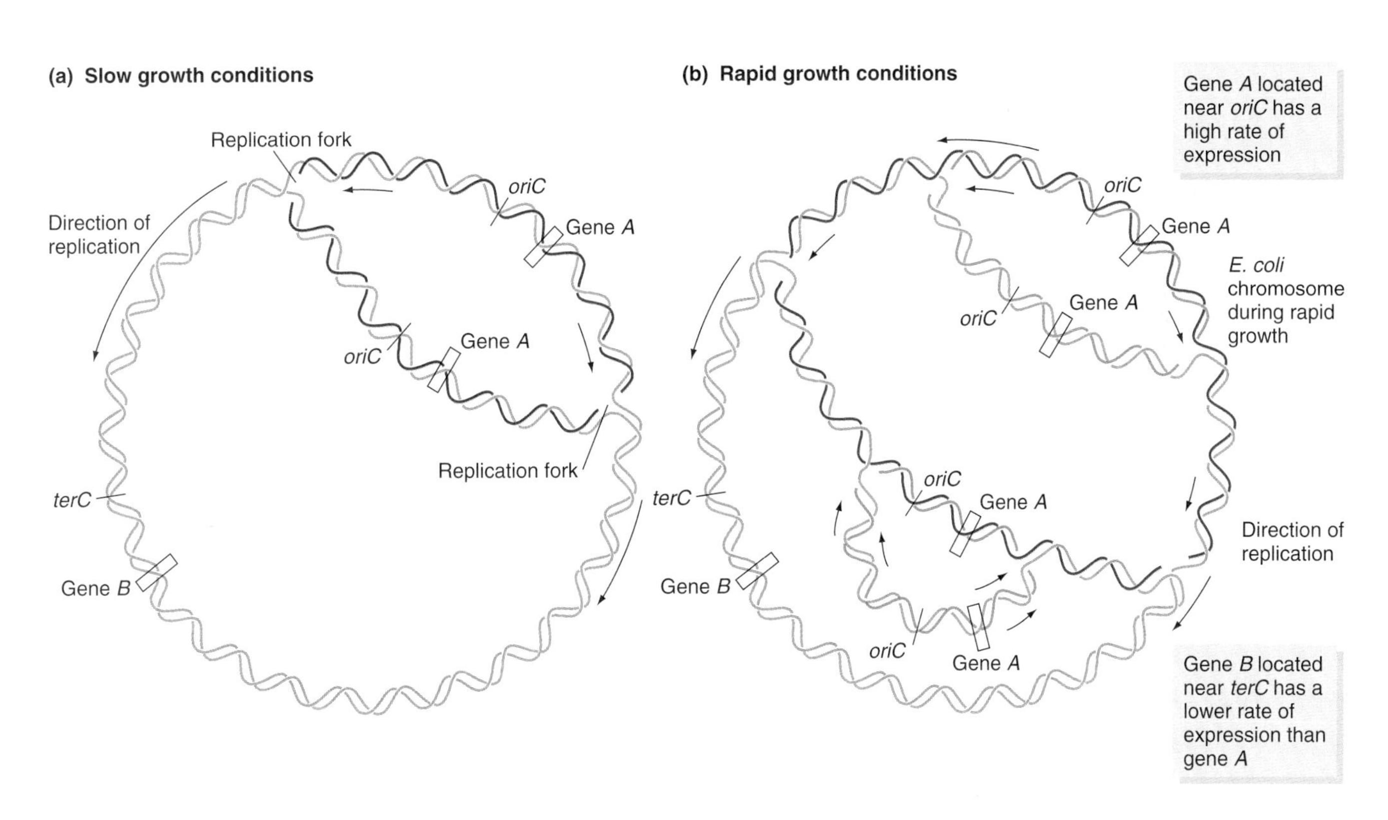

(a) Slow growth conditions

Replication fork

oriC

Gene *A*

Direction of
replication

oriC

Gene *A*

Replication fork

terC

Gene *B*

(b) Rapid growth conditions

oriC

Gene *A*

oriC

Gene *A*

oriC

Gene *A*

terC

Gene *B*

oriC

Gene *A*

Direction of
replication

Gene *A* located
near *oriC* has a
high rate of
expression

E. coli
chromosome
during rapid
growth

Gene *B* located
near *terC* has a
lower rate of
expression than
gene *A*

Replication of the *E. coli* chromosome
Figure 14.7

(b)

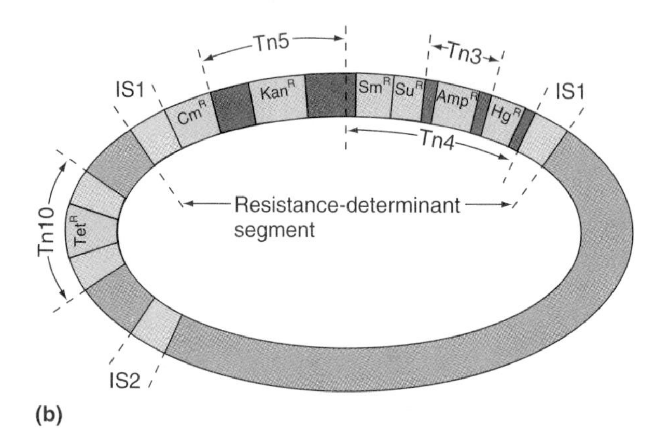

Plasmids are small circles of double-stranded DNA
Figure 14.8

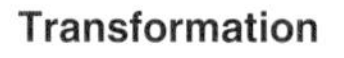

Transformation

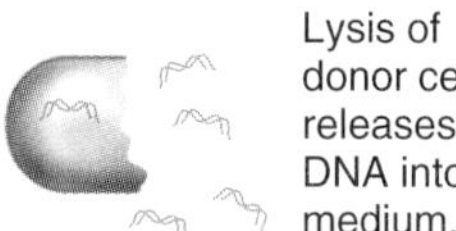

Lysis of donor cell releases DNA into medium.

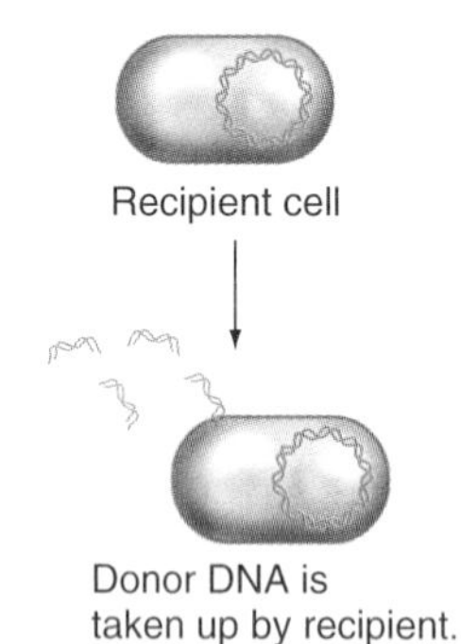

Recipient cell

↓

Donor DNA is taken up by recipient.

Conjugation

Donor cell   Recipient cell

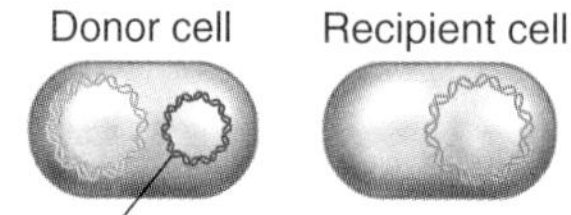

Donor cell plasmid

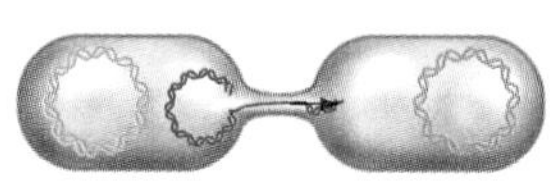

Donor DNA is transferred directly to recipient through a connecting tube. Contact and transfer are promoted by a specialized plasmid in the donor cell.

Transduction

Donor cell   Recipient cell

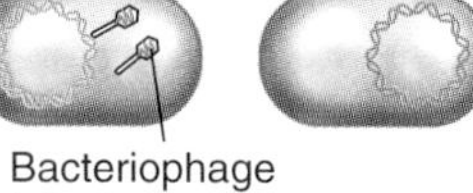

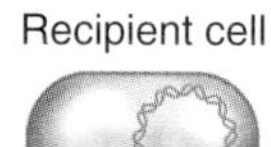

Bacteriophage infects a cell.

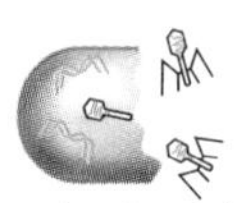

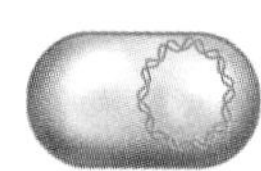

Lysis of donor cell. Donor DNA is packaged in released bacteriophage.

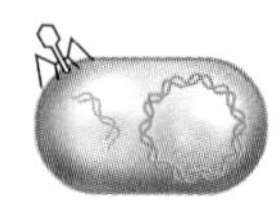

Donor DNA is transferred when phage particle infects recipient cell.

Three mechanisms of gene transfer in bacteria: An overview
Figure 14.9

(a) Donor and recipient genomes

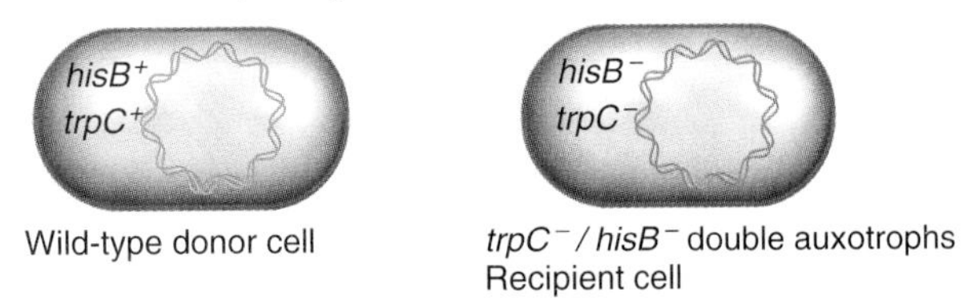

Wild-type donor cell

$trpC^-$ / $hisB^-$ double auxotrophs
Recipient cell

(b) Mechanism of natural transformation

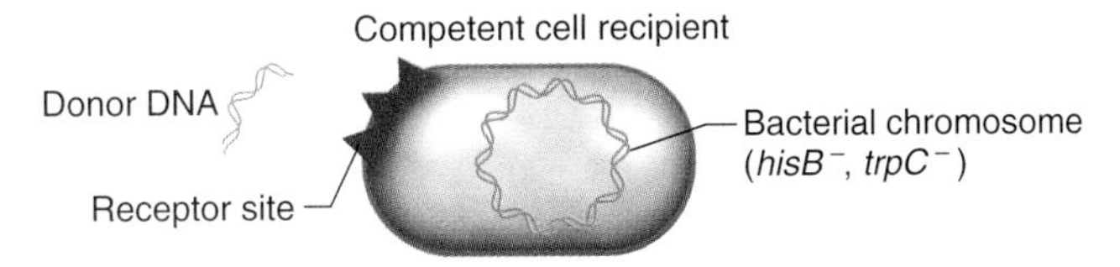

Competent cell recipient

Donor DNA

Bacterial chromosome
($hisB^-$, $trpC^-$)

Receptor site

Donor DNA binds to recipient cell at receptor site.

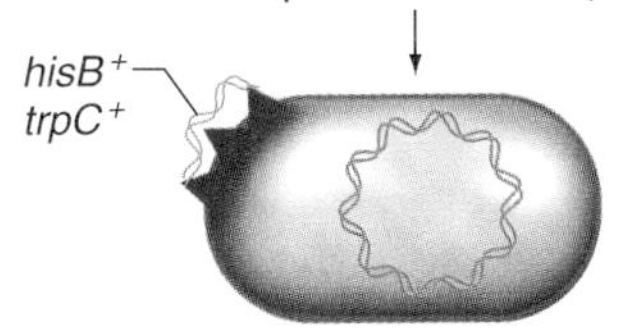

$hisB^+$
$trpC^+$

One donor strand is degraded. Admitted donor strand pairs with homologous region of bacterial chromosome.
Replaced strand is degraded.

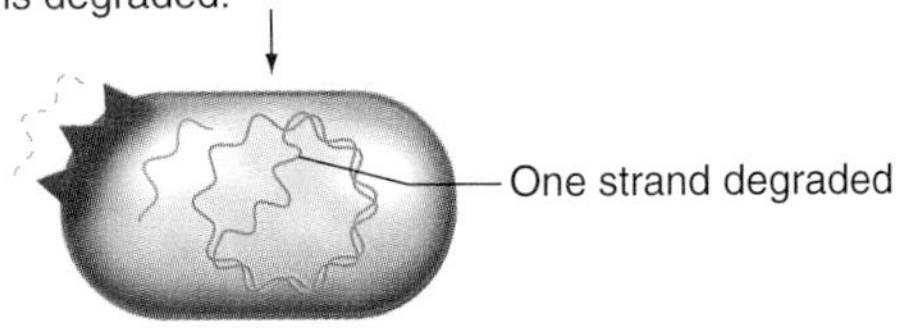

One strand degraded

Donor strand is integrated into bacterial chromosome.

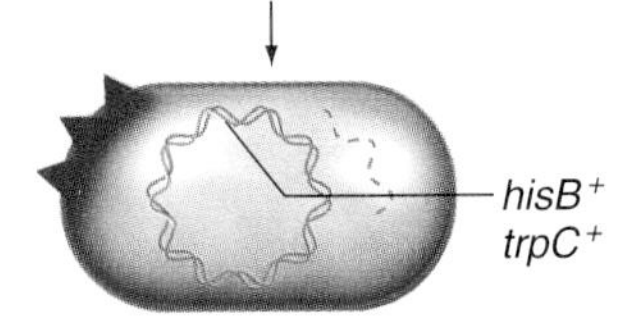

$hisB^+$
$trpC^+$

After cell replication, one cell is identical to original recipient; the other carries the mutant gene.

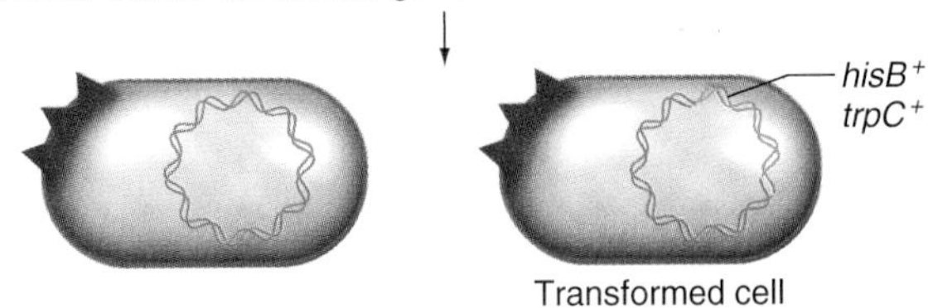

$hisB^+$
$trpC^+$

Transformed cell

Natural transformation in *B. subtilis*
Figure 14.10

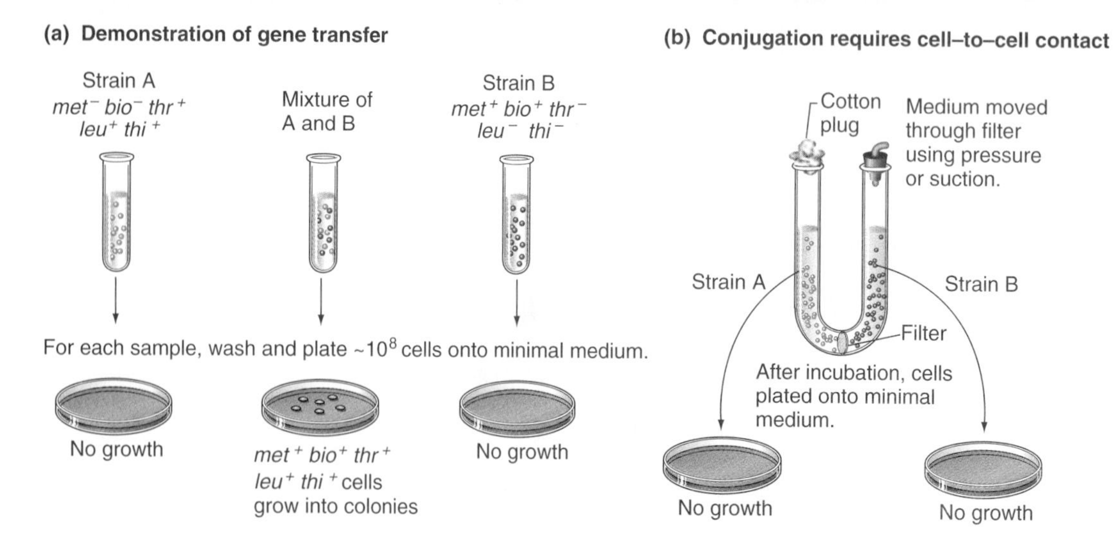

(a) Demonstration of gene transfer

Strain A
met⁻ bio⁻ thr⁺ leu⁺ thi⁺

Mixture of A and B

Strain B
met⁺ bio⁺ thr⁻ leu⁻ thi⁻

For each sample, wash and plate ~10^8 cells onto minimal medium.

No growth

met⁺ bio⁺ thr⁺ leu⁺ thi⁺ cells grow into colonies

No growth

(b) Conjugation requires cell–to–cell contact

Cotton plug

Medium moved through filter using pressure or suction.

Strain A

Strain B

Filter

After incubation, cells plated onto minimal medium.

No growth

No growth

Conjugation: A type of gene transfer requiring cell-to-cell contact
Figure 14.11

(a) Creation of Hfr chromosome.

F plasmid

Target site for single-stranded DNase

IS3

Same IS3's on plasmid and chromosome

Chromosome in F⁺ cell   IS3

IS3   IS3

Chromosome in Hfr cell

(b) Many different Hfr strains can form.

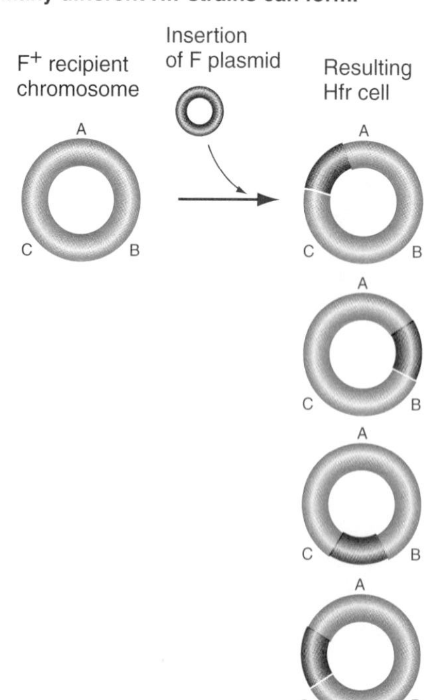

F⁺ recipient chromosome

Insertion of F plasmid

Resulting Hfr cell

Integration of the F plasmid into the bacterial chromosome forms an Hfr bacterium
Figure 14.13

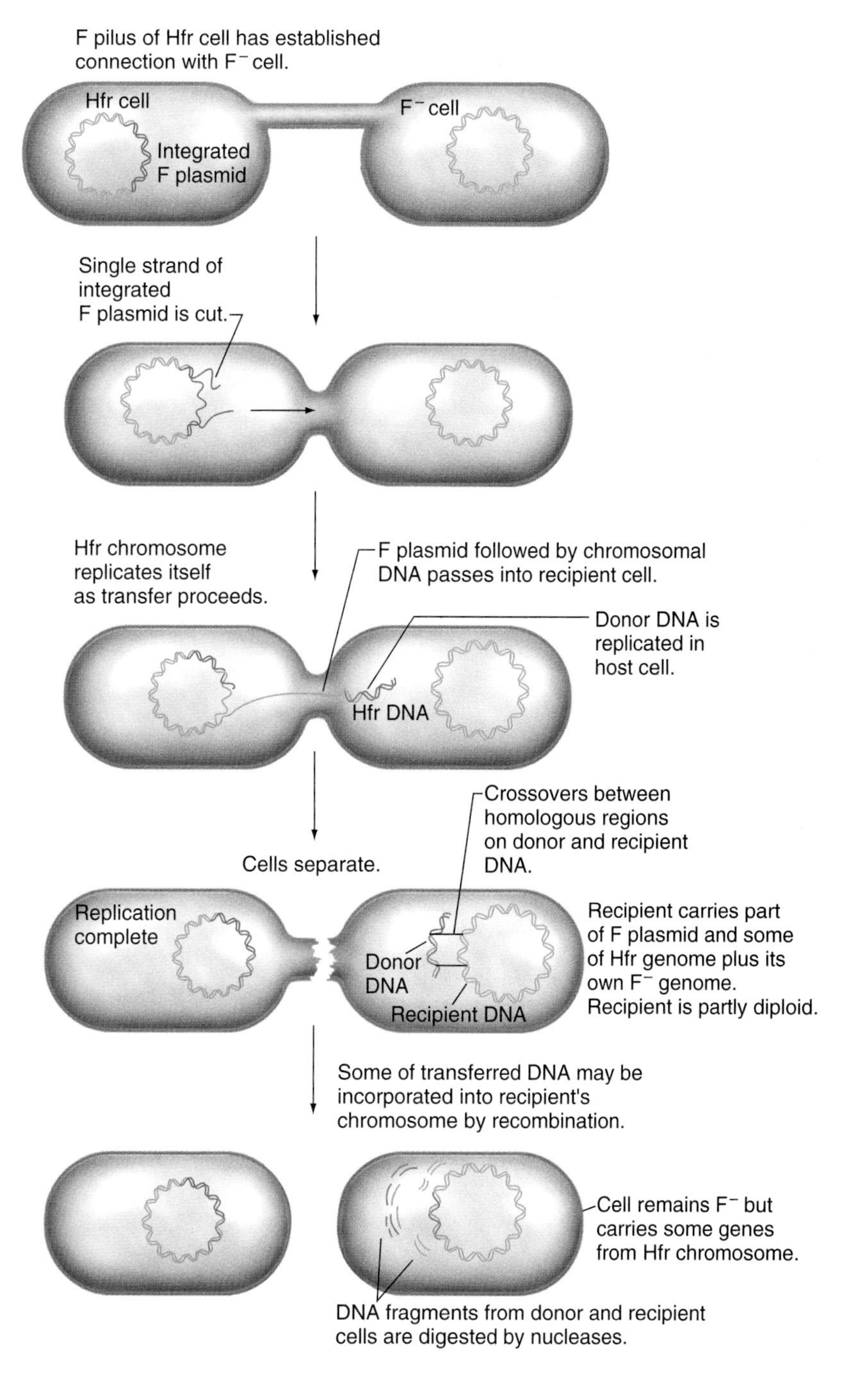

F pilus of Hfr cell has established connection with F⁻ cell.

Hfr cell

Integrated F plasmid

F⁻ cell

Single strand of integrated F plasmid is cut.

Hfr chromosome replicates itself as transfer proceeds.

F plasmid followed by chromosomal DNA passes into recipient cell.

Donor DNA is replicated in host cell.

Hfr DNA

Crossovers between homologous regions on donor and recipient DNA.

Cells separate.

Replication complete

Donor DNA

Recipient DNA

Recipient carries part of F plasmid and some of Hfr genome plus its own F⁻ genome. Recipient is partly diploid.

Some of transferred DNA may be incorporated into recipient's chromosome by recombination.

Cell remains F⁻ but carries some genes from Hfr chromosome.

DNA fragments from donor and recipient cells are digested by nucleases.

Gene transfer in a mating between Hfr donors and F⁻ recipients
Figure 14.14

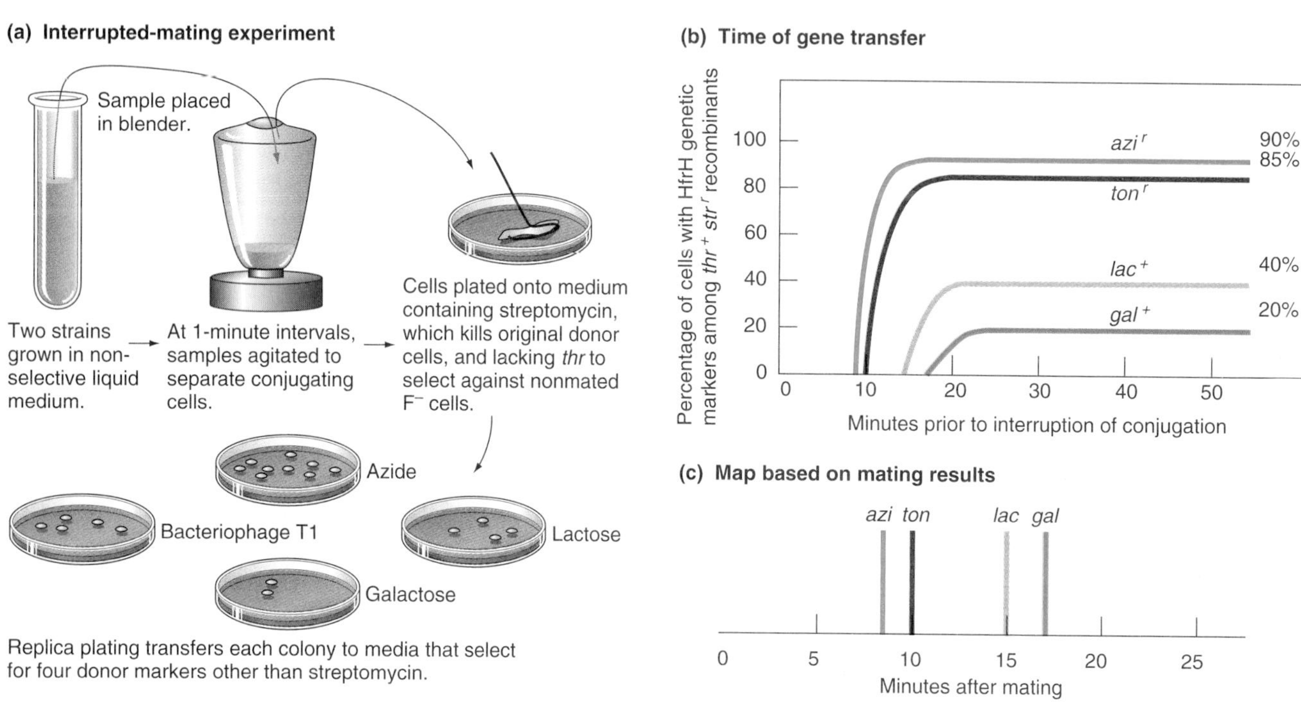

(a) Interrupted-mating experiment

Sample placed in blender.

Cells plated onto medium containing streptomycin, which kills original donor cells, and lacking *thr* to select against nonmated F⁻ cells.

Two strains grown in non-selective liquid medium.

At 1-minute intervals, samples agitated to separate conjugating cells.

Azide

Bacteriophage T1

Lactose

Galactose

Replica plating transfers each colony to media that select for four donor markers other than streptomycin.

(b) Time of gene transfer

Percentage of cells with HfrH genetic markers among *thr⁺ str⁻* recombinants

*azi*ʳ — 90%
*ton*ʳ — 85%
lac⁺ — 40%
gal⁺ — 20%

Minutes prior to interruption of conjugation

(c) Map based on mating results

azi ton   *lac gal*

0   5   10   15   20   25

Minutes after mating

Mapping genes by interrupted-mating experiments
Figure 14.15

(a) Gene transfer in different Hfrs

Hfr strain	Order of transfer ⟶
H	*thr* **azi** **ton** *lac* *pur* **gal** *his* *gly* *thi*
1	*thr* *thi* *gly* *his* **gal** *pur* *lac* **ton** **azi**
2	*lac* *pur* **gal** *his* *gly* *thi* *thr* **azi** **ton**
3	**gal** *pur* **lac** **ton** **azi** *thr* *thi* *gly* *his*

(b) Data interpretation

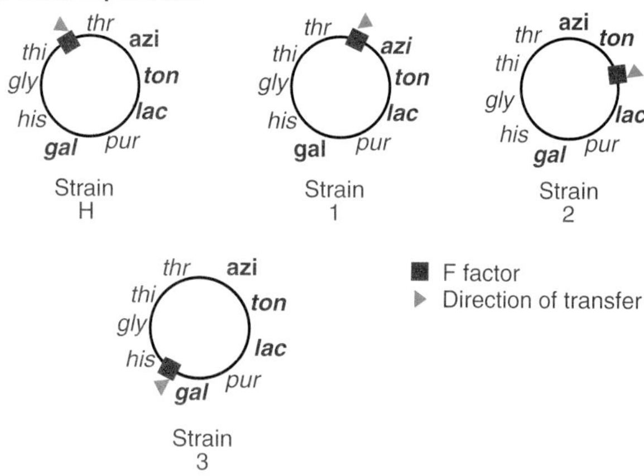

Strain H   Strain 1   Strain 2

■ F factor
▶ Direction of transfer

Strain 3

(c) The circular *E.coli* chromosome

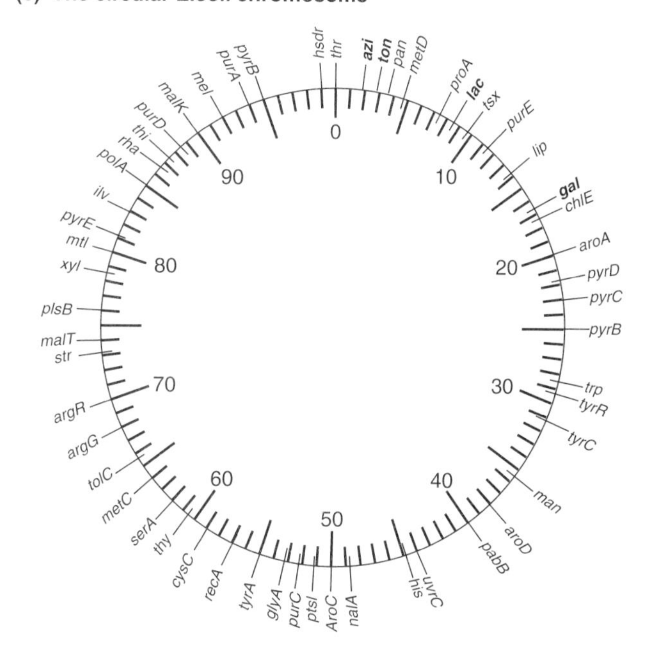

Order of gene transfers in different Hfr strains

Figure 14.16

(a) Hfr mapping: *mal⁺* **transfers last**

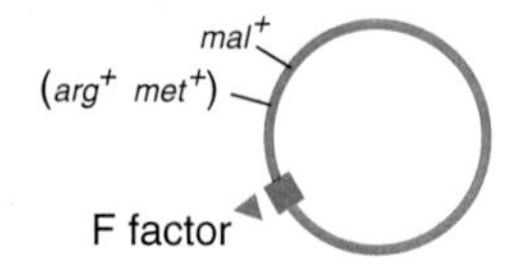

(b) Transfer of genes in *mal⁺* **exconjugants**

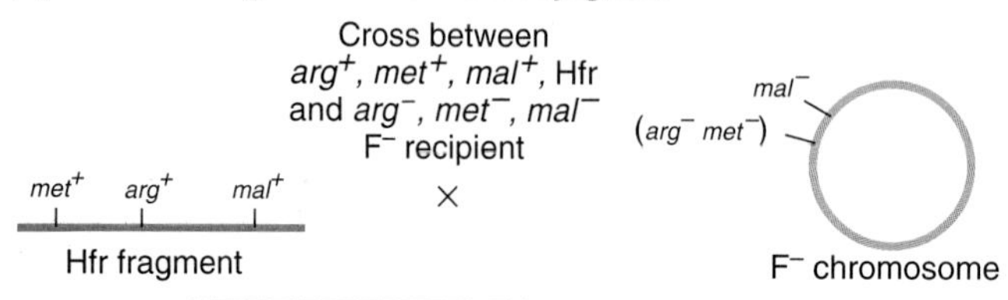

Exconjugant phenotype	Number
met⁺ arg⁺ mal⁺	375
met⁻ arg⁻ mal⁺	149
met⁻ arg⁺ mal⁺	32
met⁺ arg⁻ mal⁺	2

(c) Types of crossover events

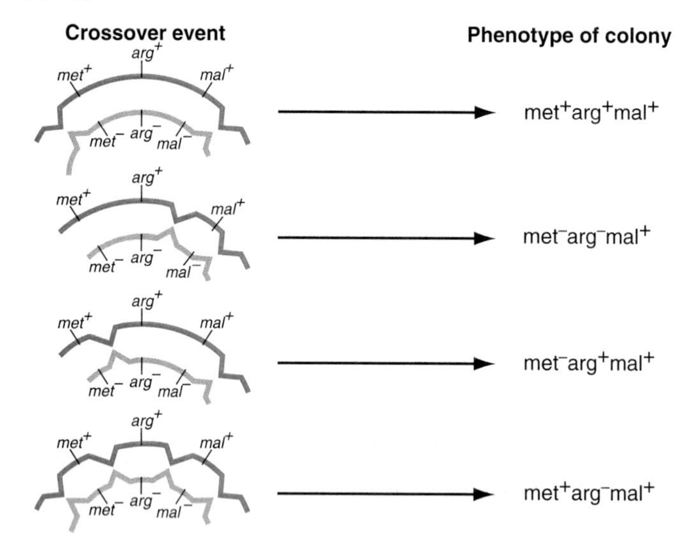

Mapping genes using a three-point cross
Figure 14.17

(a) F' plasmid formation

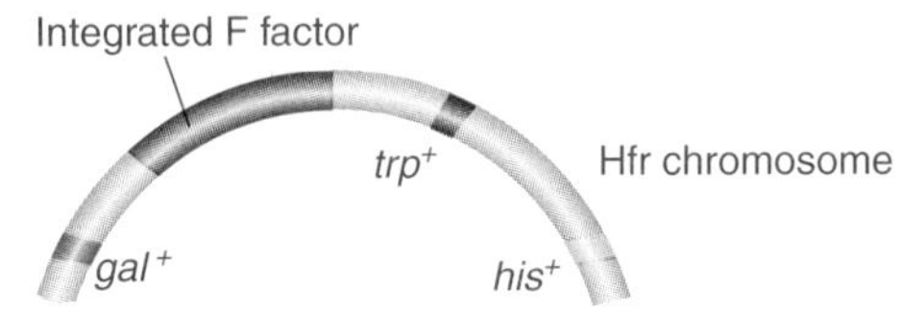

Integrated F factor

trp^+

Hfr chromosome

gal^+ his^+

A rare recombination event between regions of limited sequence homology permits out-looping of F factor including trp^+ locus.

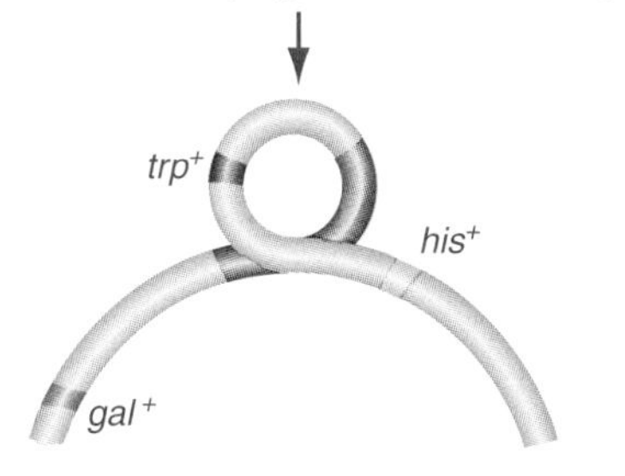

trp^+

his^+

gal^+

Separation of F creates F' trp^+ plasmid and a chromosome deleted for the trp genes.

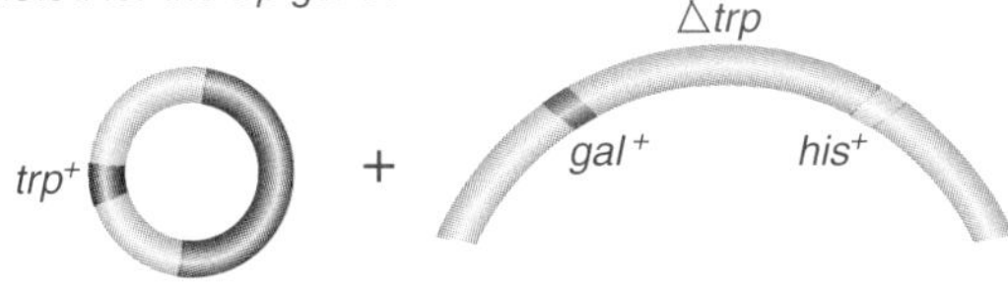

$\triangle trp$

trp^+ + gal^+ his^+

F' plasmids
Figure 14.18

(b) F' plasmid transfer

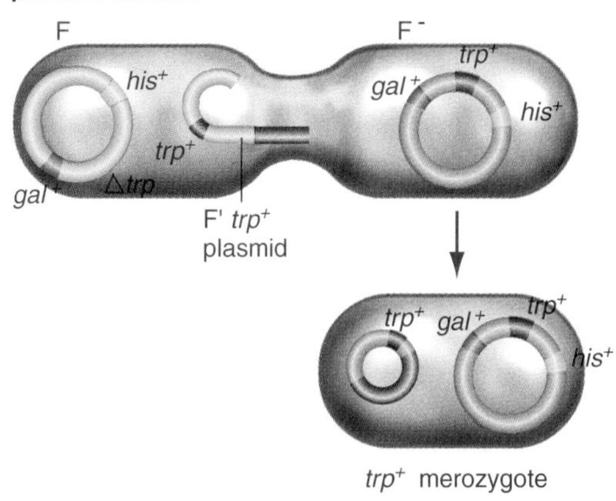

F   F⁻

his^+ gal^+ trp^+

trp^+ his^+

gal $\triangle trp$

F' trp^+
plasmid

trp^+ gal^+ trp^+

his^+

trp^+ merozygote

(c) Complementation testing using F' plasmids

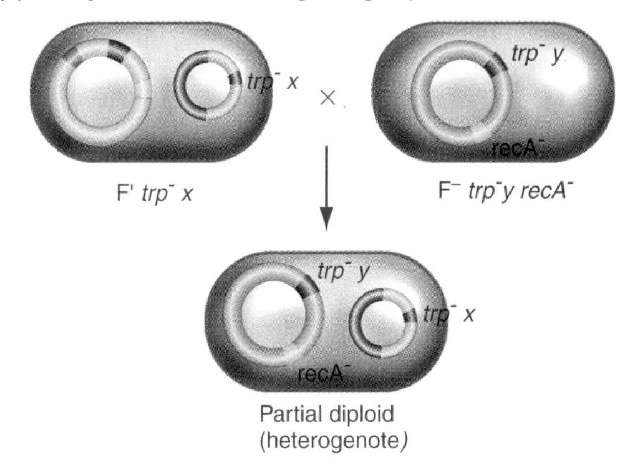

$trp^-\ x$ $trp^-\ y$

$\times$

$recA^-$

F' $trp^-\ x$ F⁻ $trp^-y\ recA^-$

$trp^-\ y$

$trp^-\ x$

$recA^-$

Partial diploid
(heterogenote)

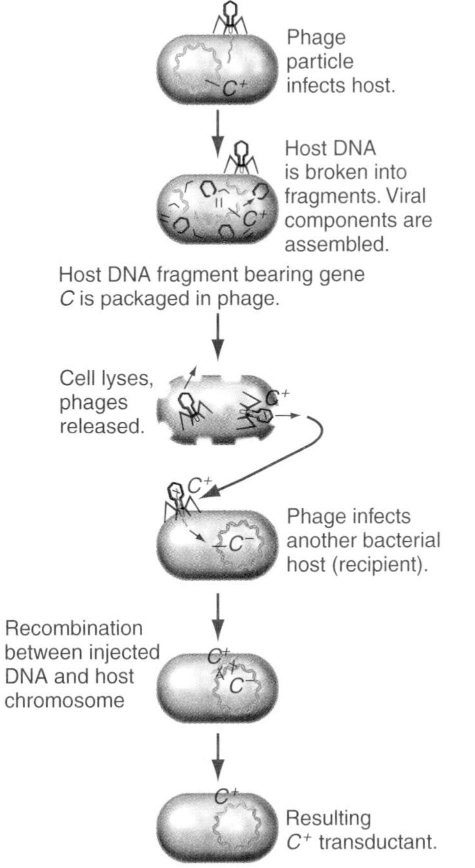

Phage particle infects host.

Host DNA is broken into fragments. Viral components are assembled.

Host DNA fragment bearing gene C is packaged in phage.

Cell lyses, phages released.

Phage infects another bacterial host (recipient).

Recombination between injected DNA and host chromosome

Resulting C^+ transductant.

Generalized transduction
Figure 14.19

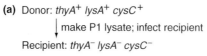

(a) Donor: *thyA⁺ lysA⁺ cysC⁺*

    ↓ make P1 lysate; infect recipient

Recipient: *thyA⁻ lysA⁻ cysC⁻*

Selected marker	Unselected marker
thy⁺	47% *lys⁺*; 2% *cys⁺*
lys⁺	50% *thy⁺*; 0% *cys⁺*

(b)

lysA   thyA   cysC

Mapping genes by cotransduction frequencies
Figure 14.20

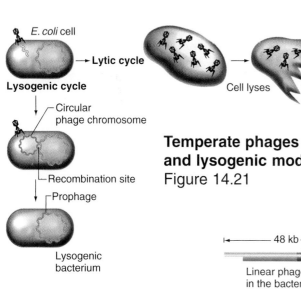

E. coli cell

→ **Lytic cycle**

Lysogenic cycle

↓

Circular phage chromosome

Recombination site

Prophage

Lysogenic bacterium

Cell lyses

Temperate phages can choose between lytic and lysogenic modes of reproduction
Figure 14.21

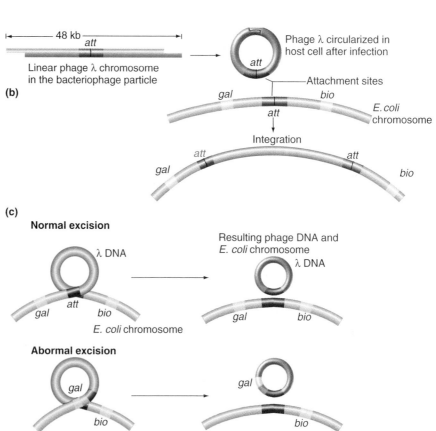

(b)

48 kb

att

Linear phage λ chromosome in the bacteriophage particle

Phage λ circularized in host cell after infection

att

Attachment sites

gal   *bio*

att

E. coli chromosome

↓ Integration

att   *att*

gal   *bio*

(c)

Normal excision

λ DNA

gal *att* *bio*

E. coli chromosome

Resulting phage DNA and E. coli chromosome

λ DNA

gal   *bio*

Abormal excision

gal

bio

gal

bio

Bacteriophage lambda and lysogeny
Figure 14.22

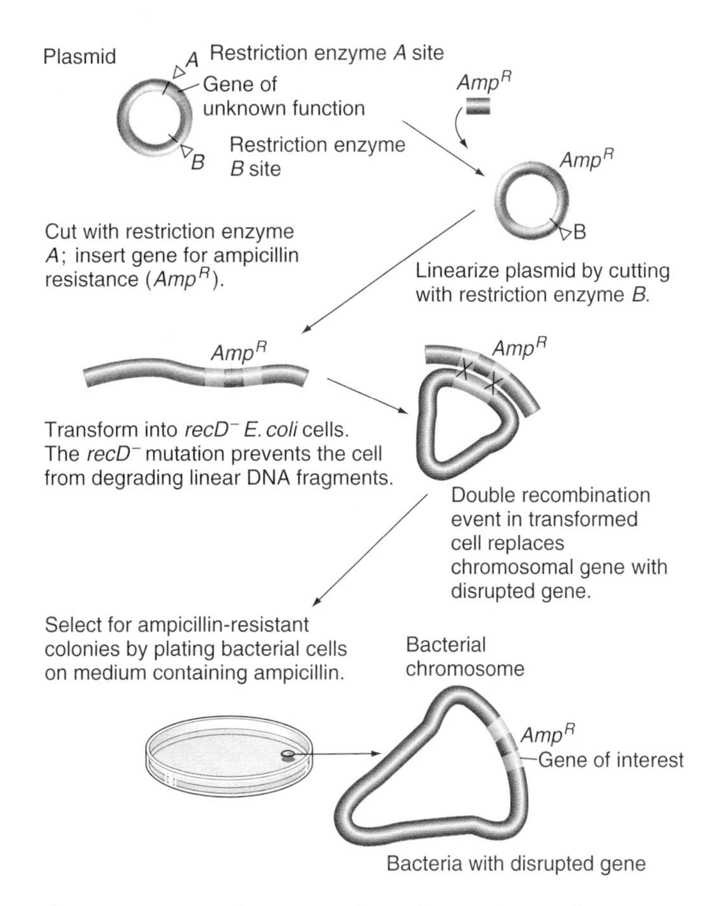

Plasmid

A Restriction enzyme A site
Gene of unknown function
B Restriction enzyme B site

Amp^R

Amp^R

Amp^R

B

Cut with restriction enzyme A; insert gene for ampicillin resistance (Amp^R).

Linearize plasmid by cutting with restriction enzyme B.

Amp^R

Amp^R

Transform into $recD^-$ E. coli cells. The $recD^-$ mutation prevents the cell from degrading linear DNA fragments.

Double recombination event in transformed cell replaces chromosomal gene with disrupted gene.

Select for ampicillin-resistant colonies by plating bacterial cells on medium containing ampicillin.

Bacterial chromosome

Amp^R
Gene of interest

Bacteria with disrupted gene

One way to characterize the roles of uncharted bacterial genes
Figure 14.23

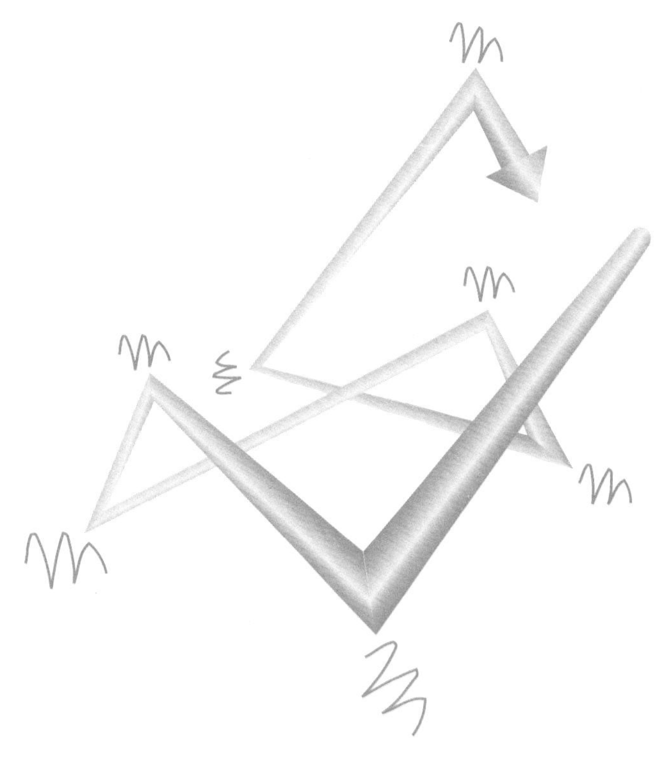

How bacteria move
Figure 14.24

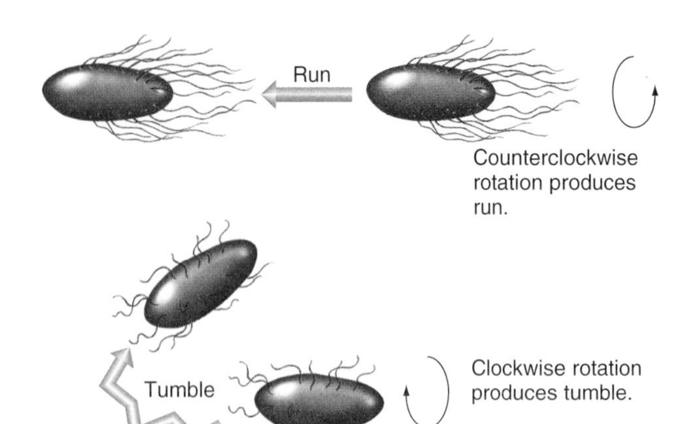

Run

Counterclockwise rotation produces run.

Tumble

Clockwise rotation produces tumble.

Counterclockwise rotations of bacterial flagella produce runs; clockwise rotations cause tumbles
Figure 14.25

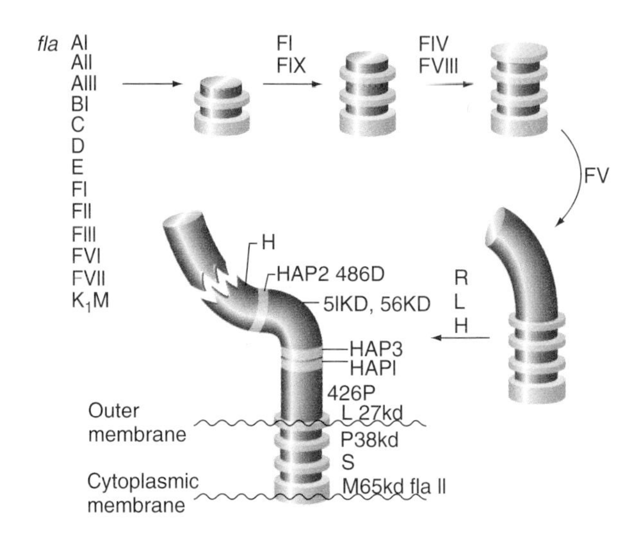

fla AI
AII
AIII
BI
C
D
E
FI
FII
FIII
FVI
FVII
K₁M

FI
FIX

FIV
FVIII

FV

R
L
H

H
HAP2 486D
5IKD, 56KD

HAP3
HAPI
426P
L 27kd
P38kd
S
M65kd *fla* II

Outer membrane

Cytoplasmic membrane

More than 20 genes are needed to generate a bacterial flagellum
Figure 14.27

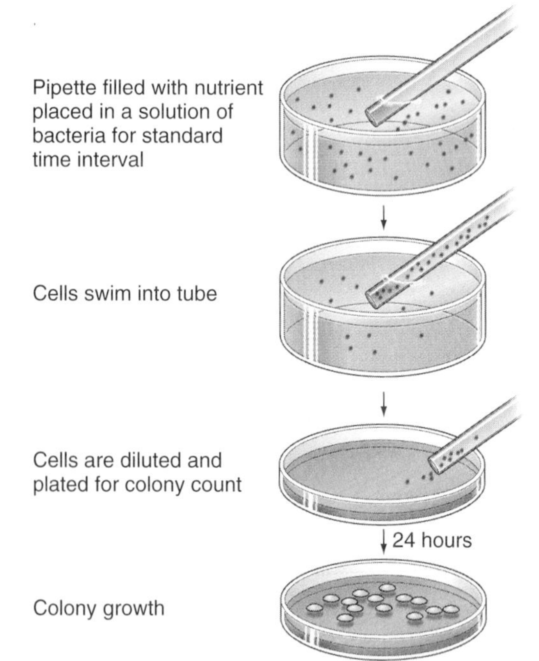

Pipette filled with nutrient placed in a solution of bacteria for standard time interval

Cells swim into tube

Cells are diluted and plated for colony count

24 hours

Colony growth

The capillary test for chemotaxis
Figure 14.28

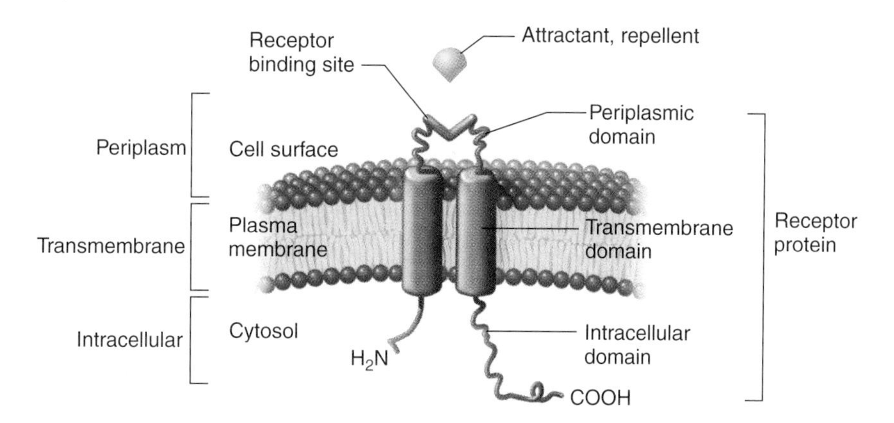

Bacteria have cell surface receptor proteins that recognize particular attractants or repellents

Figure 14.29

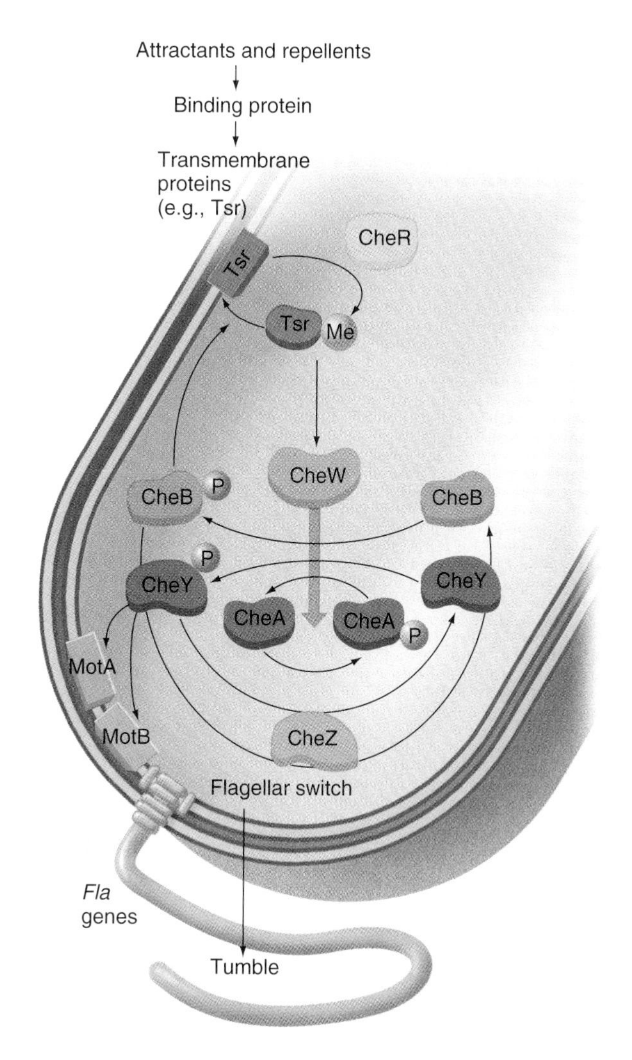

The genetic and molecular basis of bacterial chemotaxis

Figure 14.30

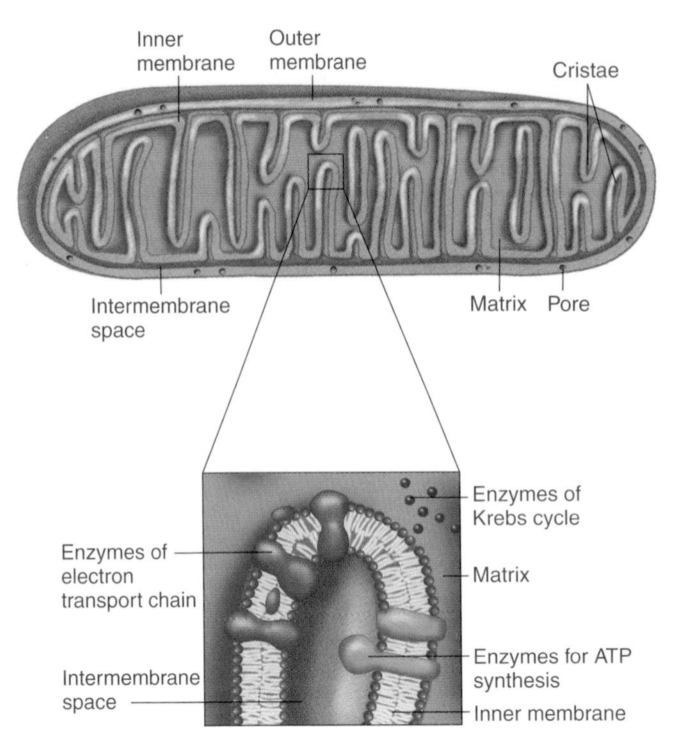

Inner membrane   Outer membrane

Cristae

Intermembrane space

Matrix   Pore

Enzymes of Krebs cycle

Enzymes of electron transport chain

Matrix

Intermembrane space

Enzymes for ATP synthesis

Inner membrane

Anatomy of a mitochondrion
Figure 15.2

Chloroplast

Thylakoid

Granum

Stroma

Outer membrane
Inner membrane

Chlorophyll and light-absorbing proteins

Proteins of photosynthetic electron transport chain

Anatomy of a chloroplast
Figure 15.3

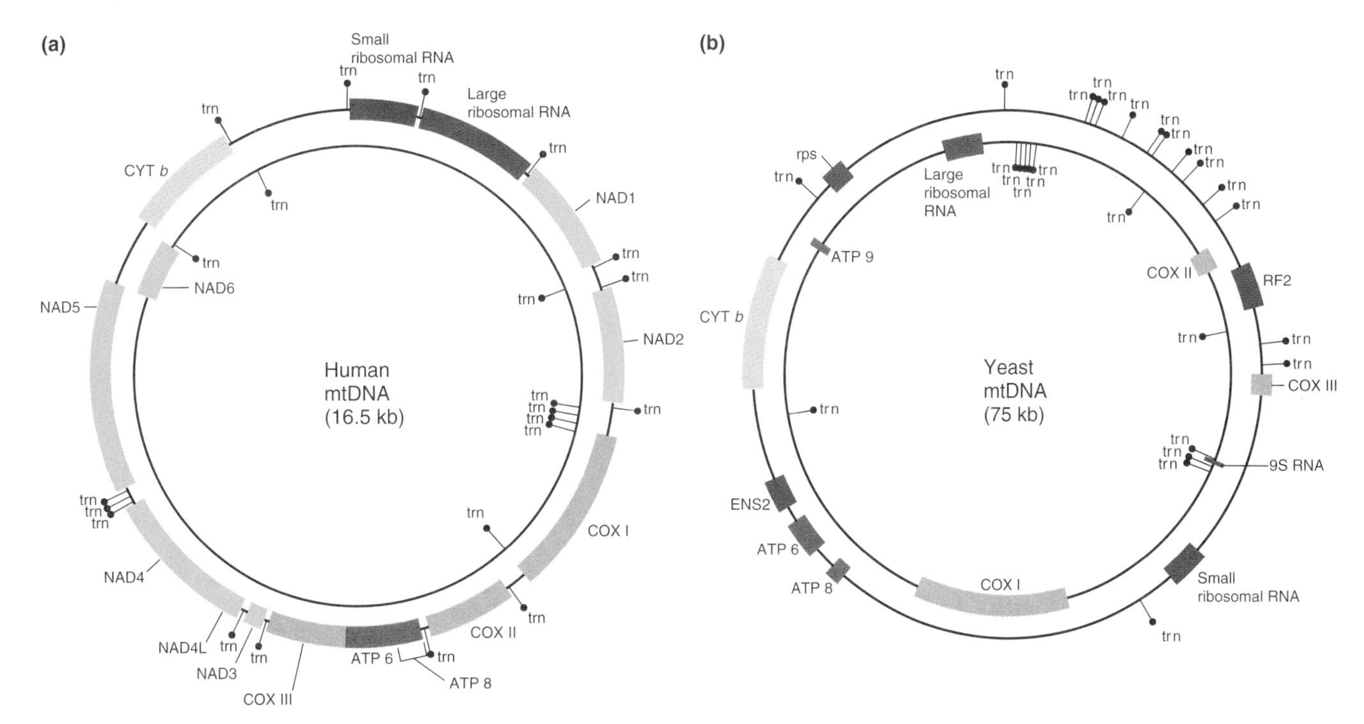

Mitochondrial genomes of three species
Figure 15.5

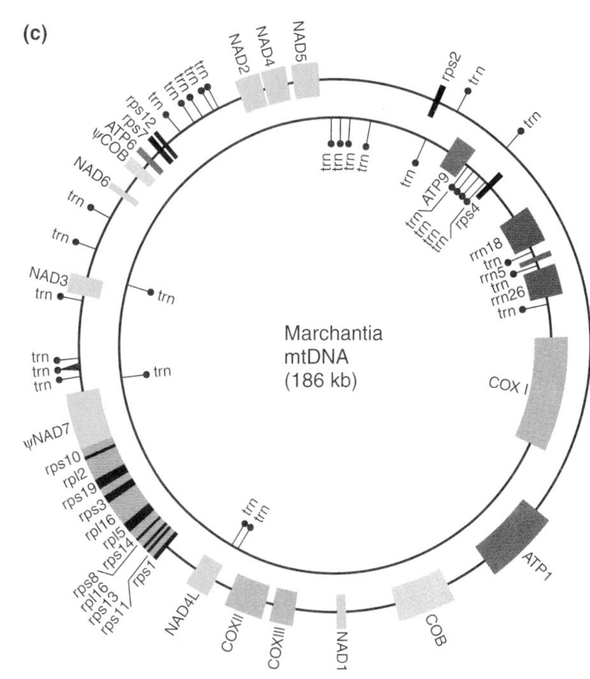

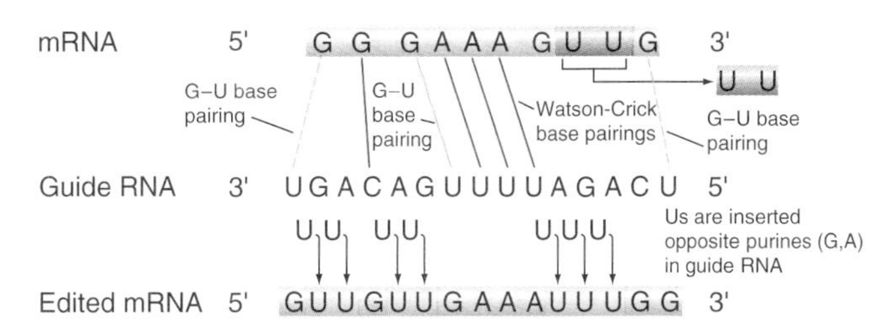

RNA editing in trypanosomes
Figure 15.6

169

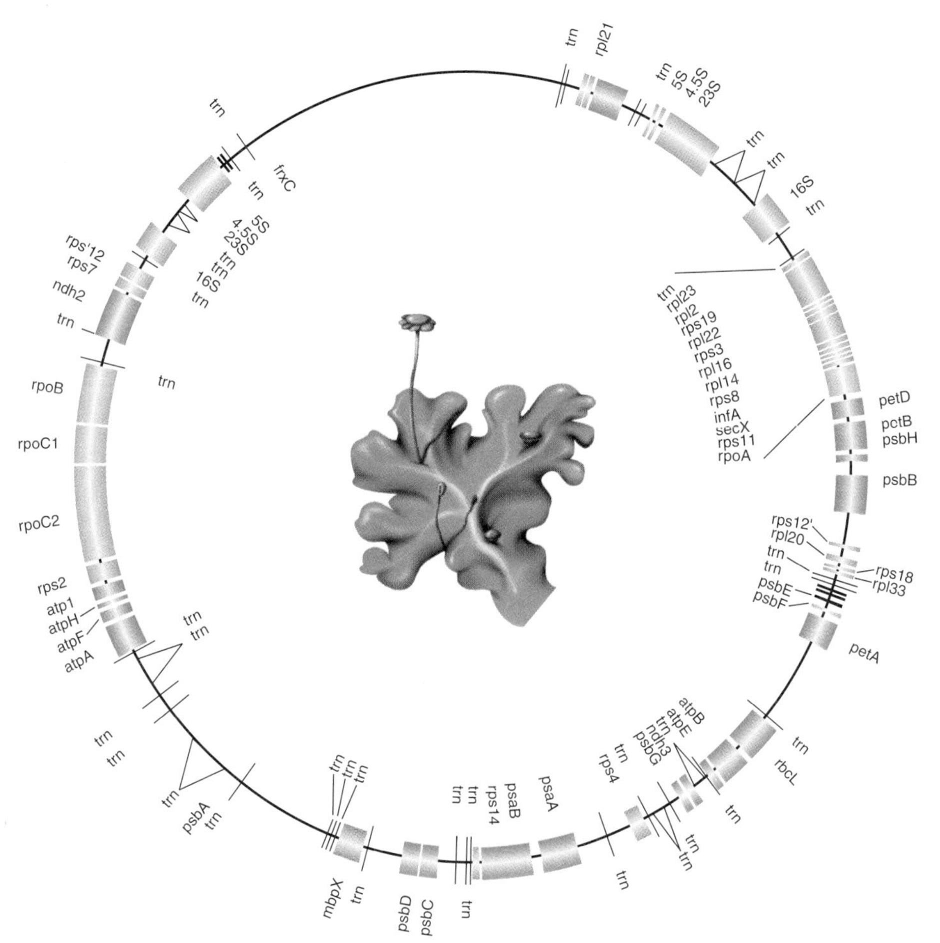

Translation

rps	30S ribosomal proteins
rpl	50S ribosomal subunit proteins
trn	tRNAs
4.5S, 5S 16S, 23S	rRNAs
infA	initiation factor
secX	50S ribosomal protein

Photosynthesis and electron transport

rbc	ribulose bisphosphate carboxylase
psa	photosystem 1
psb	photosystem 2
pet	cytochrome *b/f* complex
atp	ATP synthase
frx	iron-sulfur proteins
ndh	NAD(P)H oxidoreductase

Transcription

rpo	RNA polymerase

Miscellaneous

mbpx	permease

Chloroplast genome of the liverwort *M. polymorpha*
Figure 15.7

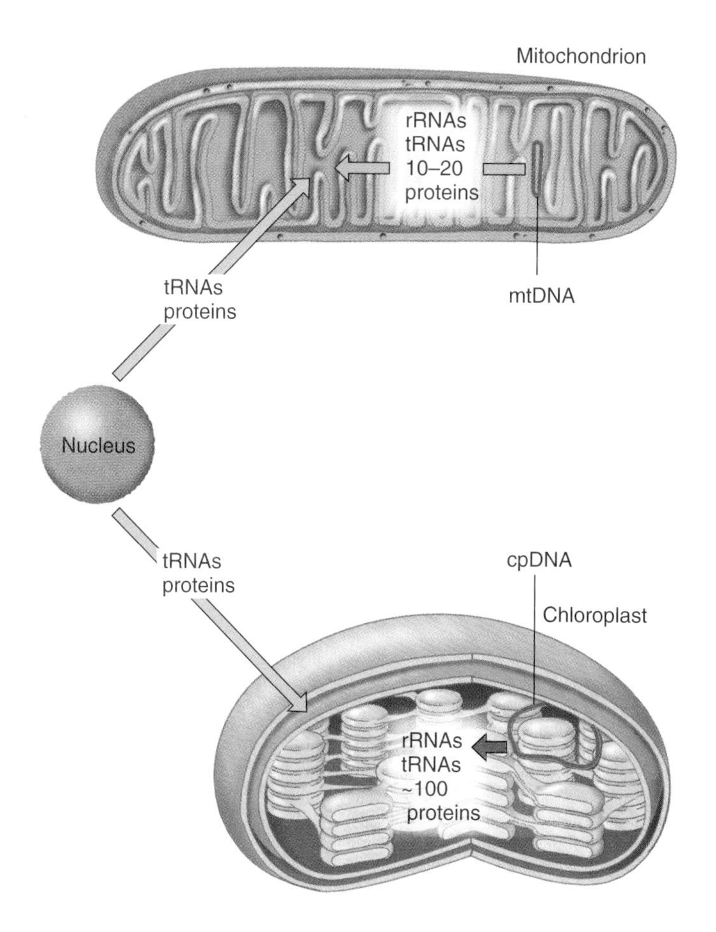

Number and genomic location of oxidative phosphorylation genes						
	Number of polypeptides					
	Electron transport chain				ATP synthase	
Genomic location	I	II	III	IV	V	Total
Mitochondrion	7	0	1	3	2	13
Nucleus	≥33	4	10	10	10	≥67
Total	≥40	4	11	13	12	≥80

Mitochondria and chloroplasts depend on gene products from the nucleus
Figure 15.8

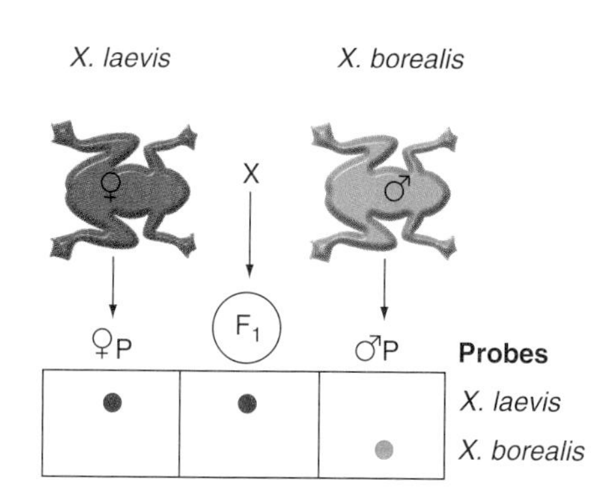

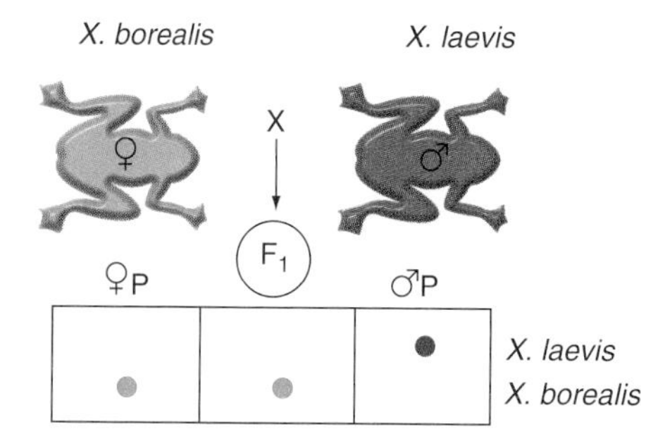

Maternal inheritance of *Xenopus* mtDNA
Figure 15.9

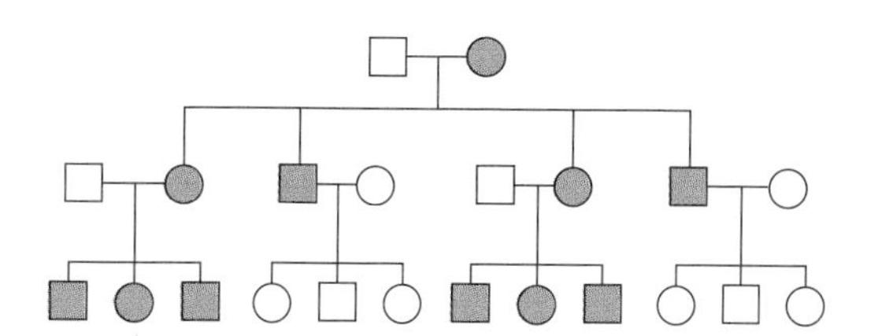

Hypothetical example of LHON pedigree
Figure 15.10

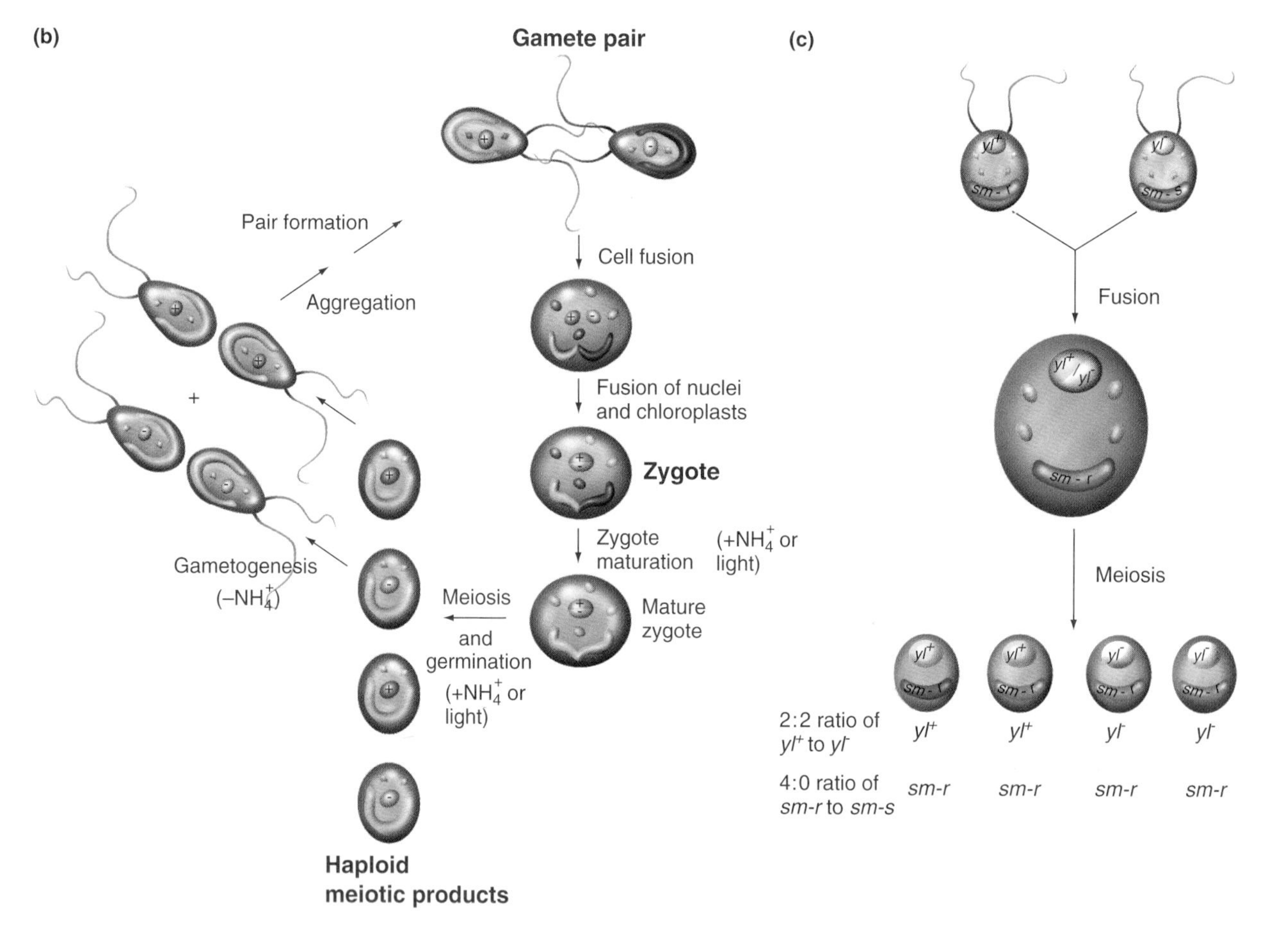

(b)

Gamete pair

Pair formation

Aggregation

Cell fusion

Fusion of nuclei
and chloroplasts

Zygote

Gametogenesis
(−NH$_4^+$)

Zygote
maturation

(+NH$_4^+$ or
light)

Meiosis
and
germination
(+NH$_4^+$ or
light)

Mature
zygote

**Haploid
meiotic products**

(c)

Fusion

yl^+/yl^-

sm-r

Meiosis

2:2 ratio of
yl^+ to yl^-

| yl^+ | yl^+ | yl^- | yl^- |

4:0 ratio of
sm-r to sm-s

| sm-r | sm-r | sm-r | sm-r |

Genes in the cpDNA of *C. reinhardtii* do not segregate at meiosis
Figure 15.11

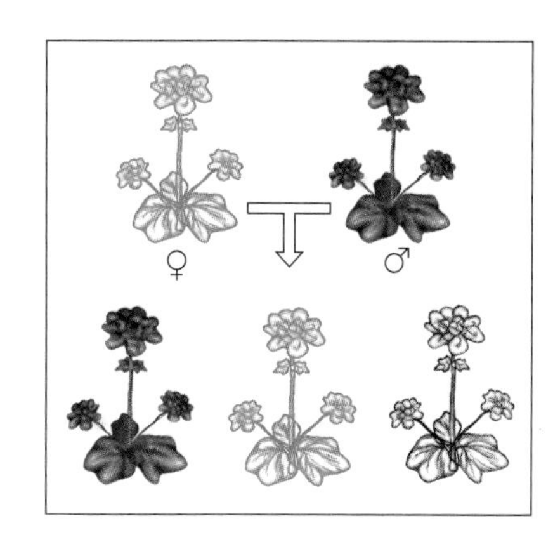

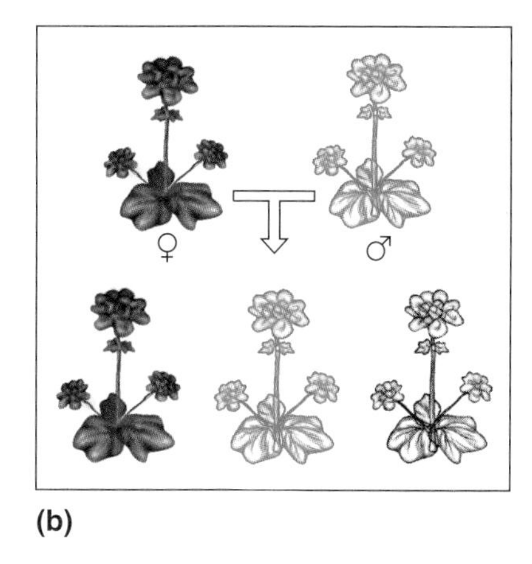

(b)

Biparental inheritance of variegation in geraniums
Figure 15.12

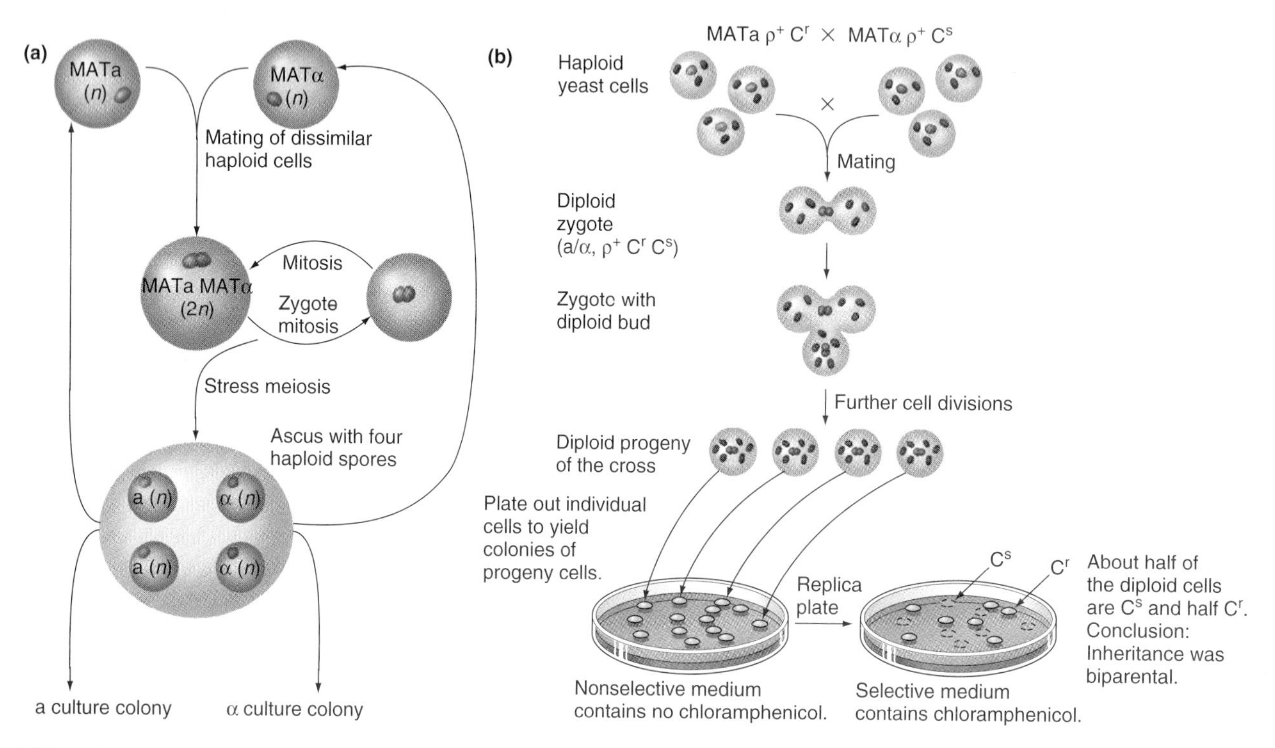

Mitochondrial transmission from the yeast *S. cerevisiae*
Figure 15.13

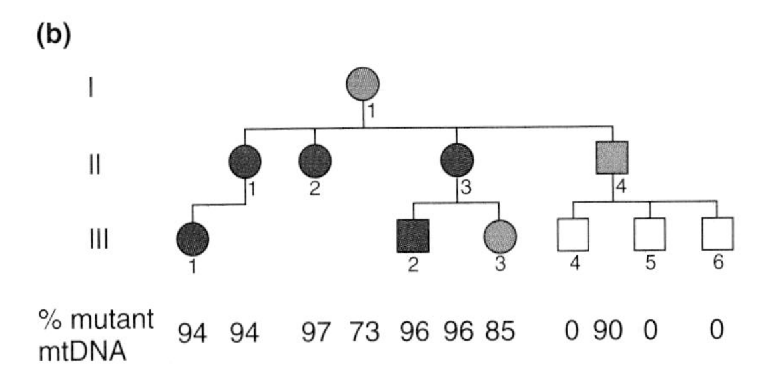

(b)

% mutant mtDNA 94 94 97 73 96 96 85 0 90 0 0

Maternal inheritance of the mitochon-drial disease MERRF

Figure 15.15

Individual mtDNA genotypes		Tissues Affected				
				Skeletal Muscle		
		Brain	Heart	Type I	Type II	Skin
I — 20% mutant mtDNAs		+	–	–	–	–
II — 40% mutant mtDNAs		+	+/–	–	–	–
III — 60% mutant mtDNAs		+	+	+	–	–
IV — 80% mutant mtDNAs		+	+	+	+/–	+/–

Disease phenotypes reflect the ratio of mutant-to-wild-type mtDNAs and the reliance of cell type on mitochondrial function

Figure 15.16

175

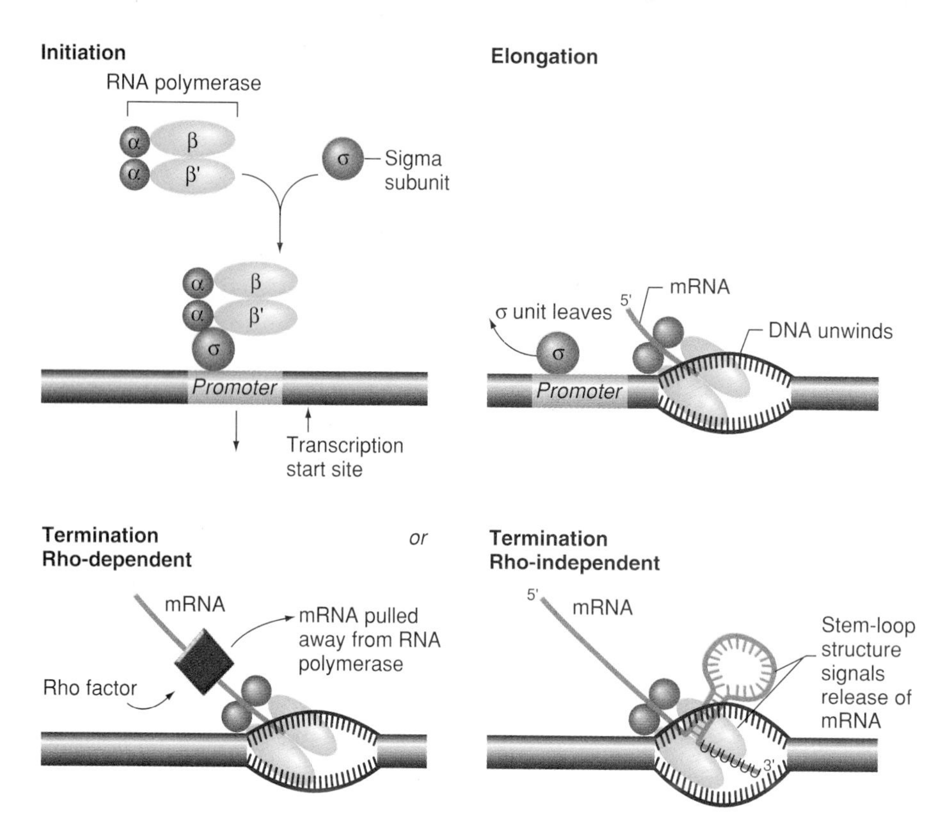

Initiation

RNA polymerase

α β
α β'

σ — Sigma subunit

α β
α β'
σ

Promoter

↓
↑ Transcription start site

Elongation

σ unit leaves

σ

5' — mRNA

DNA unwinds

Promoter

Termination
Rho-dependent

or

mRNA

→ mRNA pulled away from RNA polymerase

Rho factor →

Termination
Rho-independent

5' mRNA

Stem-loop structure signals release of mRNA

3'

Role of RNA polymerase in transcription
Figure 16.2

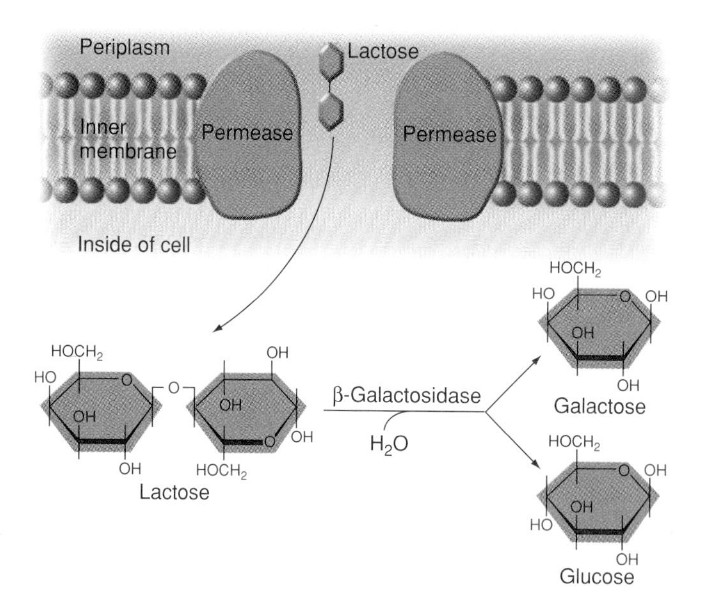

Periplasm
Lactose

Inner membrane
Permease
Permease

Inside of cell

HOCH₂
HO — O — O
OH
OH

OH
OH
HOCH₂

Lactose

β-Galactosidase

H₂O

HOCH₂
HO — O — OH
OH
OH

Galactose

HOCH₂
O — OH
OH
HO
OH

Glucose

Lactose utilization in an *E. coli* cell
Figure 16.3

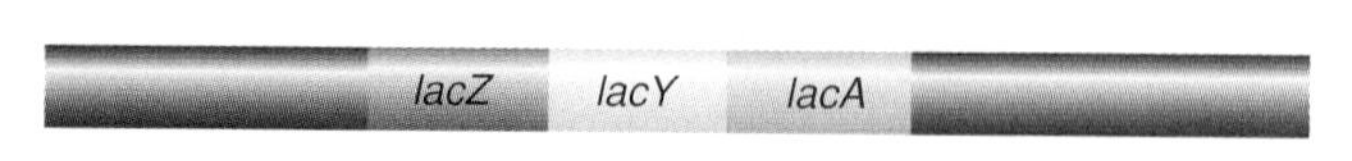

lacZ   *lacY*   *lacA*

Lactose utilization genes in *E. coli*
Figure 16.5

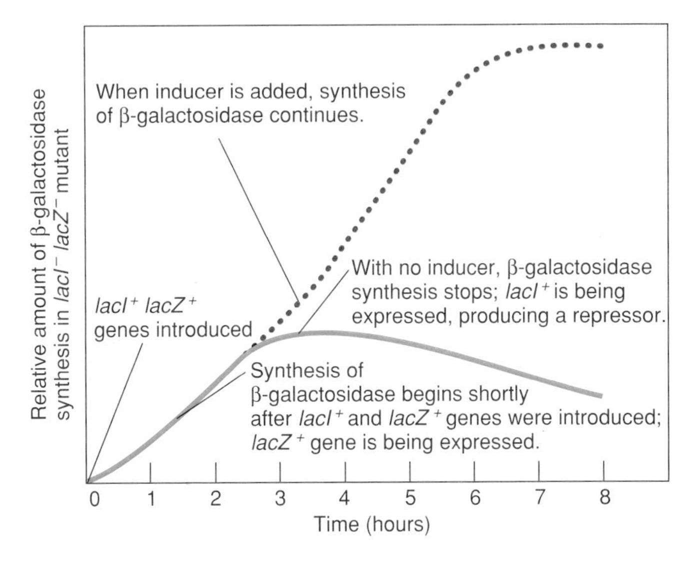

The PaJaMo experiment

Figure 16.6

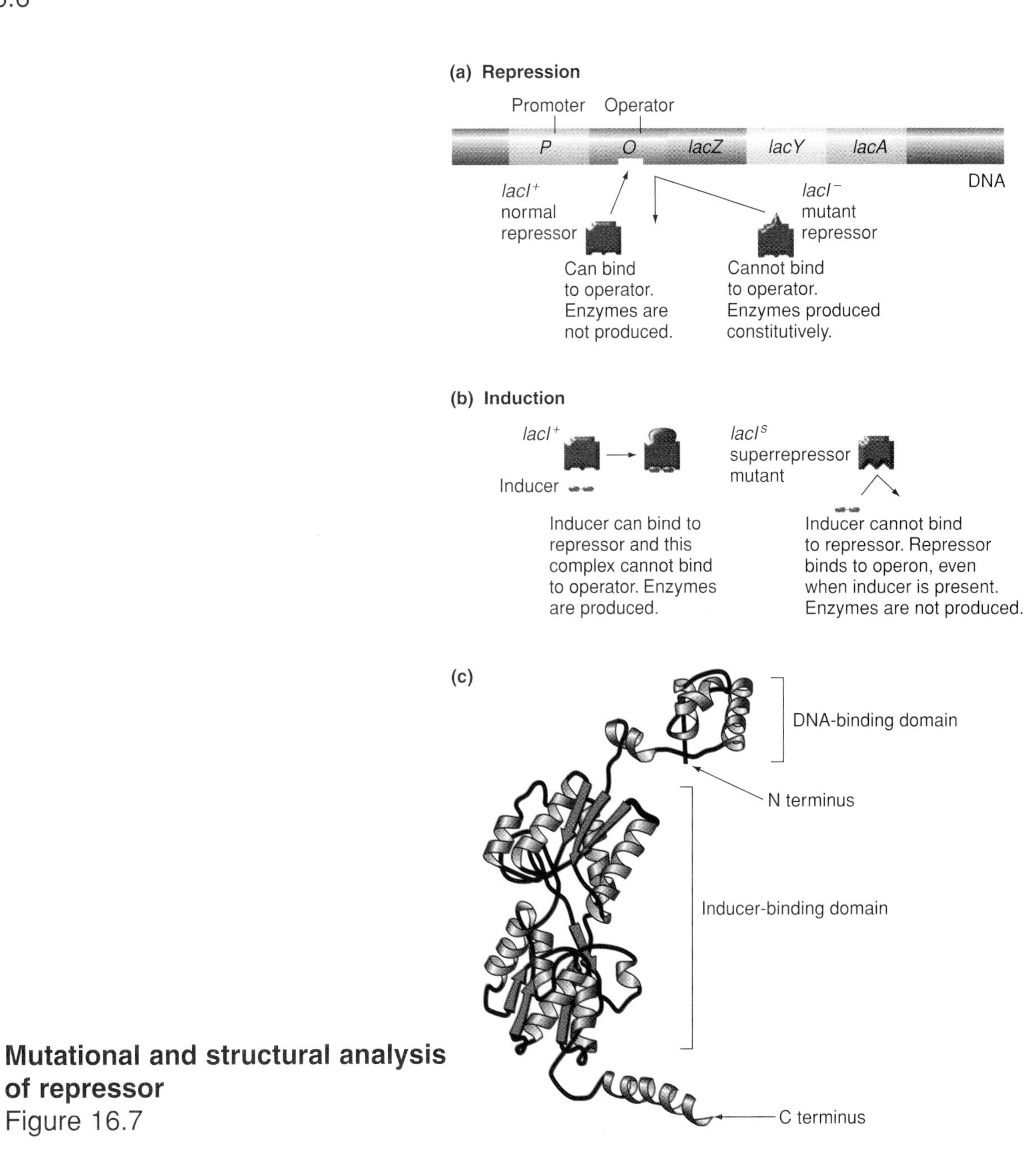

Mutational and structural analysis of repressor

Figure 16.7

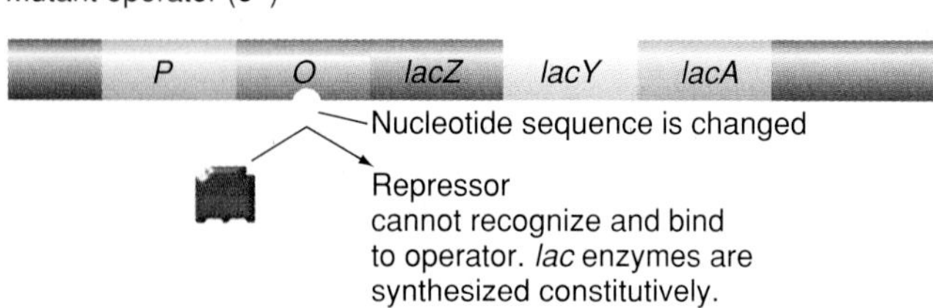

Mutant operator (o^c)

Nucleotide sequence is changed

Repressor
cannot recognize and bind
to operator. *lac* enzymes are
synthesized constitutively.

Operator mutants
Figure 16.8

(a) F' *lacI⁺ o⁺ Z⁻* plasmid in *lacI⁻ o⁺ Z⁺* bacteria

lacI⁺ gene encodes a diffusible element that acts in *trans*.

Inducible synthesis

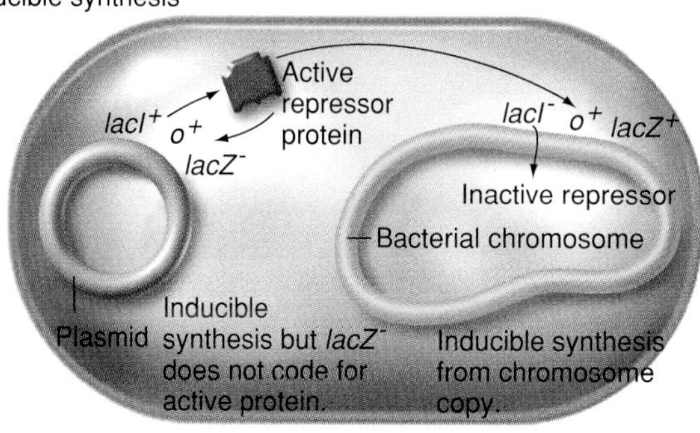

(b) F' *lacIˢ Z⁻* plasmid in *lacI⁺ Z⁺* bacteria

Noninducible—all o^+ sites eventually occupied by superrepressor.

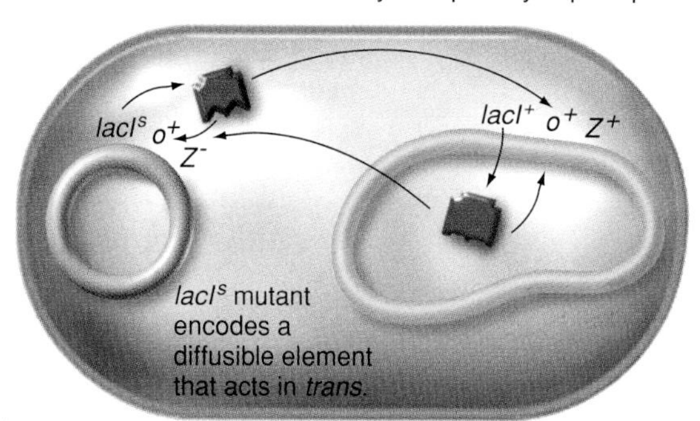

(c) F' *lacI⁺ o⁺ lacZ⁻* plasmid in *lacI⁺ oᶜ lacZ⁺* bacteria

Constitutive—presence of o^+ in plasmid has no effect on expression
of *lacZ⁺* gene in bacterial chromosome.

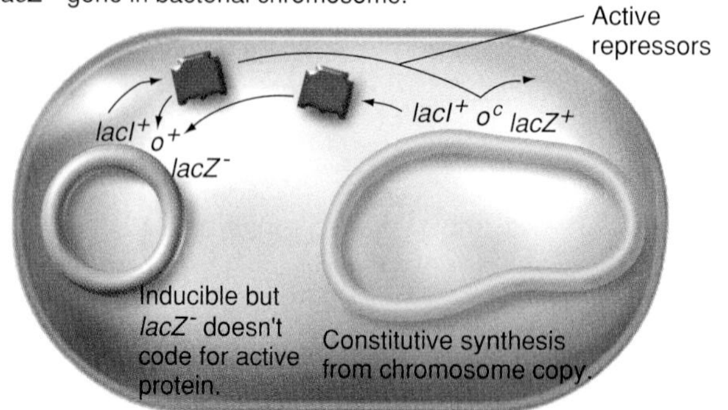

Trans-acting proteins and *cis*-acting sites
Figure 16.9

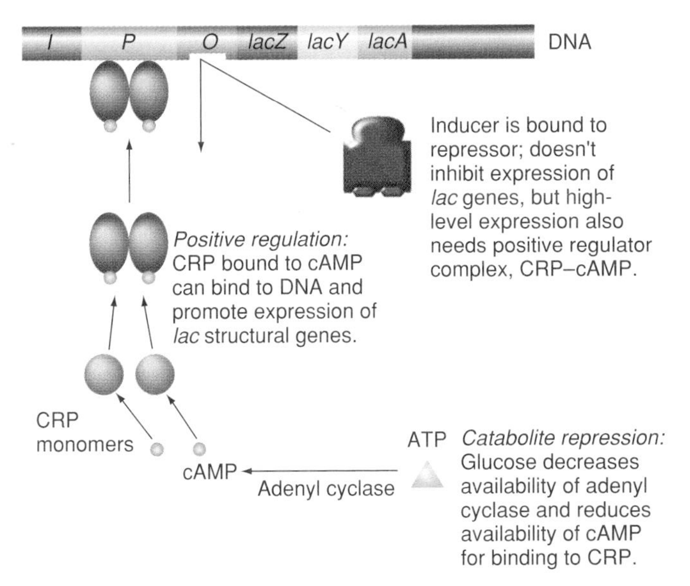

Positive regulation of the *lac* operon by CRP–cAMP
Figure 16.11

(a) **araC⁺**

araO

RNA polymerase

Ara C · Ara C

araI₁ araI₂ Promoter araB araA araC

Transcription of araBAD

(b) **araC⁻**

Ara C⁻ Ara C⁻

Promoter araB araA araC

No transcription of araBAD

Loss-of-function mutations show that AraC is a positive regulator
Figure 16.12

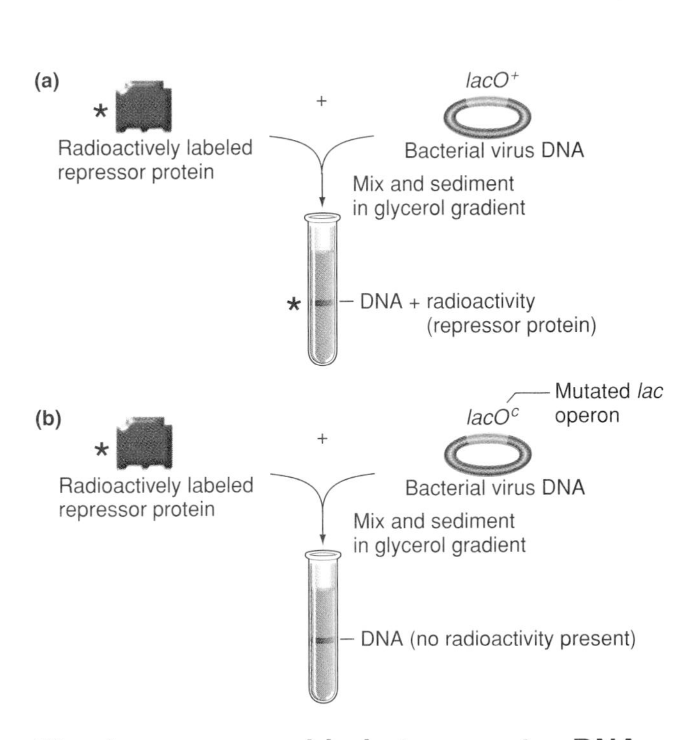

The *lac* repressor binds to operator DNA
Figure 16.13

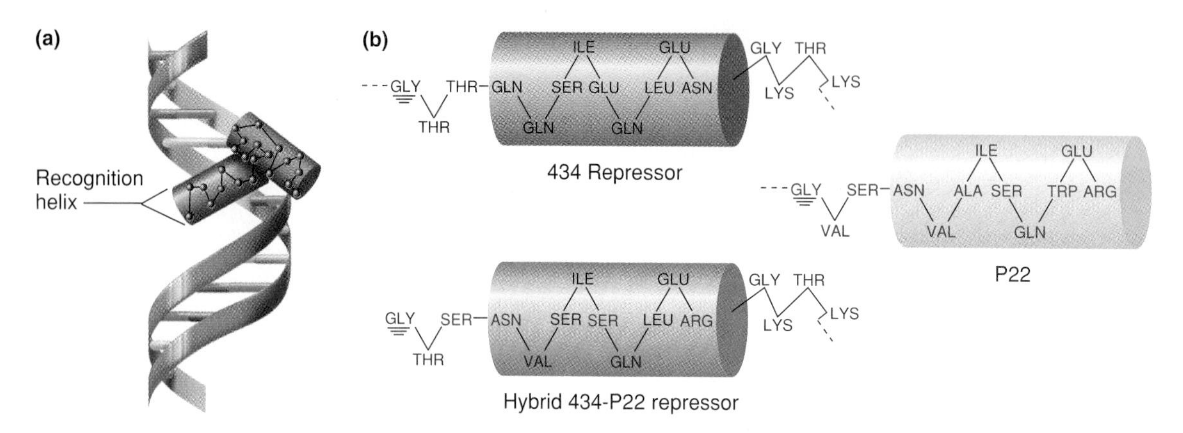

DNA recognition sequences in a helix-turn-helix section of a protein
Figure 16.14

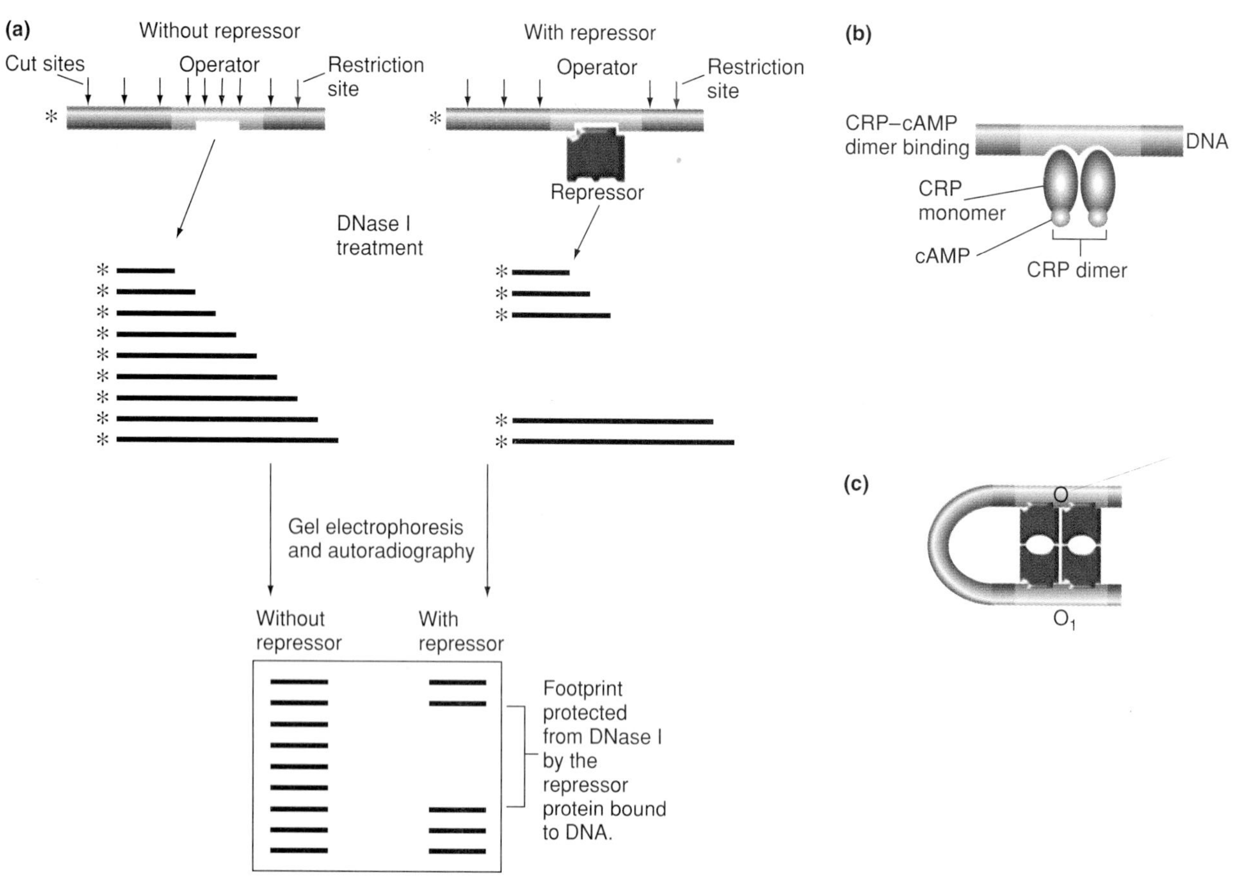

Proteins binding to DNA
Figure 16.15

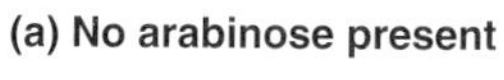

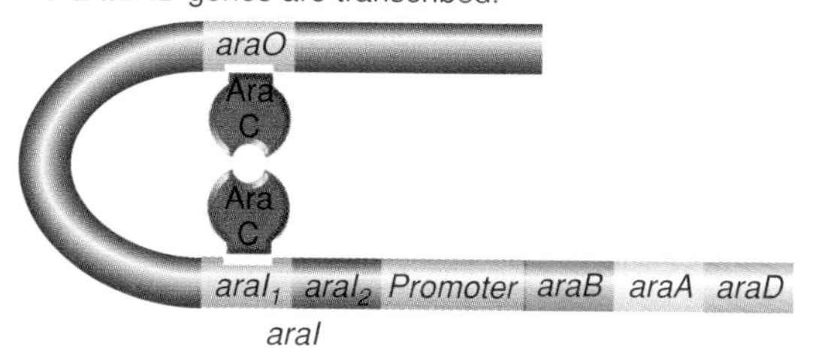

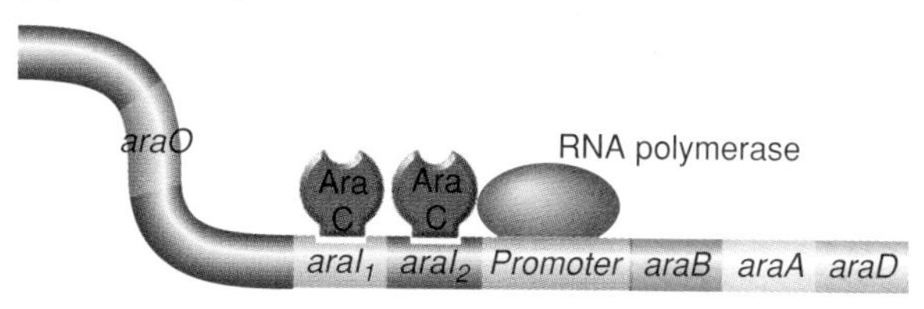

How AraC acts as both a repressor and an activator
Figure 16.16

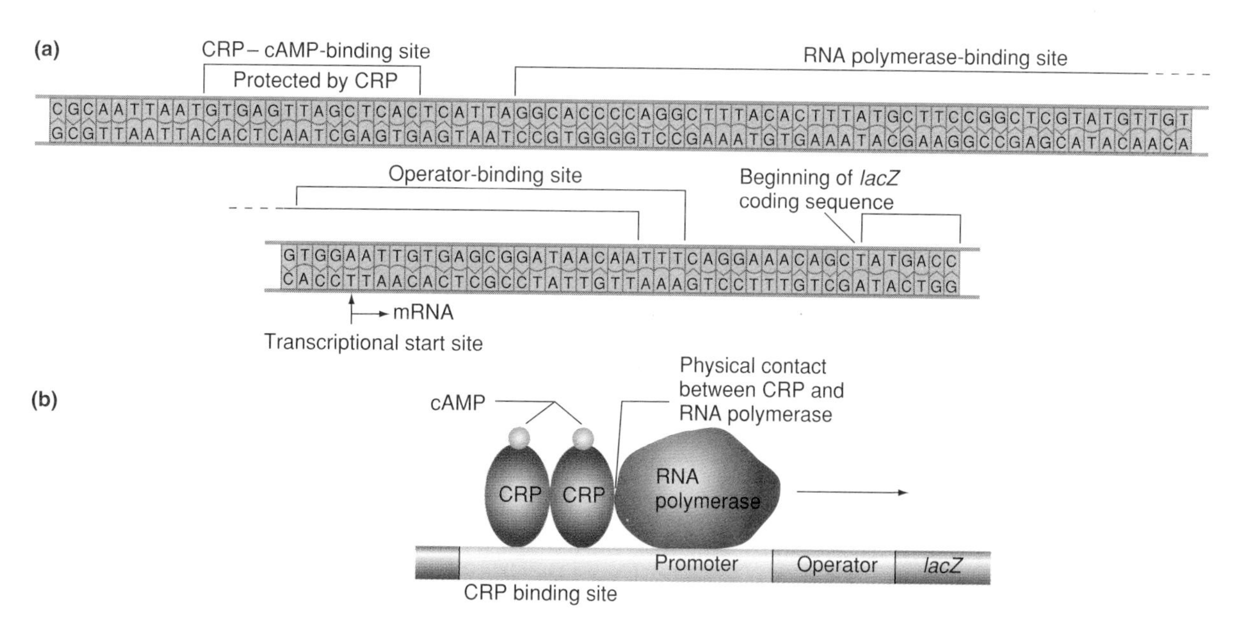

How regulatory proteins interact with RNA polymerase
Figure 16.17

(a) Fusion used to perform genetic studies of the regulatory region of gene *X*.

cis-acting gene *X* regulatory region

lacZ (Reporter gene)

Transcription ────────────►

│ Translation
▼

lacZ expression is under control of gene *X* regulatory region.

β-galactosidase

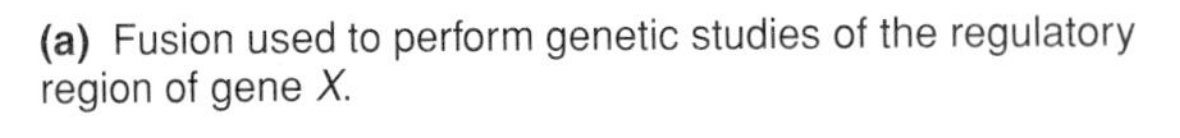

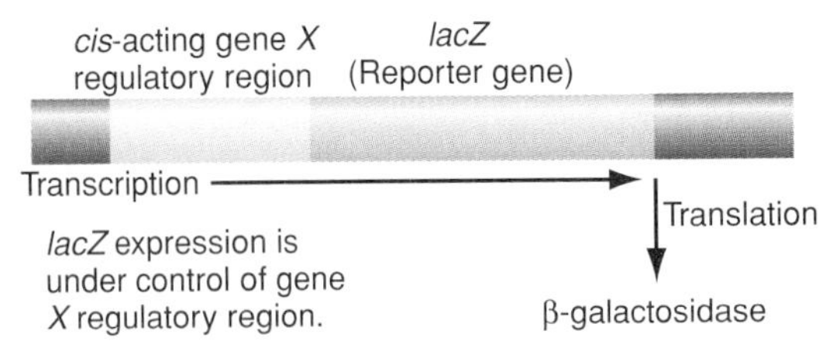

(b) Bacteriophage Mu-*lacZ* infect a population of *E. coli* cells

Mu

(c) Fusion used to obtain clone that produces large quantities of gene *X* product.

lac regulatory system

Gene *X*

◄──────── Transcription

Expression of gene *X* under control of *lac* regulatory system (inducible by lactose or analogs)

Collection (library) of *E. coli* cells each with Mu-*lacZ* inserted at a different chromosome location (and therefore fused to regulatory regions of different genes).

Human growth hormone gene (DNA fragment)

Plasmid vector

lac control region

Ligate DNA fragment and vector

Ampicillin resistance

Origin

Treat population of *E. coli* cells with UV, then assay for β-galactosidase activity.

Transform *E. coli*

Growth hormone gene

lac control region

This cell exhibits expression of β-galactosidase after UV treatment. Gene to which *lacZ* is fused is regulated by UV light.

Induce *lac* expression

Transcription

Translation

Growth hormone collects in cells

Use of the *lac* gene fusions to study gene regulation
Figure 16.18

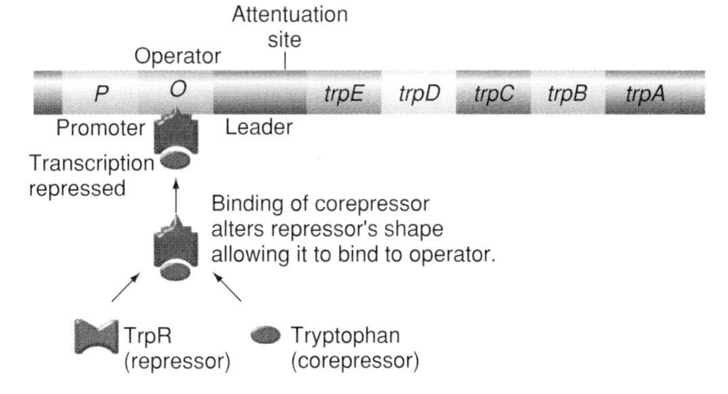

(a) Tryptophan present

Attenuation site

Operator

P O trpE trpD trpC trpB trpA

Promoter Leader

Transcription repressed

Binding of corepressor alters repressor's shape allowing it to bind to operator.

TrpR (repressor) Tryptophan (corepressor)

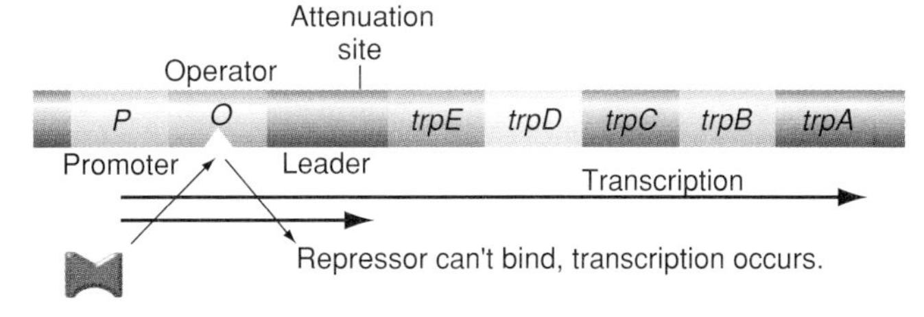

(b) Tryptophan not present

Attenuation site

Operator

P O trpE trpD trpC trpB trpA

Promoter Leader Transcription

Repressor can't bind, transcription occurs.

Tryptophan acts as a corepressor
Figure 16.19

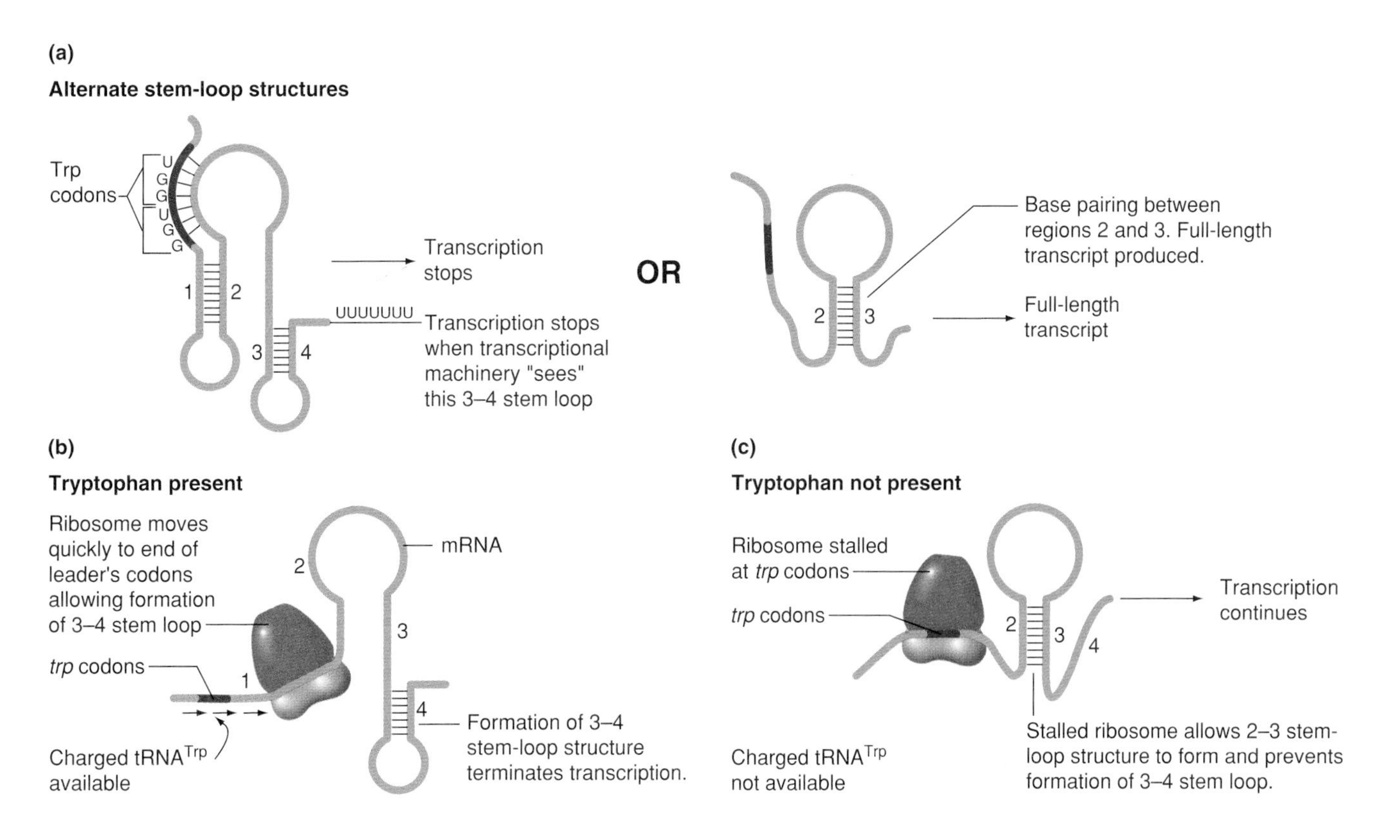

(a)

Alternate stem-loop structures

Trp codons

Transcription stops

OR

Transcription stops when transcriptional machinery "sees" this 3–4 stem loop

UUUUUUU

Base pairing between regions 2 and 3. Full-length transcript produced.

Full-length transcript

(b)

Tryptophan present

Ribosome moves quickly to end of leader's codons allowing formation of 3–4 stem loop

mRNA

trp codons

Charged tRNA^{Trp} available

Formation of 3–4 stem-loop structure terminates transcription.

(c)

Tryptophan not present

Ribosome stalled at trp codons

trp codons

Transcription continues

Charged tRNA^{Trp} not available

Stalled ribosome allows 2–3 stem-loop structure to form and prevents formation of 3–4 stem loop.

Attenuation in the tryptophan operon of *E. coli*
Figure 16.20

(a)

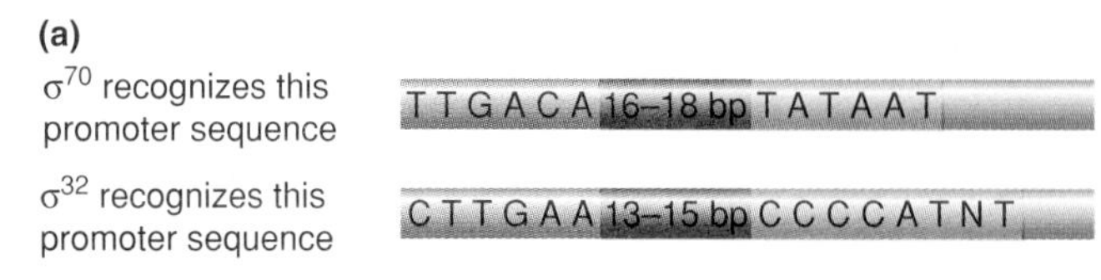

σ⁷⁰ recognizes this
promoter sequence

T T G A C A ▓16–18 bp▓ T A T A A T

σ³² recognizes this
promoter sequence

C T T G A A ▓13–15 bp▓ C C C C A T N T

(b) At high temperatures σ²⁴ recognizes different promoter sequence
on *rpoH* gene and σ³² is transcribed

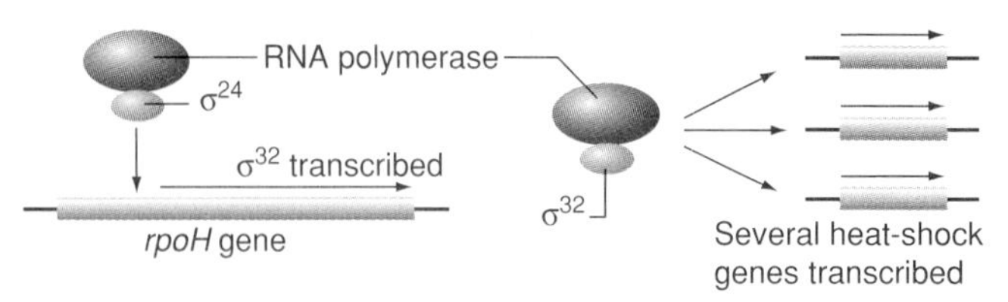

RNA polymerase

σ²⁴

σ³² transcribed

rpoH gene

σ³²

Several heat-shock
genes transcribed

Alternate sigma factors
Figure 16.21

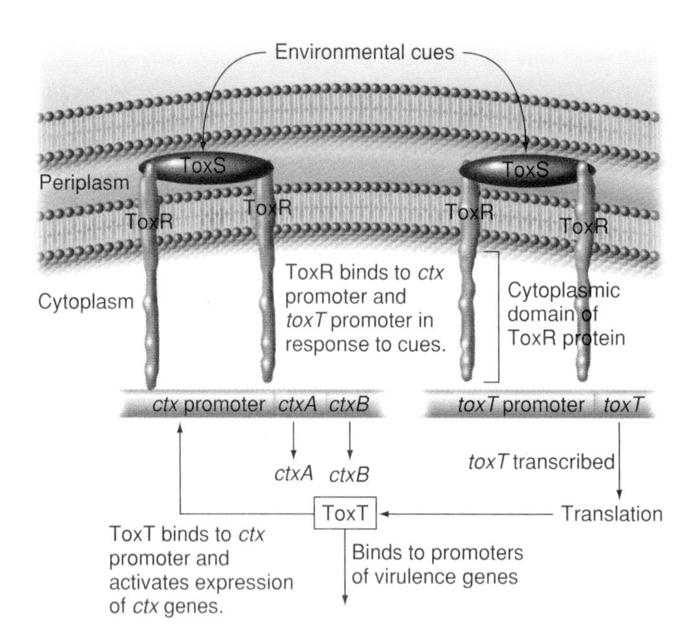

Environmental cues

ToxS

Periplasm

ToxS

ToxR   ToxR   ToxR   ToxR

Cytoplasm

ToxR binds to *ctx*
promoter and
toxT promoter in
response to cues.

Cytoplasmic
domain of
ToxR protein

ctx promoter *ctxA ctxB*   *toxT* promoter *toxT*

ctxA   ctxB   *toxT* transcribed

ToxT   Translation

ToxT binds to *ctx*
promoter and
activates expression
of *ctx* genes.

Binds to promoters
of virulence genes

Model for how *V. cholerae* regulates genes for virulence
Figure 16.22

(a) *cis*-acting elements

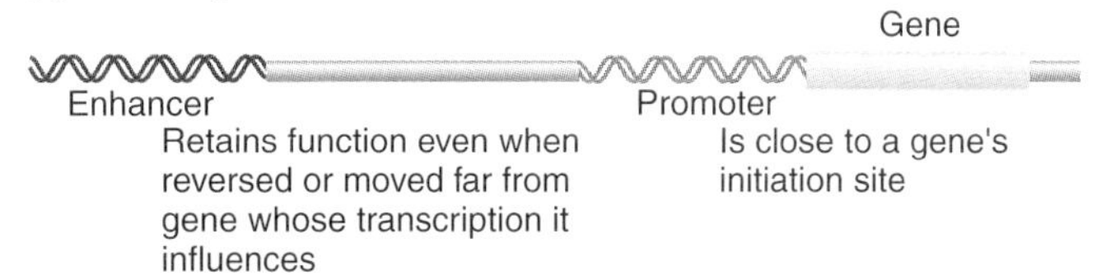

Gene

Enhancer
Retains function even when
reversed or moved far from
gene whose transcription it
influences

Promoter
Is close to a gene's
initiation site

(b) *trans*-acting gene products interact with *cis*-acting elements

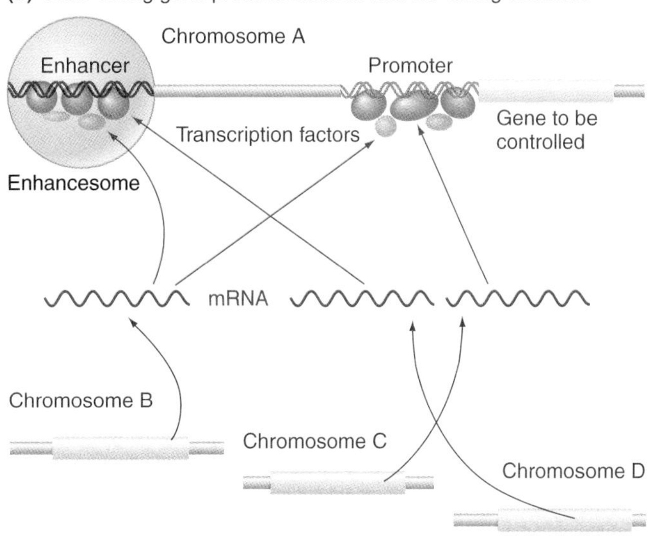

Chromosome A

Enhancer

Promoter

Transcription factors

Gene to be
controlled

Enhancesome

mRNA

Chromosome B

Chromosome C

Chromosome D

trans-acting genetic elements

How *cis*-acting and *trans*-acting elements influence transcription
Figure 17.1

(a) Tandem repeats of rRNA genes are transcribed by RNA polymerase I

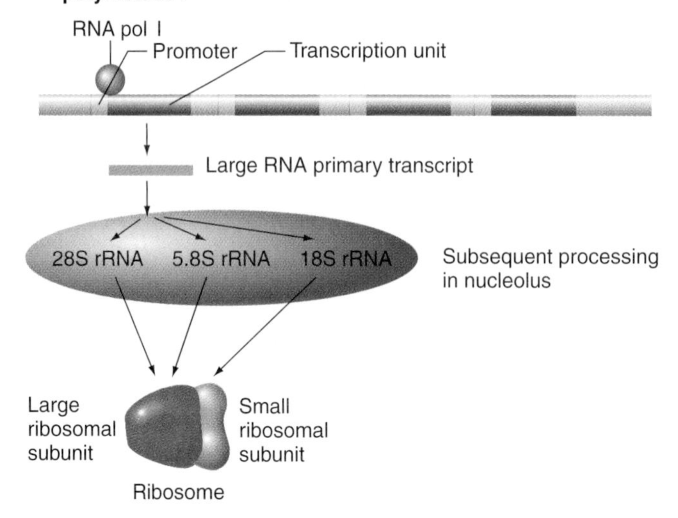

RNA pol I
Promoter — Transcription unit

Large RNA primary transcript

28S rRNA   5.8S rRNA   18S rRNA

Subsequent processing
in nucleolus

Large
ribosomal
subunit

Small
ribosomal
subunit

Ribosome

(b) Small RNA genes are transcribed by RNA pol III

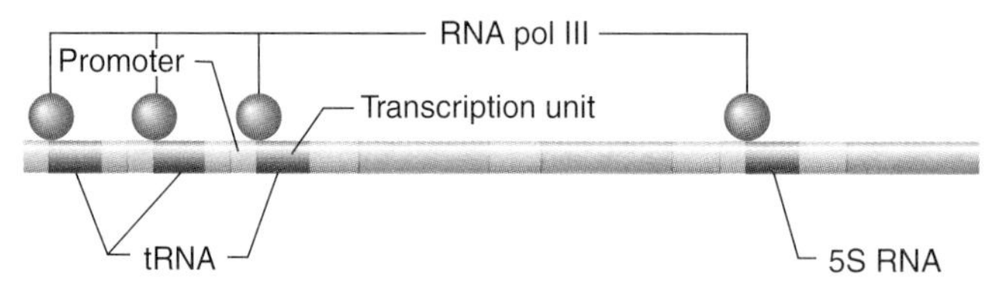

Promoter
RNA pol III
Transcription unit

tRNA

5S RNA

(c) Protein-encoding genes are transcribed by RNA pol II

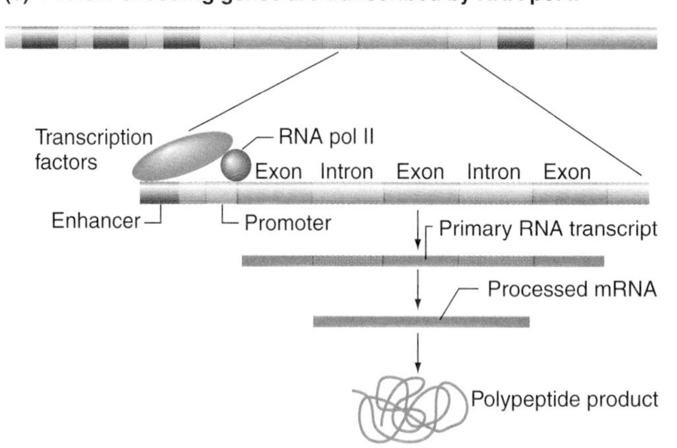

Transcription
factors
RNA pol II
Exon Intron Exon Intron Exon

Enhancer
Promoter
Primary RNA transcript

Processed mRNA

Polypeptide product

The three RNA polymerases of eukaryotic cells have different functions and recognize different promoters
Figure 17.2

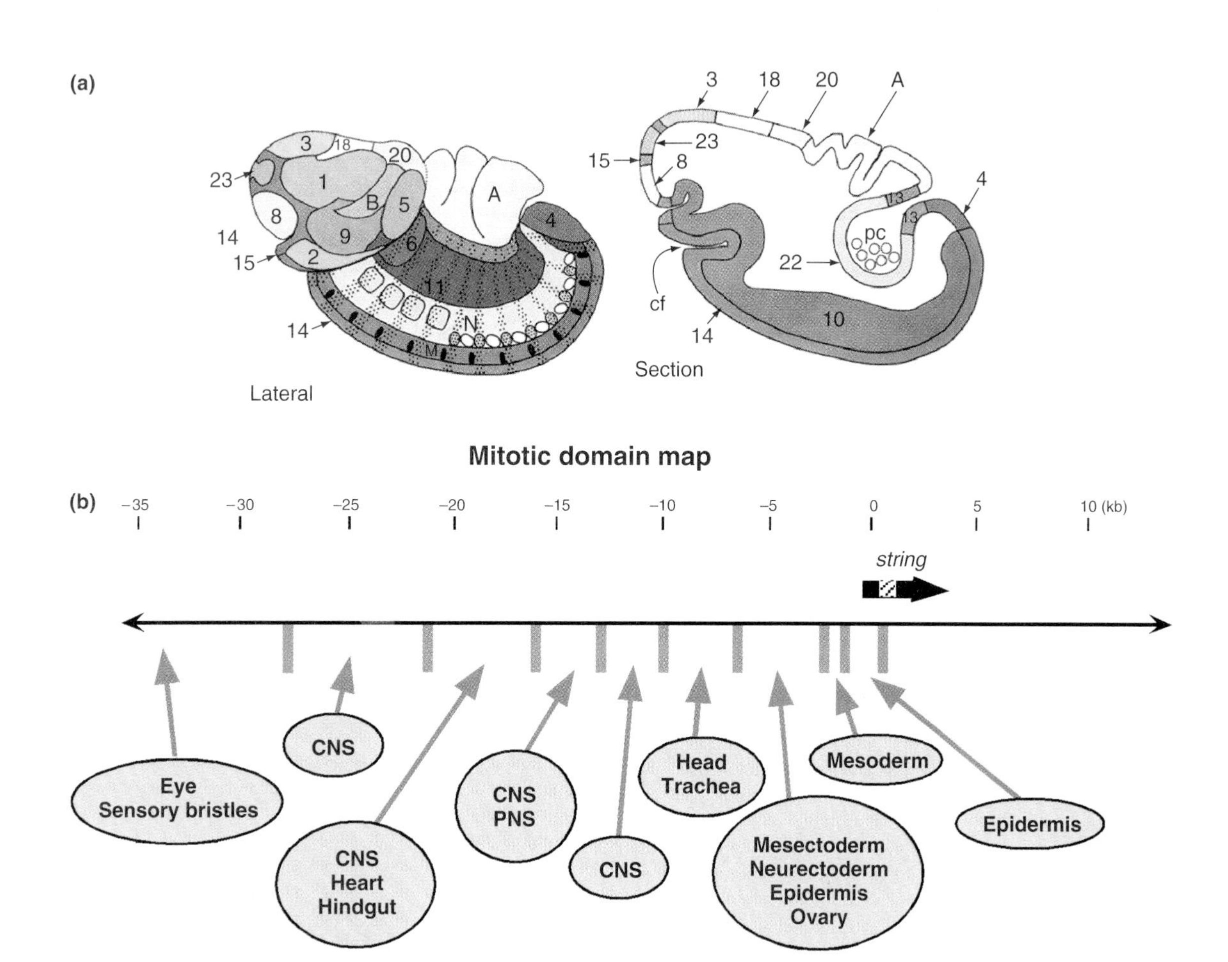

Mitotic domain map

The large enhancer region of the *Drosophila string* gene helps create discrete mitotic domains during embryogenesis
Figure 17.3

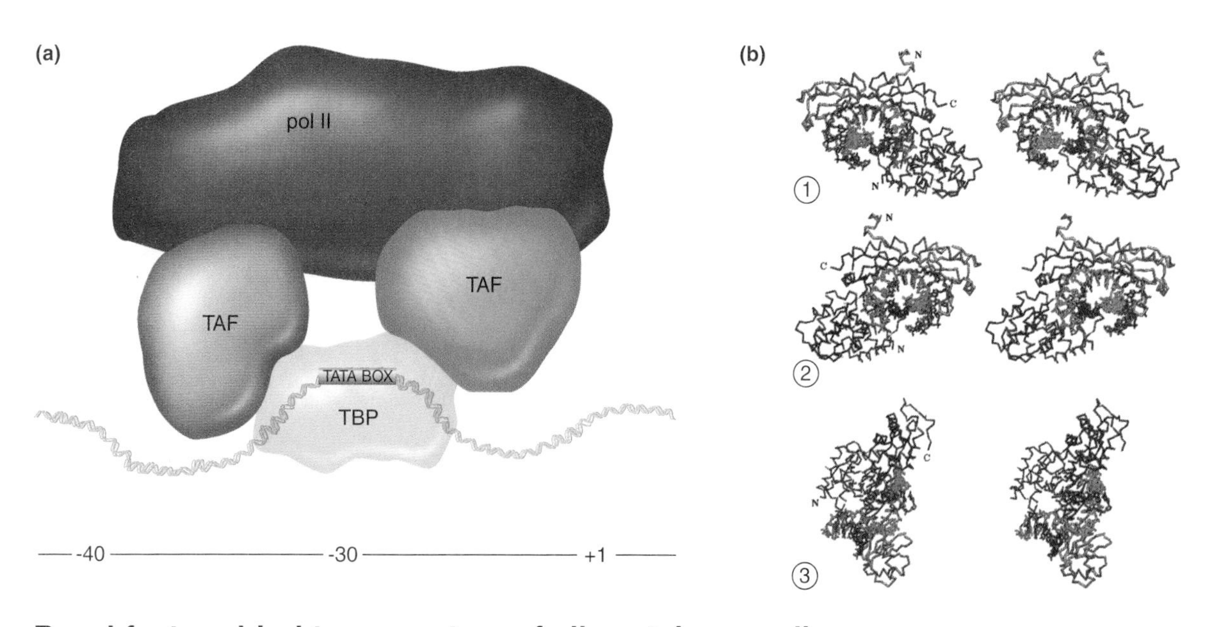

Basal factors bind to promoters of all protein-encoding genes
Figure 17.4

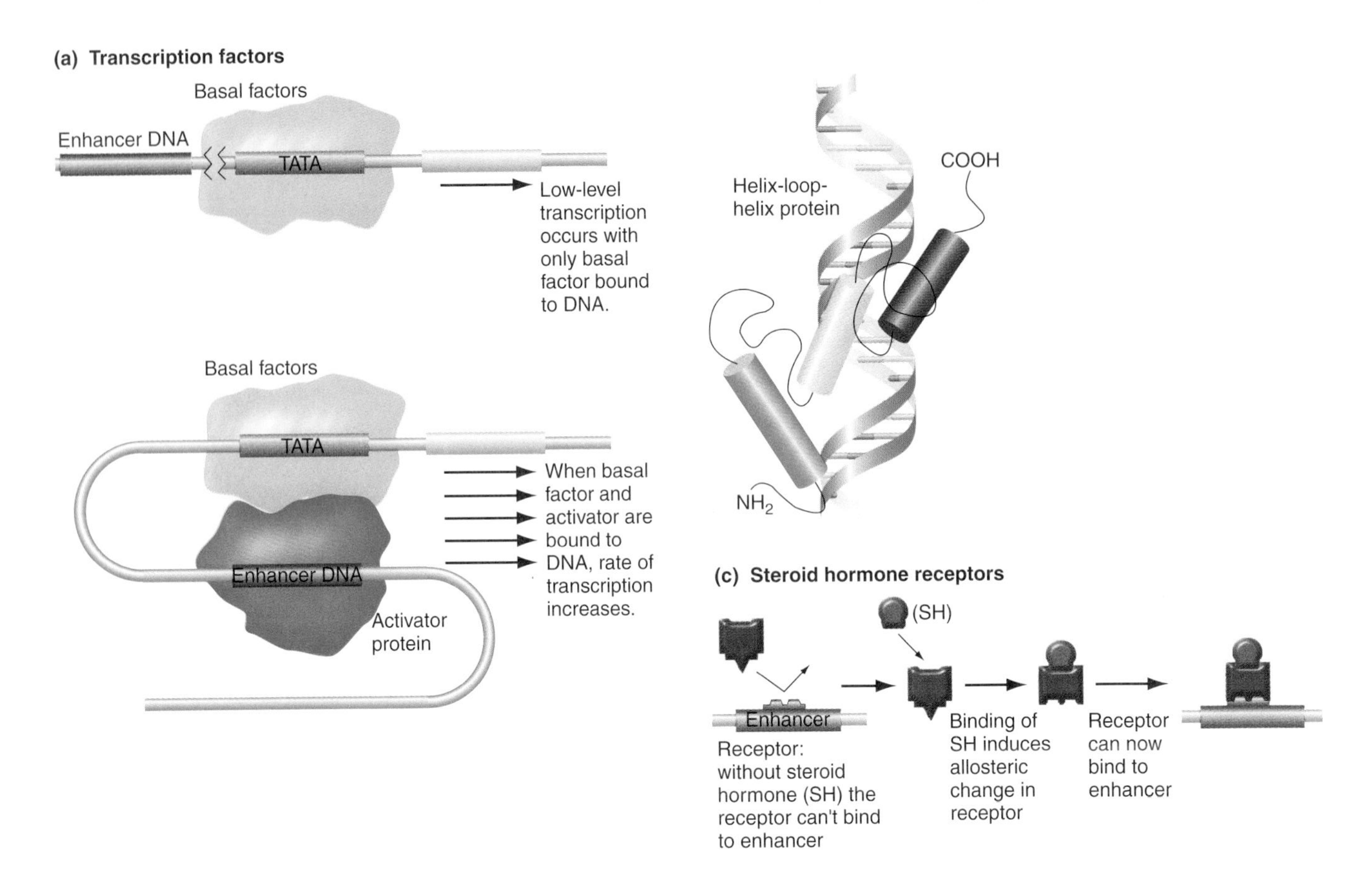

(a) Transcription factors

Basal factors

Enhancer DNA

TATA

Low-level transcription occurs with only basal factor bound to DNA.

Basal factors

TATA

When basal factor and activator are bound to DNA, rate of transcription increases.

Enhancer DNA

Activator protein

Helix-loop-helix protein

COOH

NH₂

(c) Steroid hormone receptors

(SH)

Enhancer

Receptor: without steroid hormone (SH) the receptor can't bind to enhancer

Binding of SH induces allosteric change in receptor

Receptor can now bind to enhancer

Transcriptional activators bind to specific enhancers at specific times to increase transcriptional levels
Figure 17.5

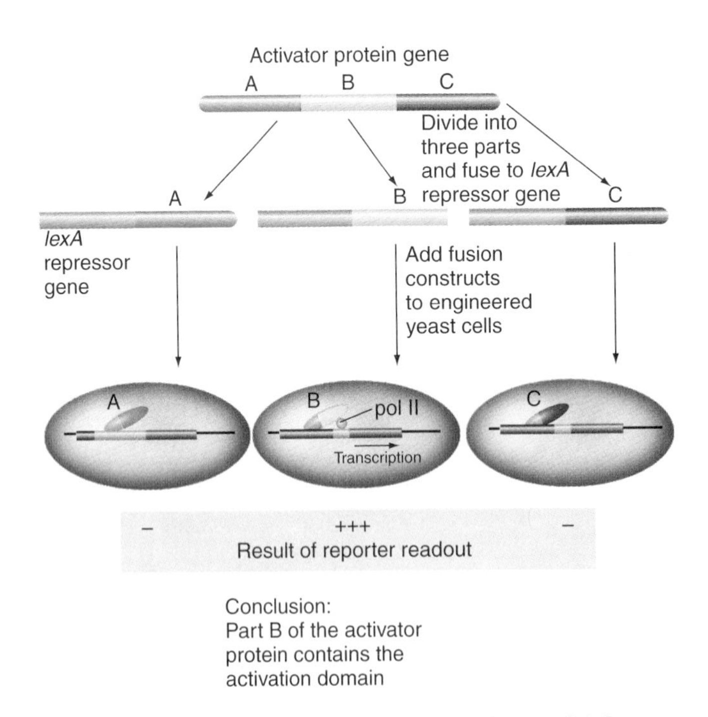

Activator protein gene

A   B   C

Divide into three parts and fuse to *lexA* repressor gene

A   B   C

lexA repressor gene

Add fusion constructs to engineered yeast cells

A   B—pol II   C

Transcription

–   +++   –
Result of reporter readout

Conclusion:
Part B of the activator protein contains the activation domain

Localization of activator domains within activator proteins can be achieved with recombinant DNA constructs
Figure 17.6

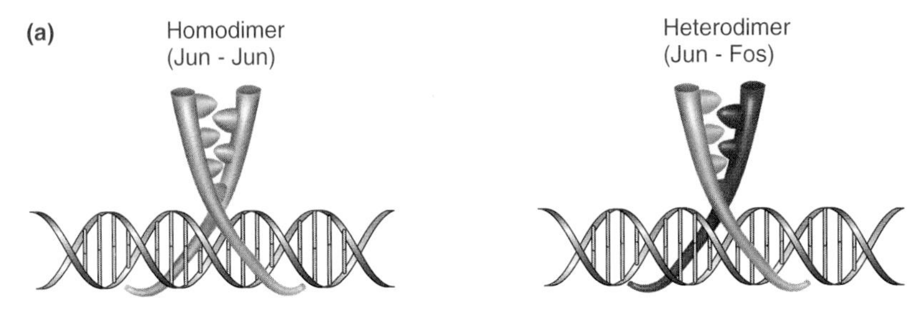

(a) Homodimer (Jun - Jun)   Heterodimer (Jun - Fos)

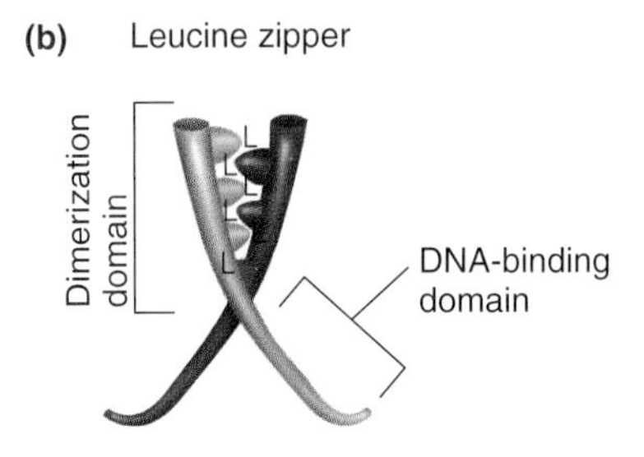

(b) Leucine zipper

Dimerization domain

DNA-binding domain

Most activator proteins function in the cell as dimers
Figure 17.7

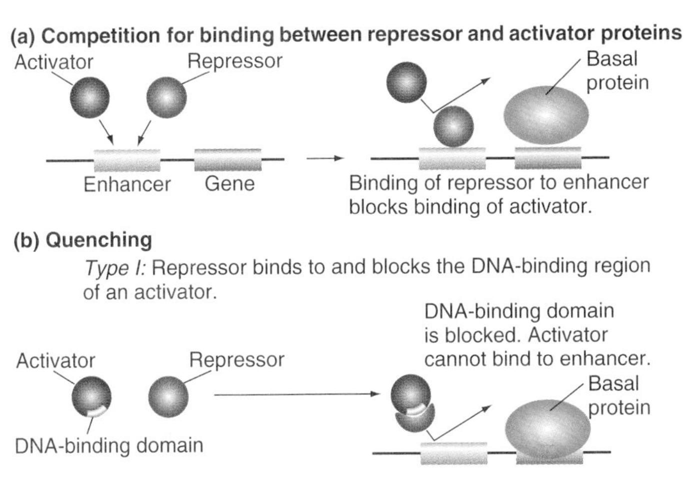

(a) Competition for binding between repressor and activator proteins

Activator   Repressor   Basal protein

Enhancer   Gene   Binding of repressor to enhancer blocks binding of activator.

(b) Quenching

Type I: Repressor binds to and blocks the DNA-binding region of an activator.

DNA-binding domain is blocked. Activator cannot bind to enhancer.

Activator   Repressor   Basal protein

DNA-binding domain

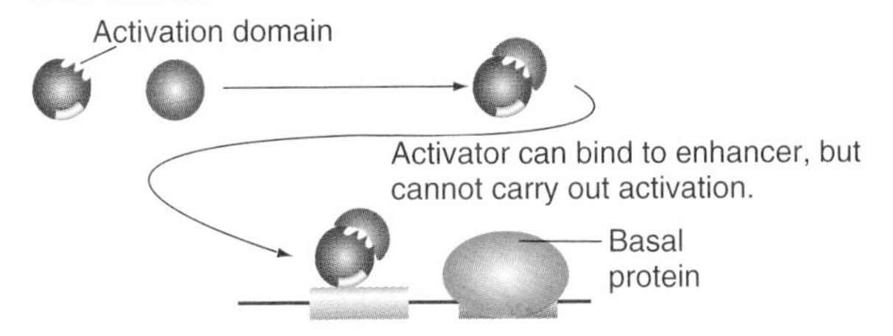

Type II: Repressor binds to and blocks the activation domain of an activator.

Activation domain

Activator can bind to enhancer, but cannot carry out activation.

Basal protein

Repressor proteins reduce transcriptional levels through competition or quenching
Figure 17.8

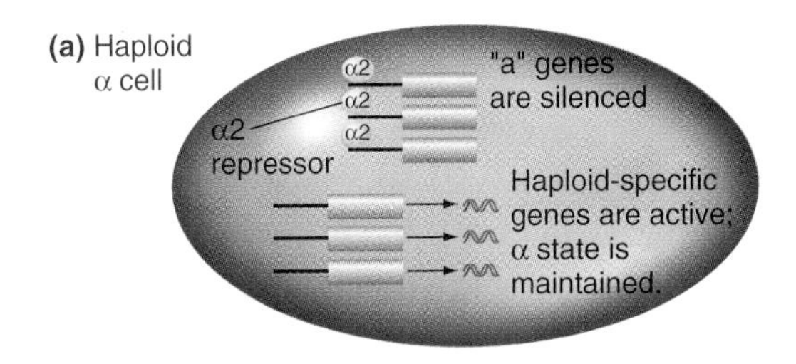

(a) Haploid α cell

α2

α2

α2

α2 repressor

"a" genes are silenced

Haploid-specific genes are active; α state is maintained.

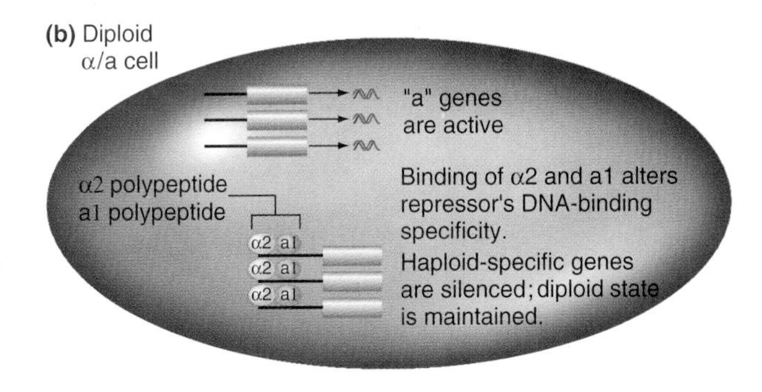

(b) Diploid α/a cell

"a" genes are active

α2 polypeptide
a1 polypeptide

α2 a1
α2 a1
α2 a1

Binding of α2 and a1 alters repressor's DNA-binding specificity.

Haploid-specific genes are silenced; diploid state is maintained.

The same transcription factors can play different roles in different cells
Figure 17.9

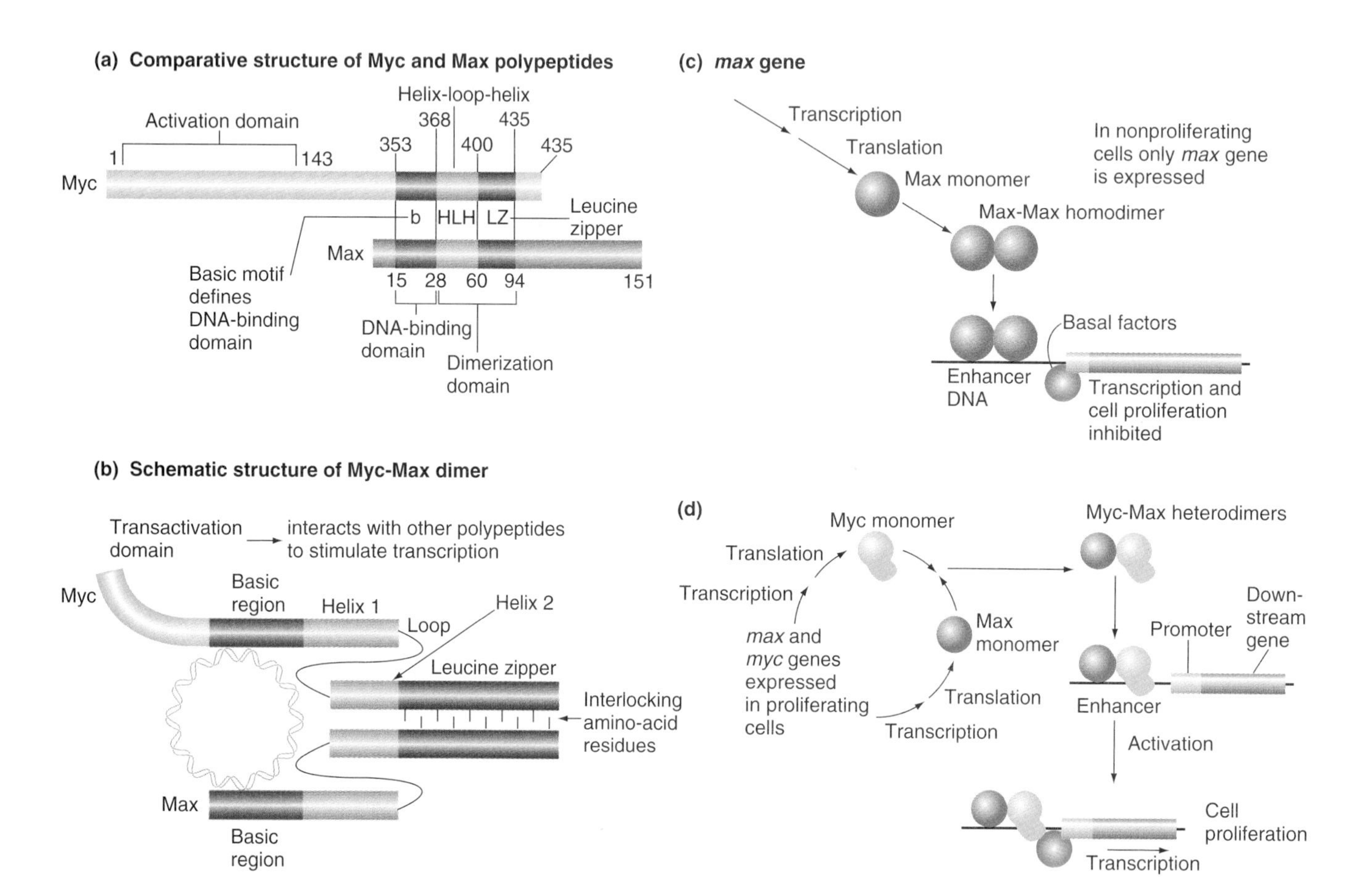

(a) Comparative structure of Myc and Max polypeptides

Activation domain

Helix-loop-helix

1   143   353   368   400   435   435

Myc

b   HLH   LZ

Leucine zipper

Basic motif defines DNA-binding domain

Max

15   28   60   94   151

DNA-binding domain

Dimerization domain

(c) *max* gene

Transcription

Translation

Max monomer

In nonproliferating cells only *max* gene is expressed

Max-Max homodimer

Basal factors

Enhancer DNA

Transcription and cell proliferation inhibited

(b) Schematic structure of Myc-Max dimer

Transactivation domain → interacts with other polypeptides to stimulate transcription

Myc

Basic region

Helix 1

Loop

Helix 2

Leucine zipper

Interlocking amino-acid residues

Max

Basic region

(d)

Myc monomer

Myc-Max heterodimers

Translation

Transcription

max and *myc* genes expressed in proliferating cells

Max monomer

Translation

Transcription

Down-stream gene

Promoter

Enhancer

Activation

Cell proliferation

Transcription

The Myc-Max system of activation and repression
Figure 17.10

The *GAL1, GAL7, GAL10* gene system is a model of complex transcriptional regulation
Figure 17.11

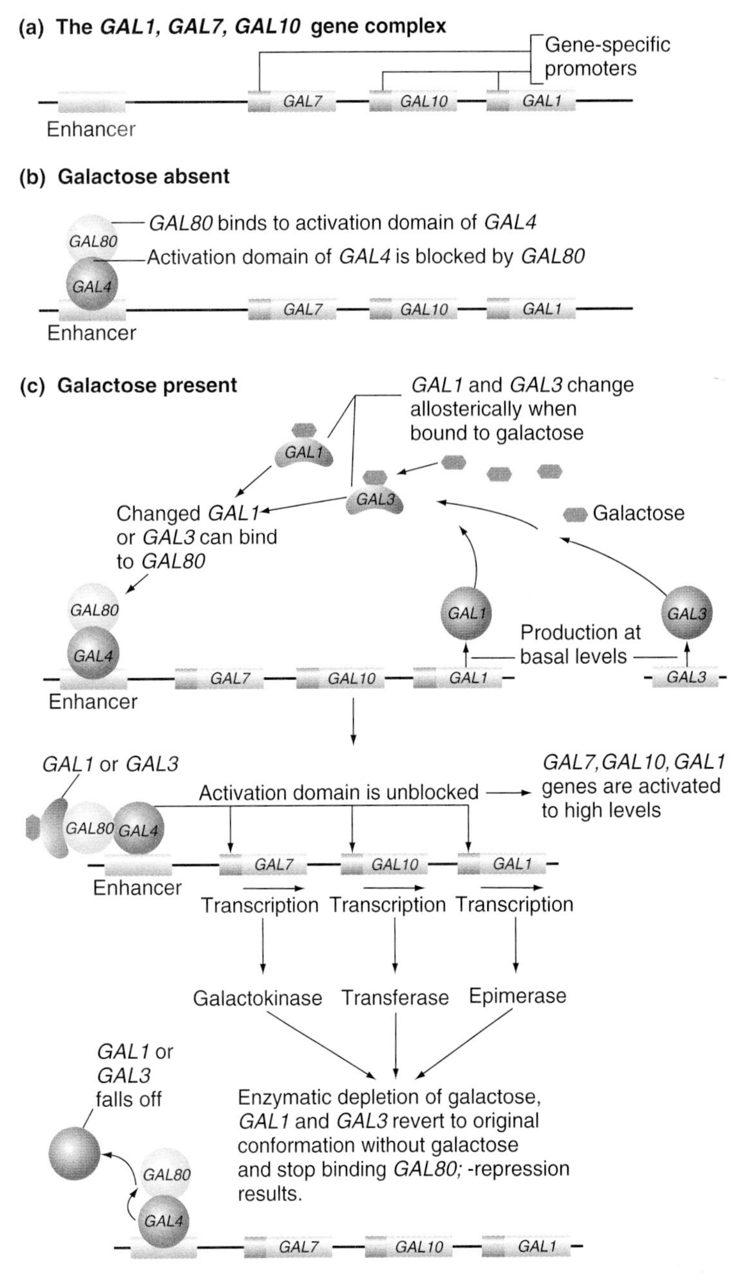

(a) The *GAL1, GAL7, GAL10* gene complex

(b) Galactose absent

(c) Galactose present

(a) Gene complex

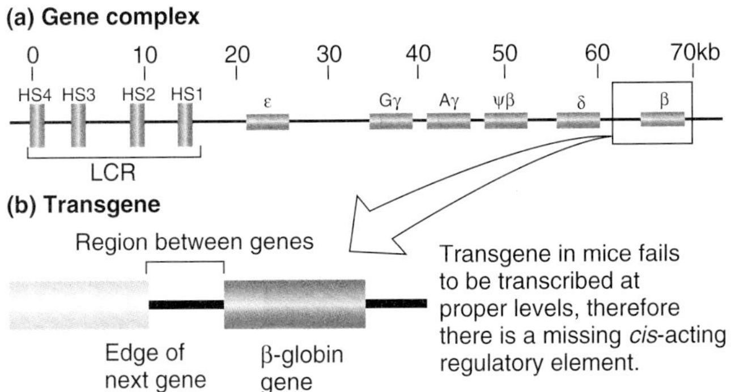

(b) Transgene

Region between genes

Transgene in mice fails to be transcribed at proper levels, therefore there is a missing *cis*-acting regulatory element.

Edge of next gene

β-globin gene

(c) Mechanism of transcriptional activation

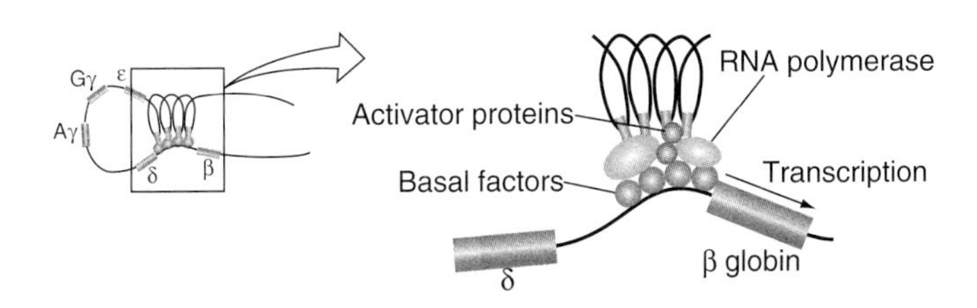

A locus control region (LCR) operates over a cluster of genes to activate transcription
Figure 17.12

(a) Naked promoter binds RNA polymerase and basal factors.

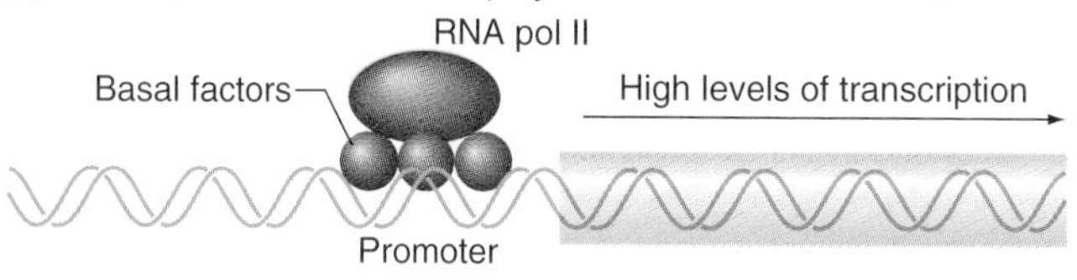

RNA pol II

Basal factors

High levels of transcription

Promoter

(b) Chromatin reduces binding to basal factors and RNA pol II to very low levels.

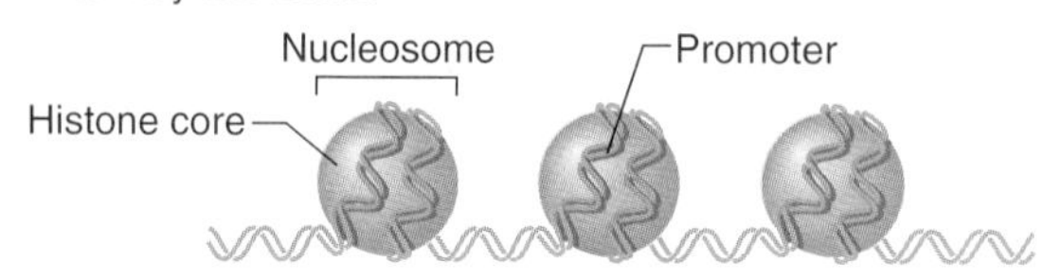

Nucleosome

Promoter

Histone core

(c) Chromatin remodeling can expose promoter.

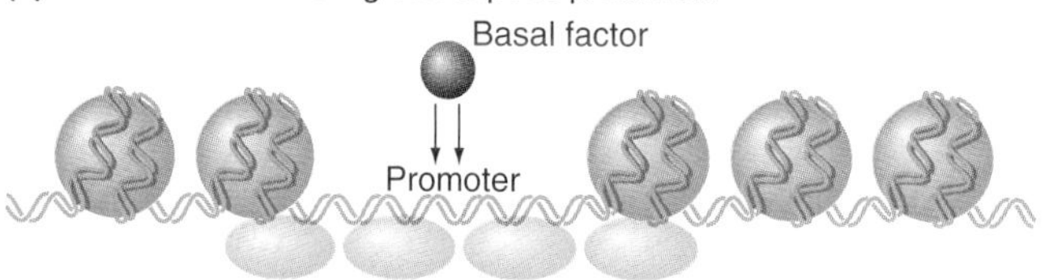

Basal factor

Promoter

Remodeling proteins

(d) DNase hypersensitive sites are at 5' ends of genes.

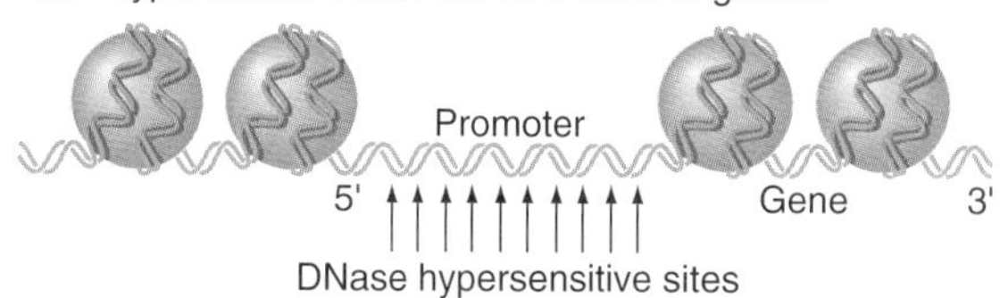

Promoter

5'

Gene

3'

DNase hypersensitive sites

(e) The SWI-SNF multisubunit complex is an example of a remodeling complex.

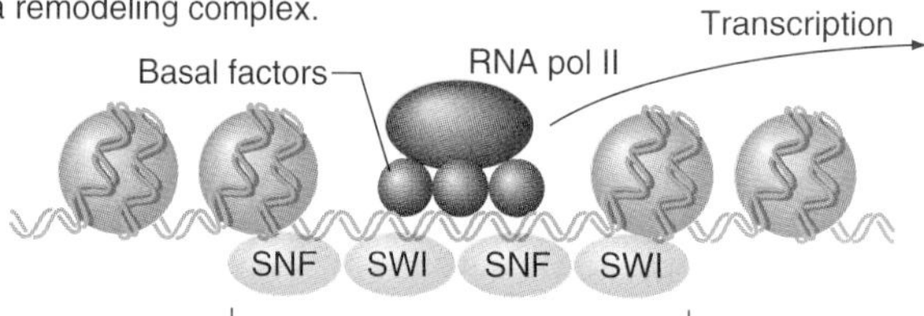

Transcription

Basal factors

RNA pol II

SNF   SWI   SNF   SWI

The SWI-SNF multisubunit destabilizes chromatin structure and gives transcription machinery access to the promoter.

Chromatin structure plays a critical role in eukaryotic gene regulation

Figure 17.13

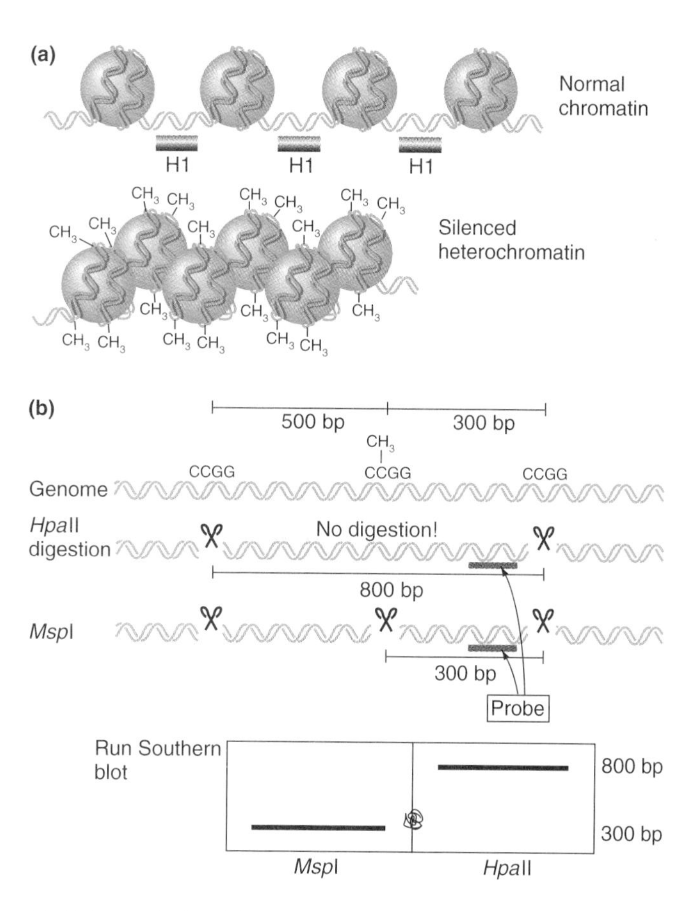

(a)

Normal chromatin

H1   H1   H1

Silenced heterochromatin

(b)

500 bp   300 bp

Genome

Hpall digestion   No digestion!

800 bp

MspI

300 bp

Probe

Run Southern blot

MspI   Hpall

800 bp

300 bp

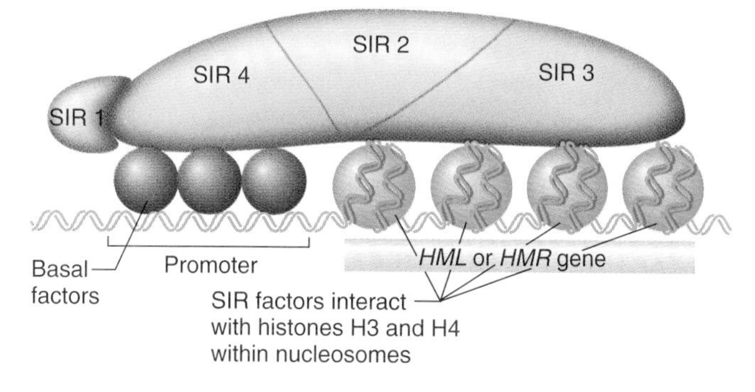

(c) SIR complex binds to basal factors and interacts with H3 and H4 components of histones

SIR 2

SIR 4   SIR 3

SIR 1

Basal factors   Promoter   HML or HMR gene

SIR factors interact with histones H3 and H4 within nucleosomes

Heterochromatin formation can lead to transcriptional silencing
Figure 17.14

(a) Early embryo

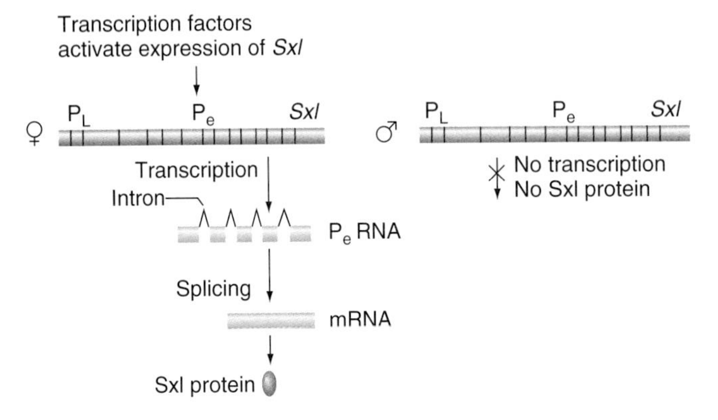

(b) Sxl protein regulates the splicing of its mRNA by blocking male acceptor site

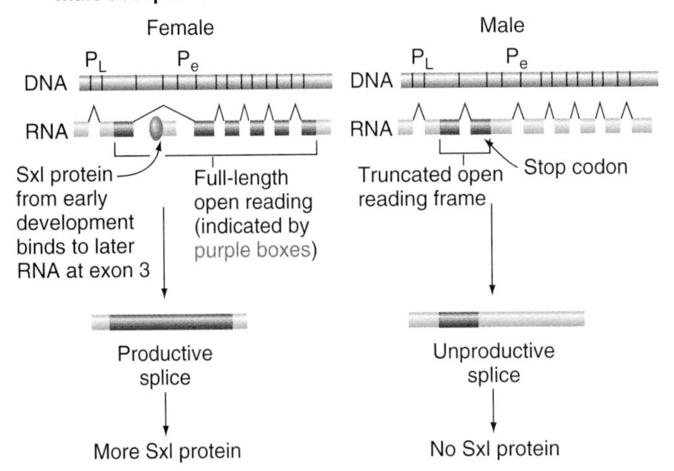

Differential RNA splicing can regulate gene expression at the polypeptide level
Figure 17.16

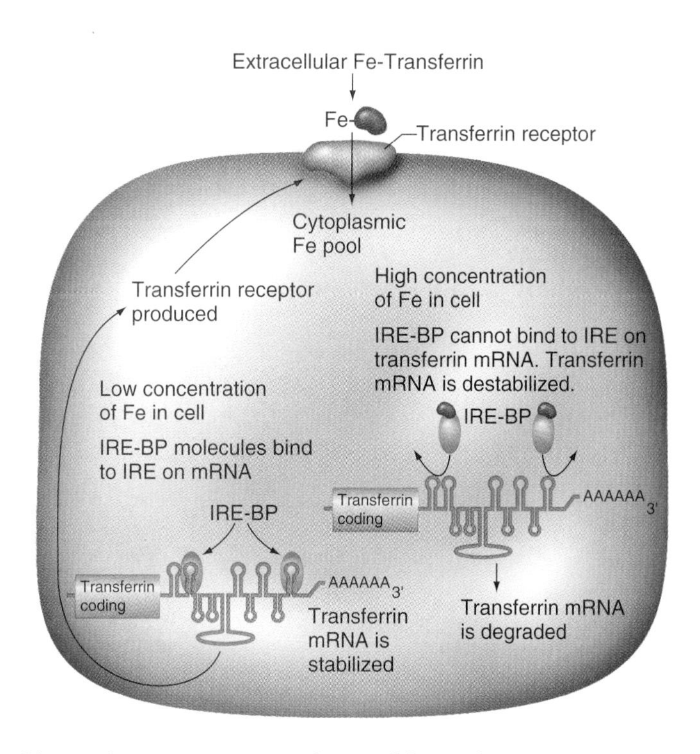

How the concentration of iron in a cell regulates transferrin receptor levels.
Figure 17.17

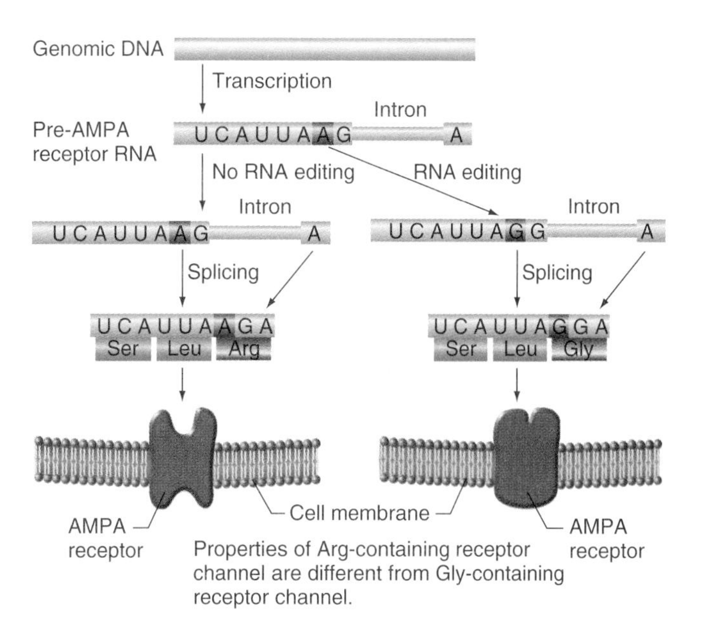

mRNA editing can regulate the function of the protein product that ensues
Figure 17.18

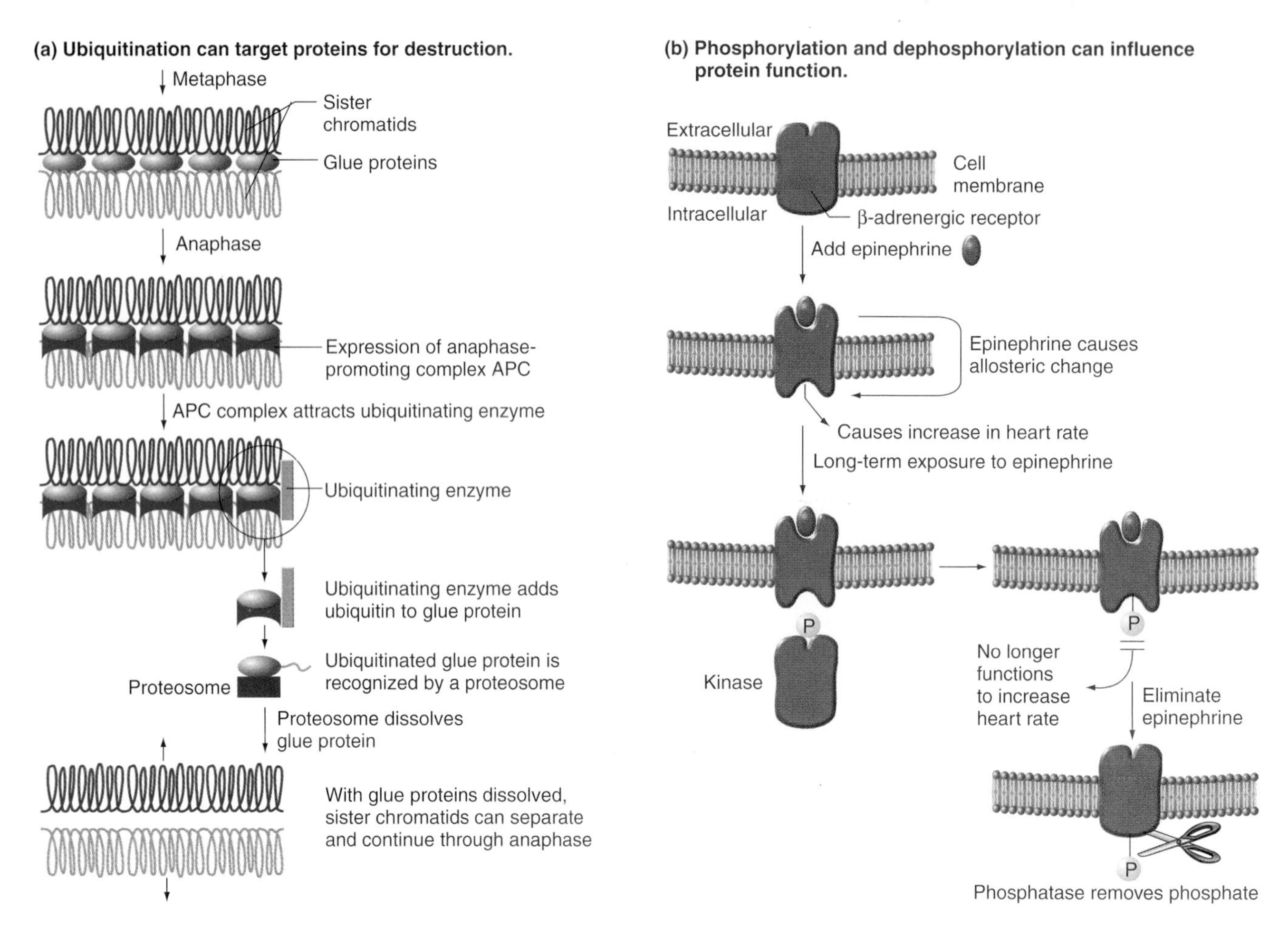

(a) Ubiquitination can target proteins for destruction.

Metaphase

Sister chromatids

Glue proteins

Anaphase

Expression of anaphase-promoting complex APC

APC complex attracts ubiquitinating enzyme

Ubiquitinating enzyme

Ubiquitinating enzyme adds ubiquitin to glue protein

Proteosome

Ubiquitinated glue protein is recognized by a proteosome

Proteosome dissolves glue protein

With glue proteins dissolved, sister chromatids can separate and continue through anaphase

(b) Phosphorylation and dephosphorylation can influence protein function.

Extracellular

Cell membrane

Intracellular

β-adrenergic receptor

Add epinephrine

Epinephrine causes allosteric change

Causes increase in heart rate

Long-term exposure to epinephrine

Kinase

P

No longer functions to increase heart rate

P

Eliminate epinephrine

P

Phosphatase removes phosphate

Protein modifications can affect the final level of gene function
Figure 17.19

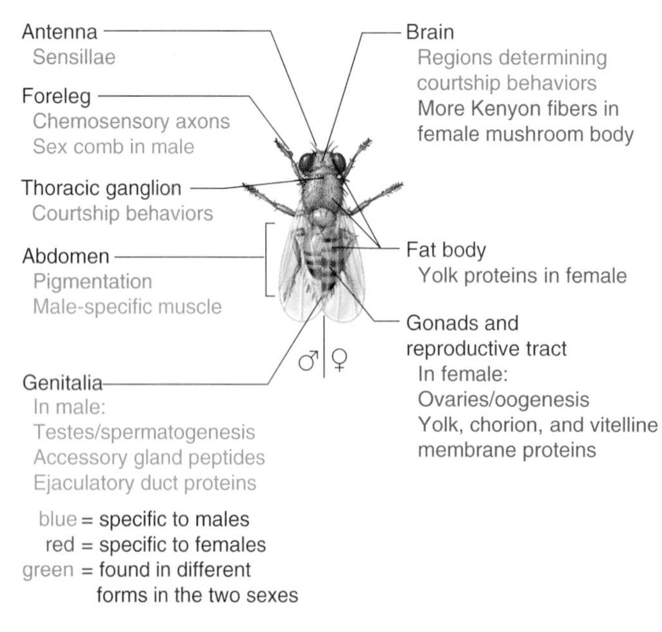

Antenna
Sensillae

Foreleg
Chemosensory axons
Sex comb in male

Thoracic ganglion
Courtship behaviors

Abdomen
Pigmentation
Male-specific muscle

Genitalia
In male:
Testes/spermatogenesis
Accessory gland peptides
Ejaculatory duct proteins

Brain
Regions determining courtship behaviors
More Kenyon fibers in female mushroom body

Fat body
Yolk proteins in female

Gonads and reproductive tract
In female:
Ovaries/oogenesis
Yolk, chorion, and vitelline membrane proteins

♂ ♀

blue = specific to males
red = specific to females
green = found in different forms in the two sexes

Sex-specific traits in *Drosophila*
Figure 17.20

Numerator ●
Denominator ●

Female
X X : A A = 1

Male
X : A A = 1/2

Gene expression

Gene products

Dimerization

Extra monomers

+

No numerator dimers

Transcription activated

Sxl

P_e
Promoter

Sxl

P_e
No activation

Sxl protein

No Sxl protein

Female
sex determination pathways

Male
sex determination pathways

The X:A ratio determines the expression of *Sxl*, the first gene in the sex-determination pathway
Figure 17.21

The expression or nonexpression of *Sxl* leads to downstream differences in expression of products that also play a role in the sex-determination pathway
Figure 17.22

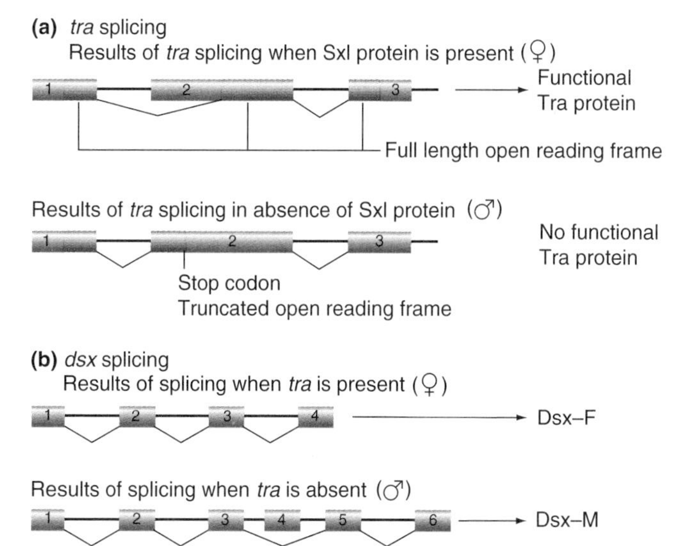

(a) *tra* splicing
Results of *tra* splicing when Sxl protein is present (♀)

1 2 3 → Functional Tra protein

Full length open reading frame

Results of *tra* splicing in absence of Sxl protein (♂)

1 2 3 → No functional Tra protein

Stop codon
Truncated open reading frame

(b) *dsx* splicing
Results of splicing when *tra* is present (♀)

1 2 3 4 → Dsx–F

Results of splicing when *tra* is absent (♂)

1 2 3 4 5 6 → Dsx–M

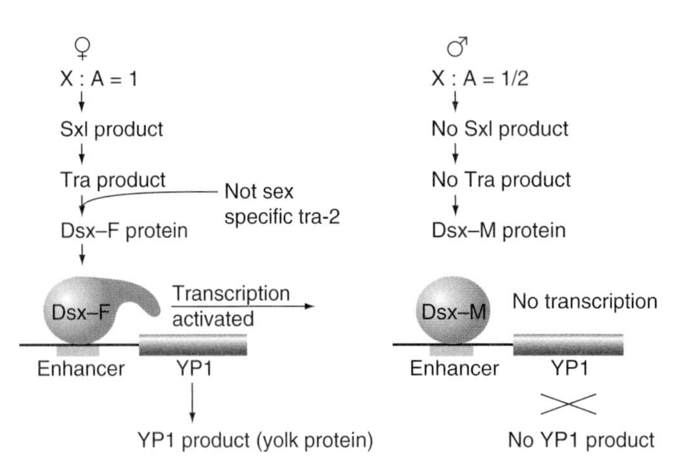

Alternative forms of the *dxs* gene product both bind to the same *YP1* enhancer, but they have opposite effects on the expression of the *YP1* gene
Figure 17.23

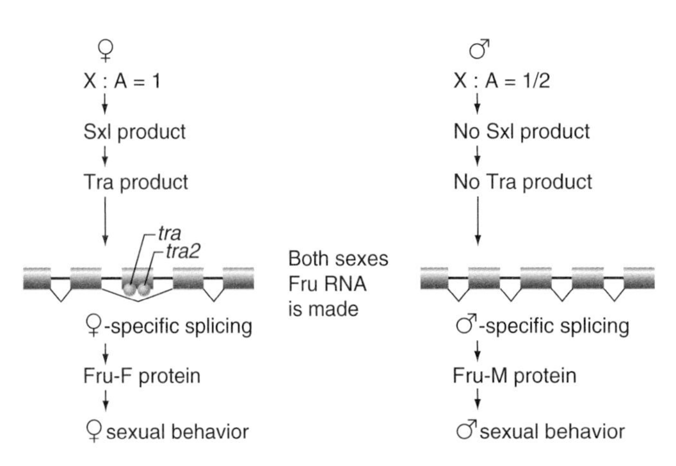

The primary *fru* RNA transcript is made in both sexes
Figure 17.24

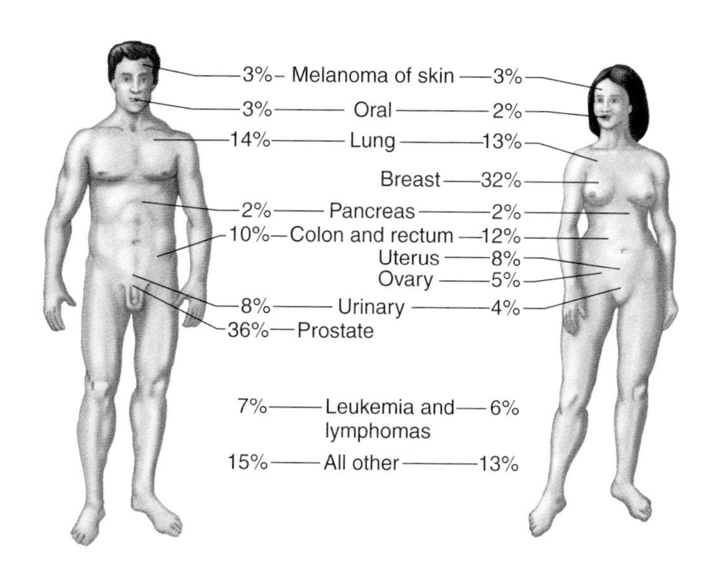

The relative percentages of new cancers in the United States that occur at different sites in the bodies of men and women
Figure 18.1

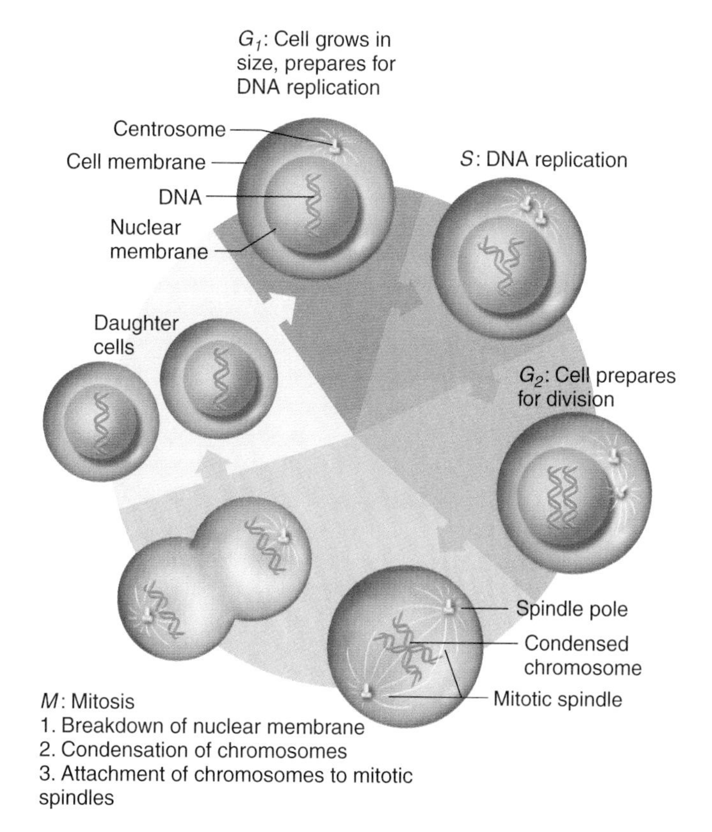

The cell cycle is the series of events that transpire between one cell division and the next
Figure 18.2

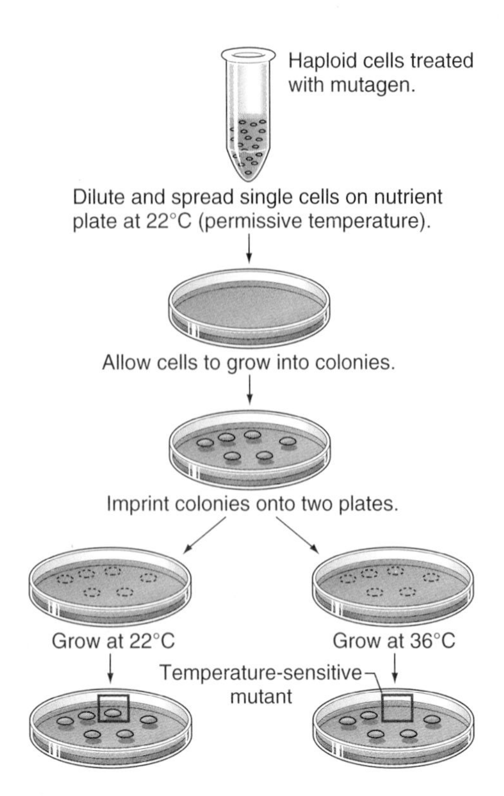

The isolation of temperature-sensitive mutants of yeast
Figure 18.3

Haploid cells treated with mutagen.

Dilute and spread single cells on nutrient plate at 22°C (permissive temperature).

Allow cells to grow into colonies.

Imprint colonies onto two plates.

Grow at 22°C   Grow at 36°C

Temperature-sensitive mutant

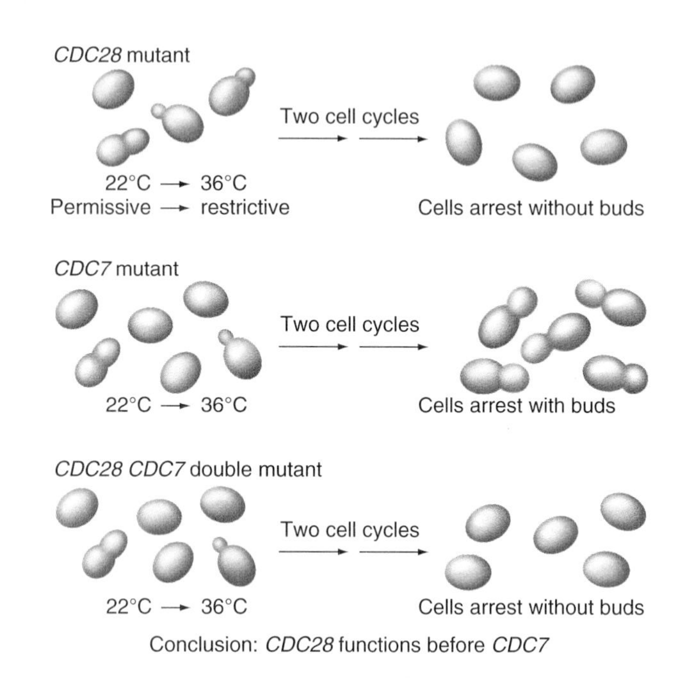

CDC28 mutant

Two cell cycles

22°C → 36°C
Permissive → restrictive

Cells arrest without buds

CDC7 mutant

Two cell cycles

22°C → 36°C

Cells arrest with buds

CDC28 CDC7 double mutant

Two cell cycles

22°C → 36°C

Cells arrest without buds

Conclusion: *CDC28* functions before *CDC7*

A double mutant reveals which gene is needed first
Figure 18.5

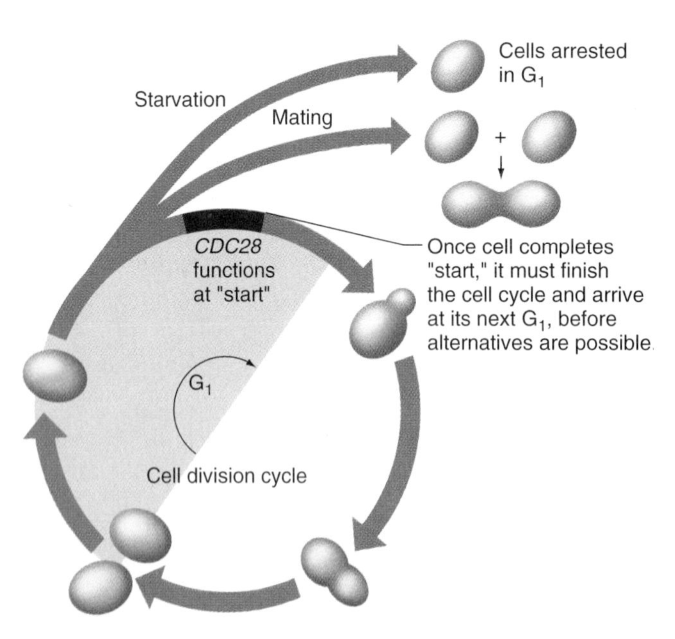

Starvation

Mating

Cells arrested in G₁

Once cell completes "start," it must finish the cell cycle and arrive at its next G₁, before alternatives are possible.

CDC28 functions at "start"

G₁

Cell division cycle

Yeast cells become committed to the cell cycle in G₁
Figure 18.6

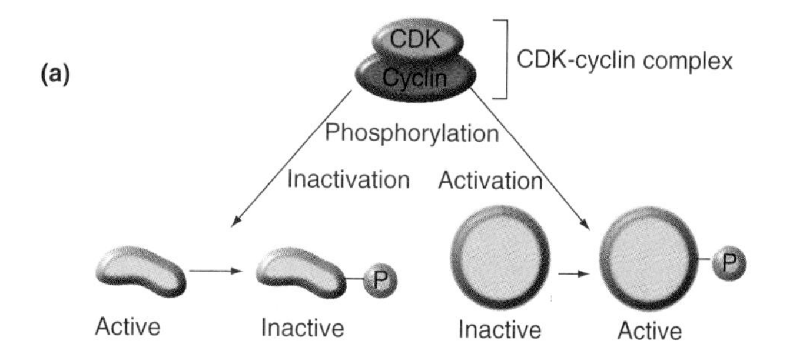

(a)

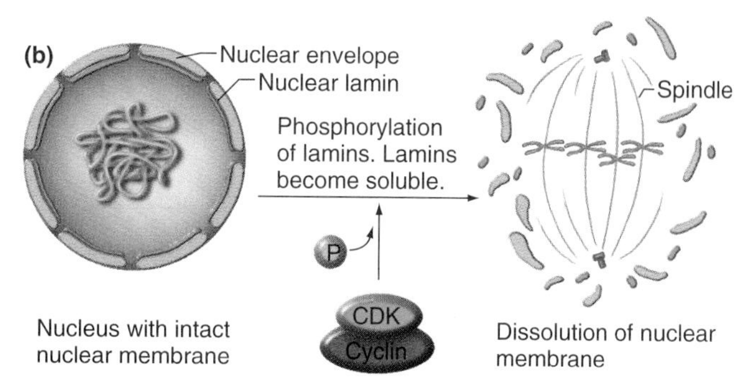

(b)

The cyclin-dependent kinases (CDK) control the cell cycle by phosphorylating other proteins
Figure 18.7

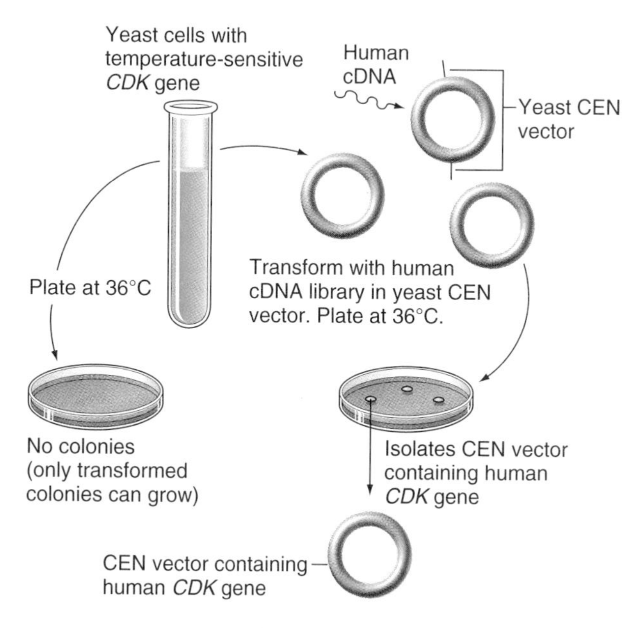

Mutant yeast permit the cloning of a human CDK gene
Figure 18.8

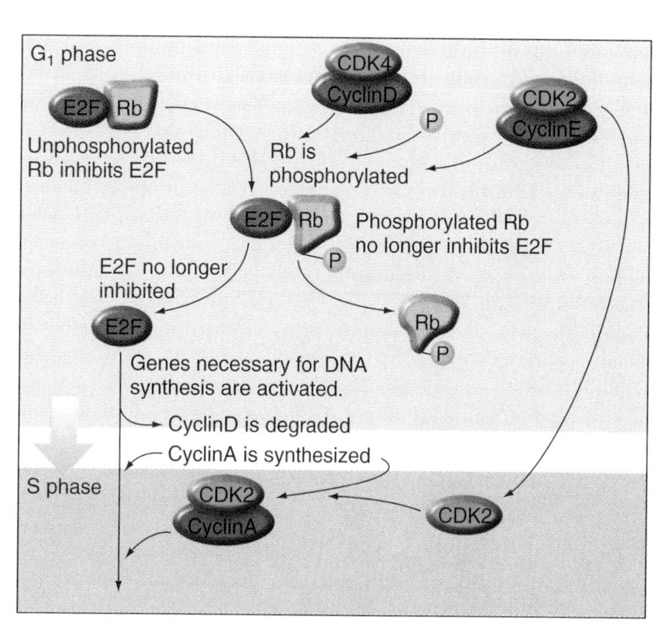

CDKs mediate the transition from the G_1 to the S phase of the cell cycle
Figure 18.9

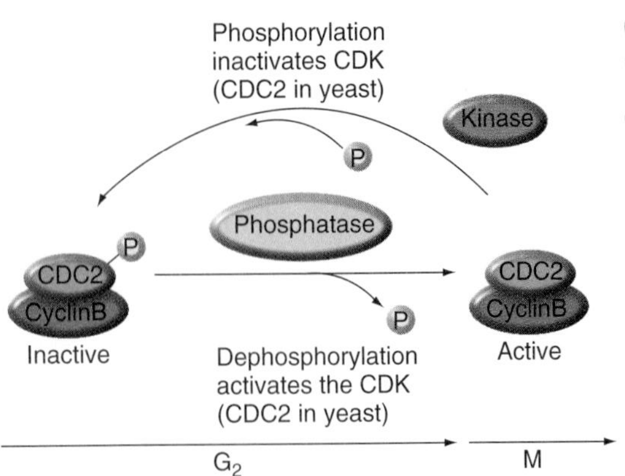

CDK activity in yeast is controlled by phosphorylation and dephosphorylation
Figure 18.10

Phosphorylation inactivates CDK (CDC2 in yeast)

Kinase

P

Phosphatase

CDC2 CyclinB
Inactive

CDC2 CyclinB
Active

Dephosphorylation activates the CDK (CDC2 in yeast)

G_2 ———— M

Cellular responses to DNA damage
Figure 18.11

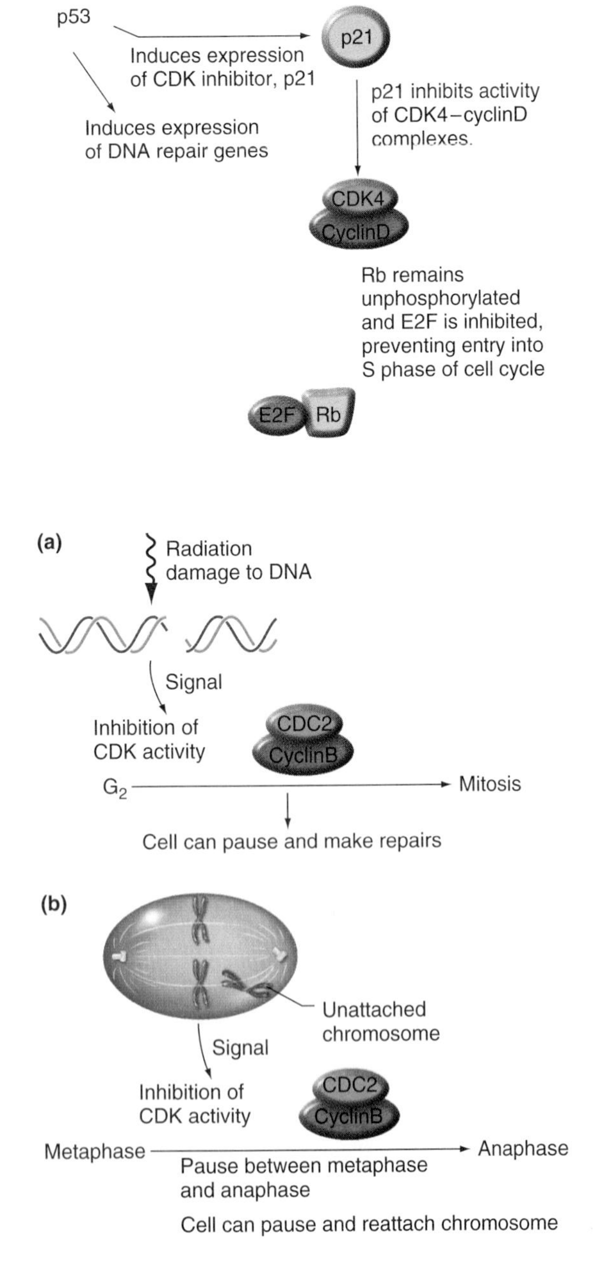

(a) Transcription factor, p53 activated by UV or ionizing radiation

p53

Induces expression of CDK inhibitor, p21

p21

Induces expression of DNA repair genes

p21 inhibits activity of CDK4–cyclinD complexes.

CDK4 CyclinD

Rb remains unphosphorylated and E2F is inhibited, preventing entry into S phase of cell cycle

E2F Rb

(a)

Radiation damage to DNA

Signal

Inhibition of CDK activity

CDC2 CyclinB

G_2 ————→ Mitosis

Cell can pause and make repairs

(b)

Unattached chromosome

Signal

Inhibition of CDK activity

CDC2 CyclinB

Metaphase ————→ Anaphase
Pause between metaphase and anaphase
Cell can pause and reattach chromosome

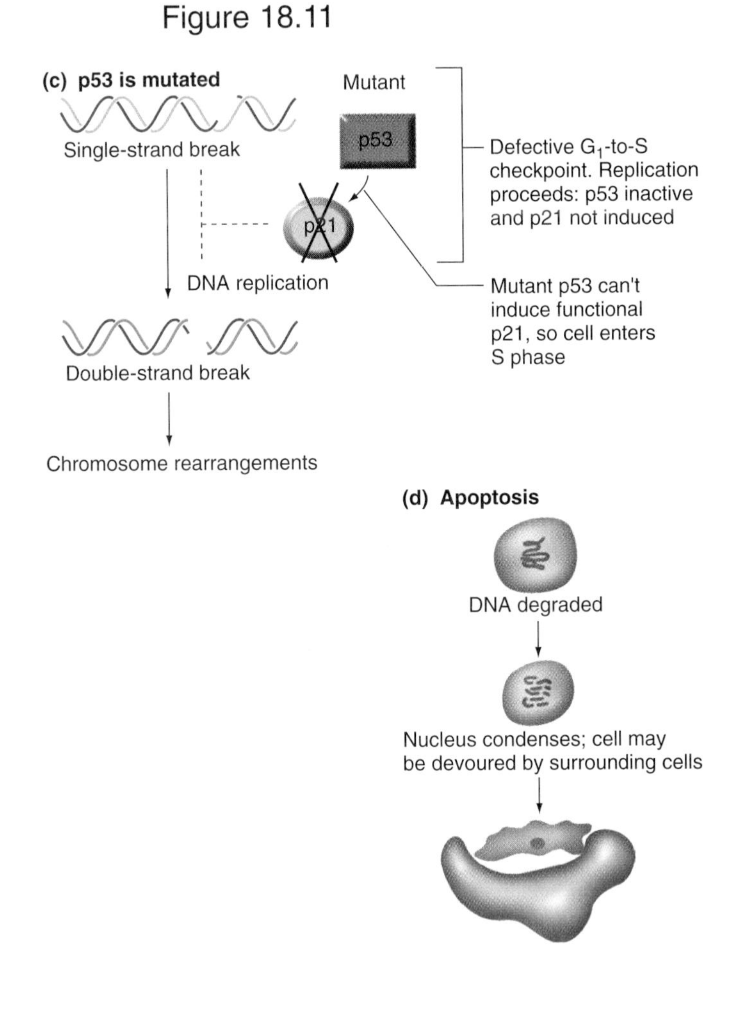

(c) p53 is mutated

Mutant

Single-strand break

p53

p21

DNA replication

Defective G_1-to-S checkpoint. Replication proceeds: p53 inactive and p21 not induced

Mutant p53 can't induce functional p21, so cell enters S phase

Double-strand break

Chromosome rearrangements

(d) Apoptosis

DNA degraded

Nucleus condenses; cell may be devoured by surrounding cells

Two checkpoints act at the G_2-to-M cell-cycle transition
Figure 18.12

Three classes of error lead to aneuploidy in tumor cells
Figure 18.13

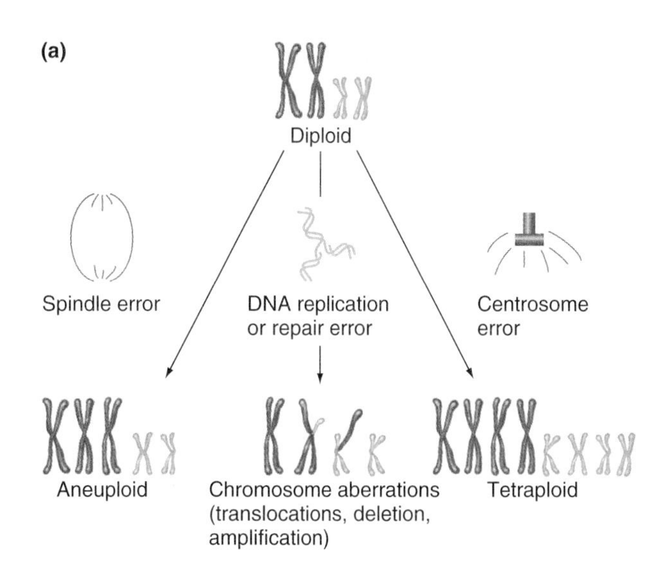

(a)

Diploid

Spindle error

DNA replication or repair error

Centrosome error

Aneuploid

Chromosome aberrations (translocations, deletion, amplification)

Tetraploid

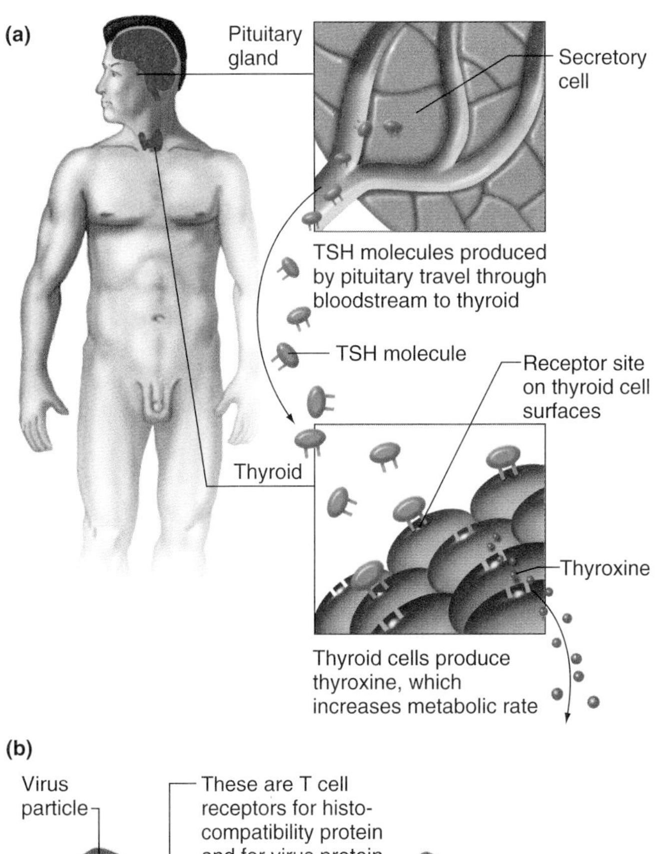

(a)

Pituitary gland

Secretory cell

TSH molecules produced by pituitary travel through bloodstream to thyroid

TSH molecule

Receptor site on thyroid cell surfaces

Thyroid

Thyroxine

Thyroid cells produce thyroxine, which increases metabolic rate

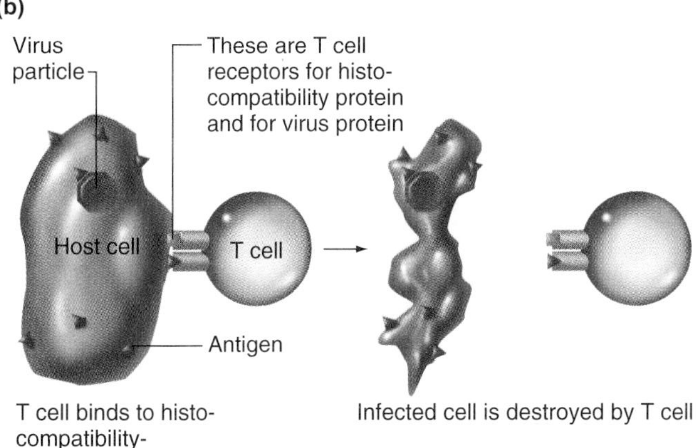

(b)

Virus particle

These are T cell receptors for histocompatibility protein and for virus protein

Host cell

T cell

Antigen

T cell binds to histocompatibility-antigen complex

Infected cell is destroyed by T cell

Extracellular signals can diffuse from one cell to another or be delivered by cell-to-cell contact
Figure 18.14

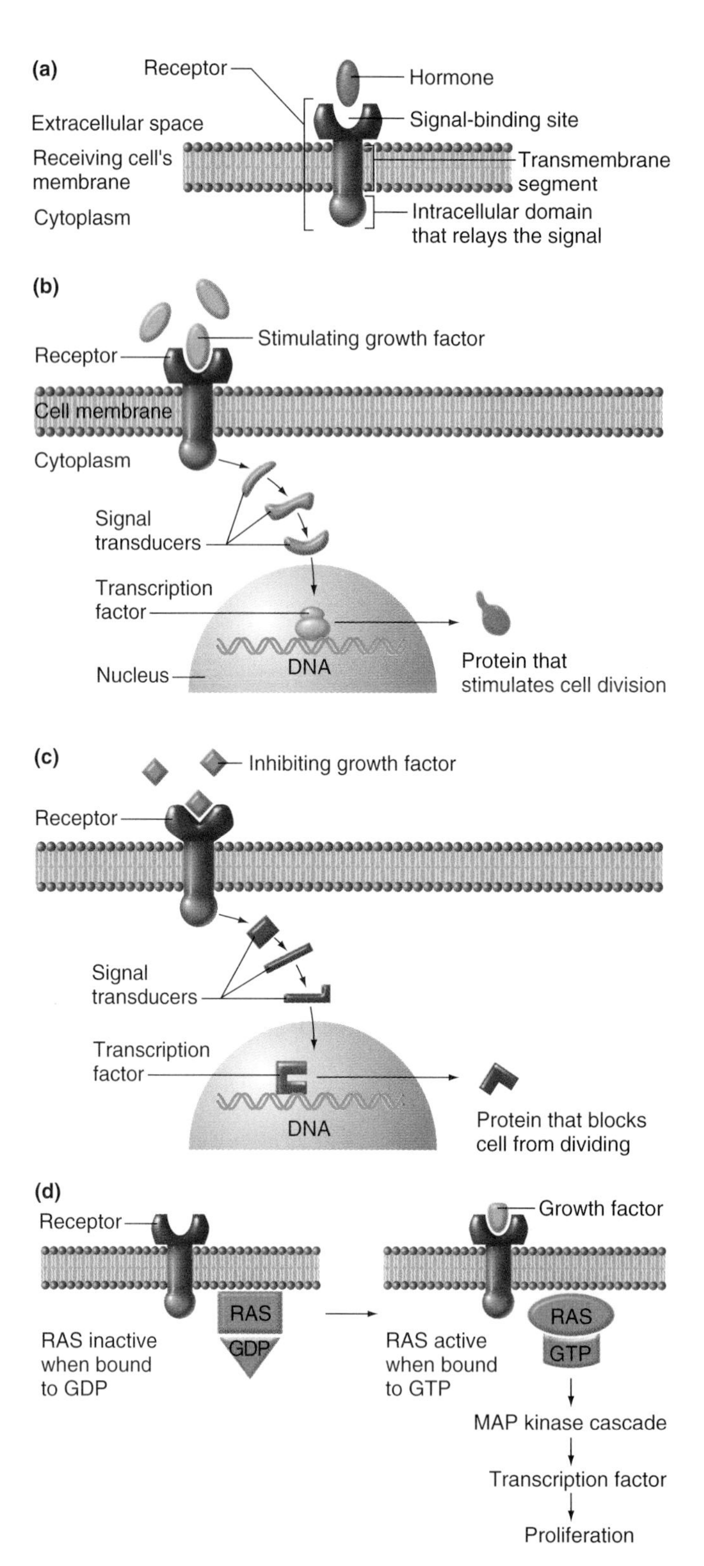

Many hormones transmit signals into cells through receptors that span the cellular membrane

Figure 18.15

The percent of mice still alive as a function of age
Figure 18.17

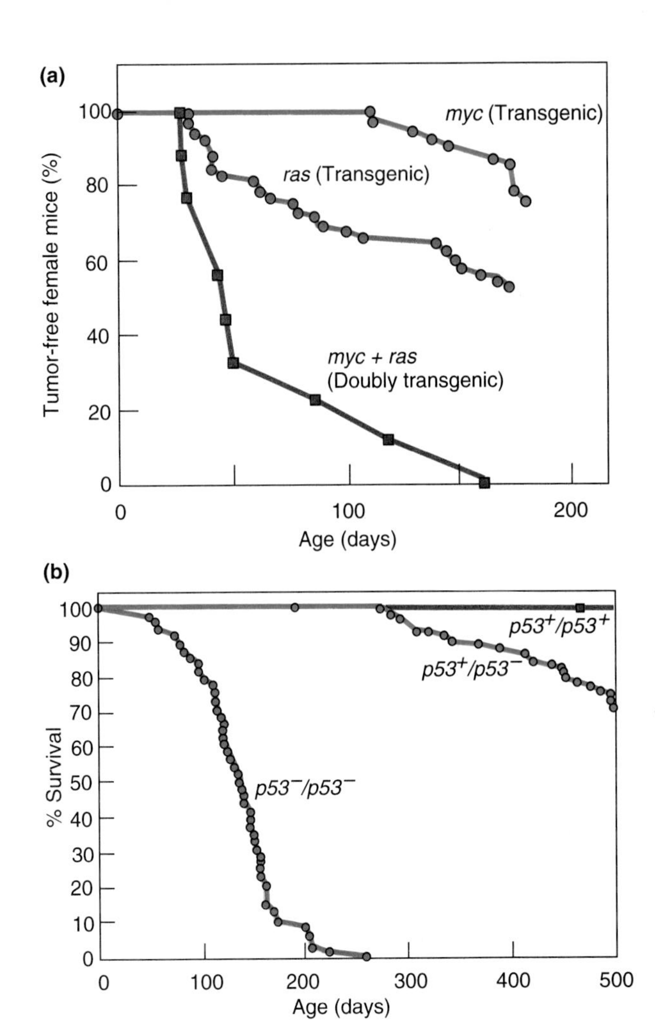

(a)

myc (Transgenic)

ras (Transgenic)

myc + ras
(Doubly transgenic)

(b)

p53+/p53+

p53+/p53−

p53−/p53−

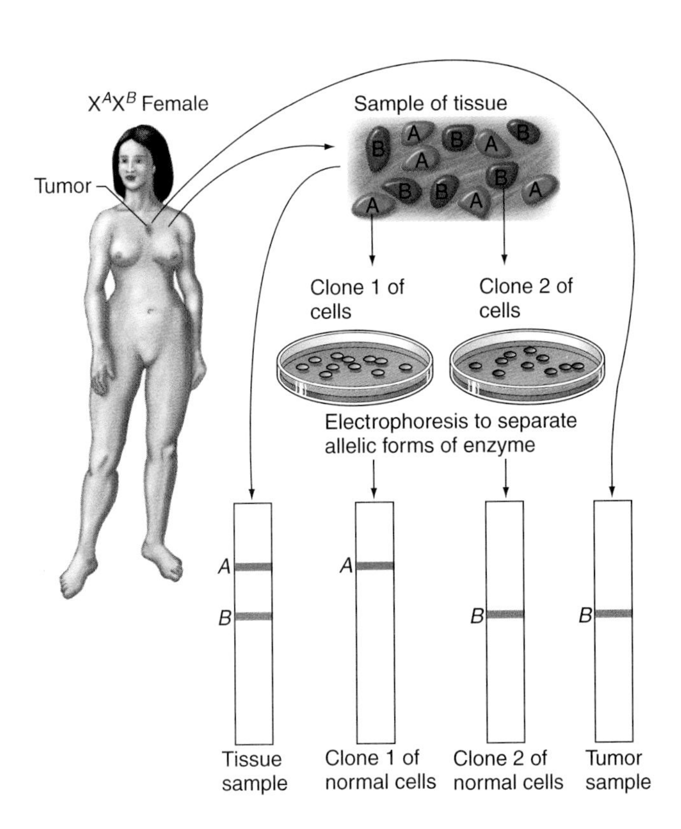

X^A X^B Female

Tumor

Sample of tissue

B A B A B
A A
A B B A A
A B B A A

Clone 1 of
cells

Clone 2 of
cells

Electrophoresis to separate
allelic forms of enzyme

A   A

B   B   B

Tissue
sample

Clone 1 of
normal cells

Clone 2 of
normal cells

Tumor
sample

Polymorphic enzymes encoded by the X chromosome reveal the clonal origin of tumors
Figure 18.18

(a) Lung cancer death rates, United States, 1930-91

MALE

FEMALE

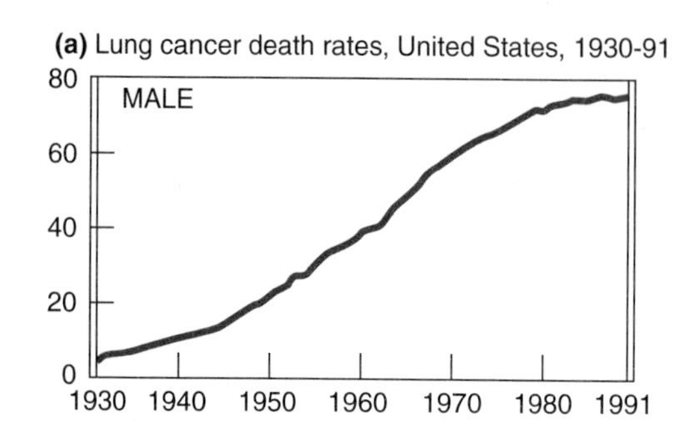

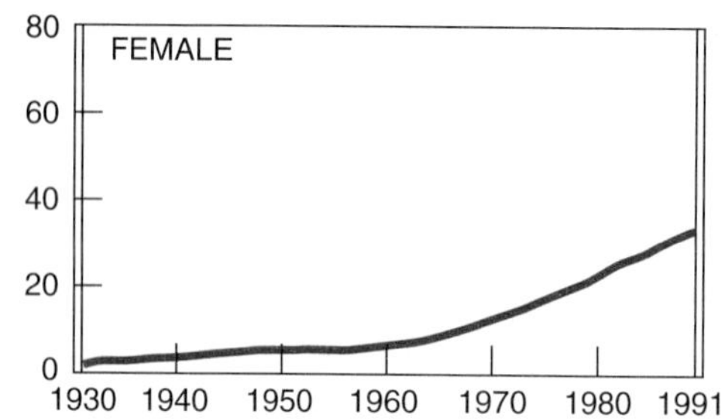

Rates are per 100,000 and are age-adjusted to the 1970 U.S. census population.

(b)

log-log plot

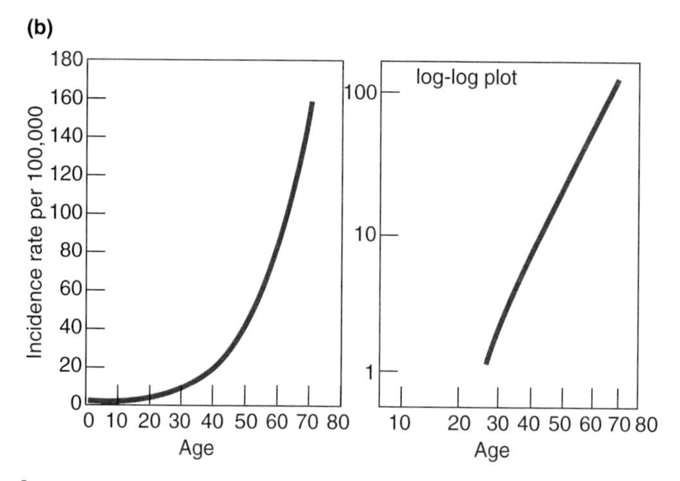

Lung cancer death rates and incidence of cancer with age
Figure 18.19

Egg   Sperm
RB^- RB^+

Cells of the retina

RB^+/RB^- cells

RB^-/RB^- cell

RB^+/RB^-
cancer-prone
individual

RB^-/RB^- cells

Tumor of the retina

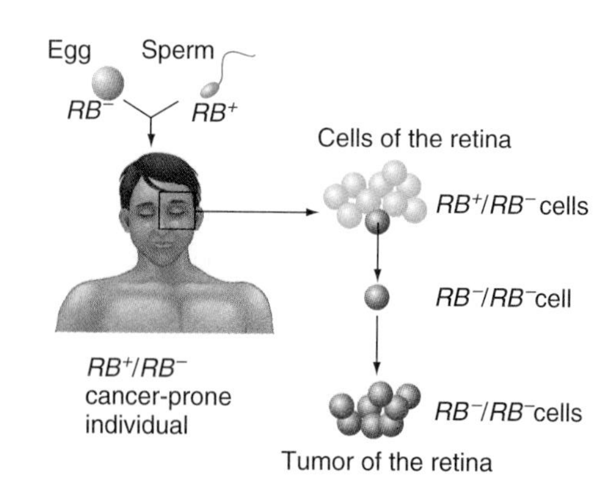

Individuals who inherit one copy of the RB^- allele are prone to cancer of the retina
Figure 18.20

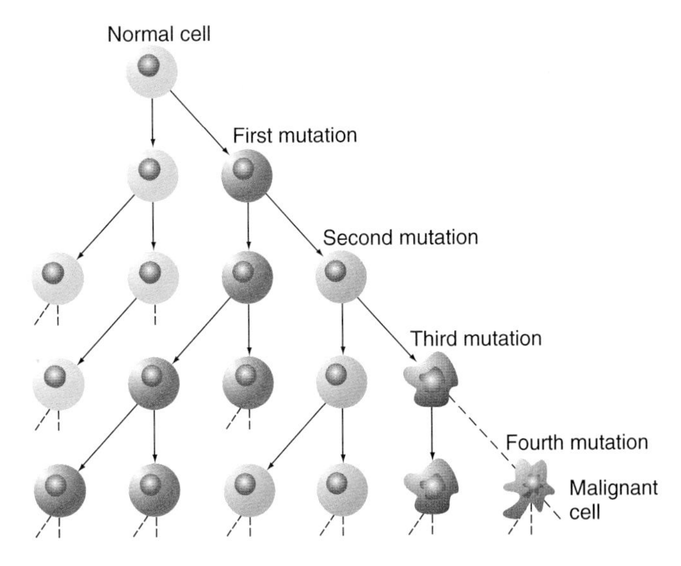

Cancer is thought to arise by successive mutations in a clone of proliferating cells
Figure 18.21

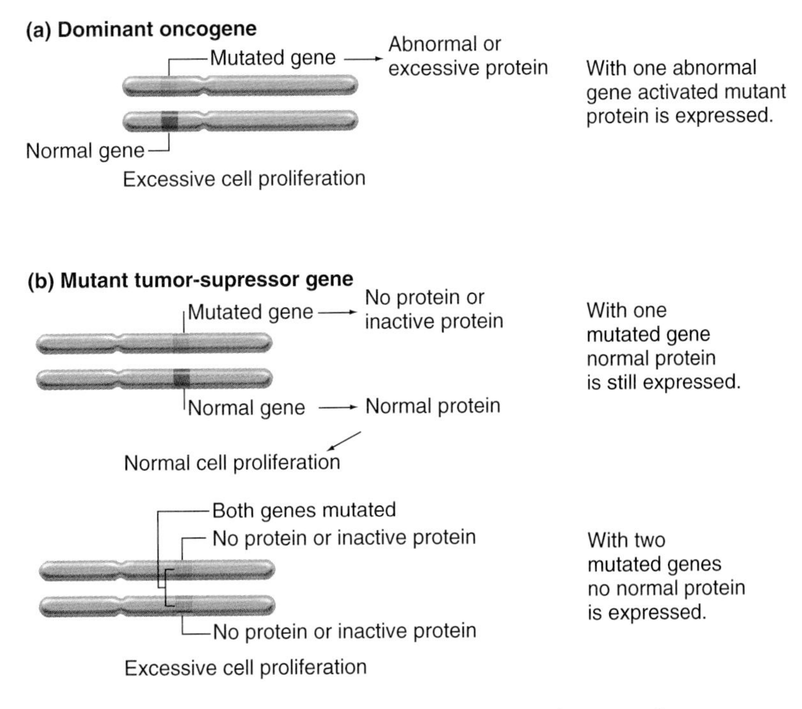

(a) Dominant oncogene

Mutated gene → Abnormal or excessive protein

Normal gene

Excessive cell proliferation

With one abnormal gene activated mutant protein is expressed.

(b) Mutant tumor-supressor gene

Mutated gene → No protein or inactive protein

Normal gene → Normal protein

Normal cell proliferation

With one mutated gene normal protein is still expressed.

Both genes mutated

No protein or inactive protein

No protein or inactive protein

Excessive cell proliferation

With two mutated genes no normal protein is expressed.

Cancer-producing mutations occur in two forms
Figure 18.22

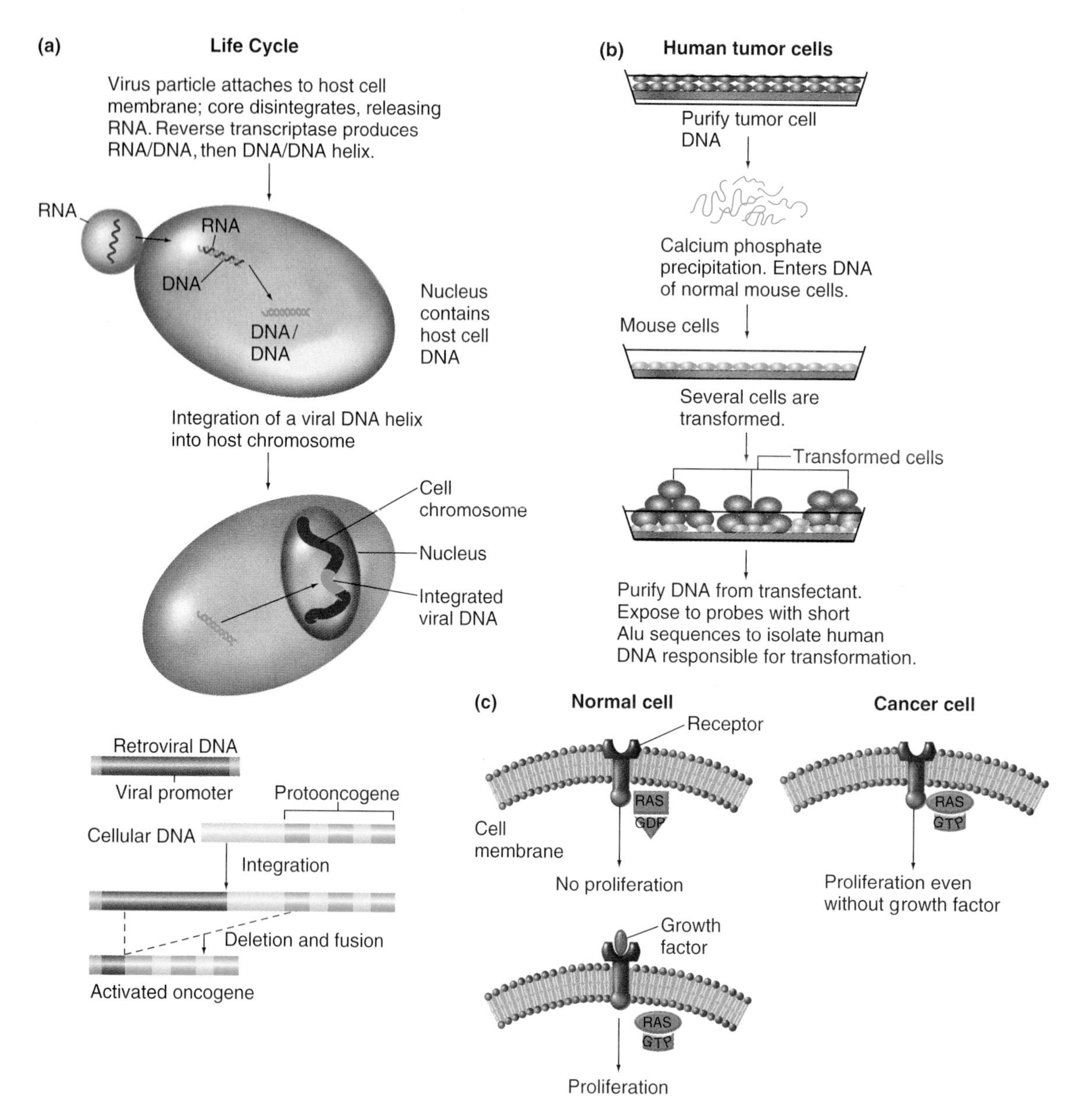

(a) **Life Cycle**

Virus particle attaches to host cell membrane; core disintegrates, releasing RNA. Reverse transcriptase produces RNA/DNA, then DNA/DNA helix.

RNA

RNA

DNA

DNA/ DNA

Nucleus contains host cell DNA

Integration of a viral DNA helix into host chromosome

Cell chromosome

Nucleus

Integrated viral DNA

Retroviral DNA

Viral promoter   Protooncogene

Cellular DNA

Integration

Deletion and fusion

Activated oncogene

(b) **Human tumor cells**

Purify tumor cell DNA

Calcium phosphate precipitation. Enters DNA of normal mouse cells.

Mouse cells

Several cells are transformed.

Transformed cells

Purify DNA from transfectant. Expose to probes with short Alu sequences to isolate human DNA responsible for transformation.

(c) **Normal cell**   **Cancer cell**

Receptor

RAS GDP

RAS GTP

Cell membrane

No proliferation

Proliferation even without growth factor

Growth factor

RAS GTP

Proliferation

Two methods to isolate oncogenes
Figure 18.23

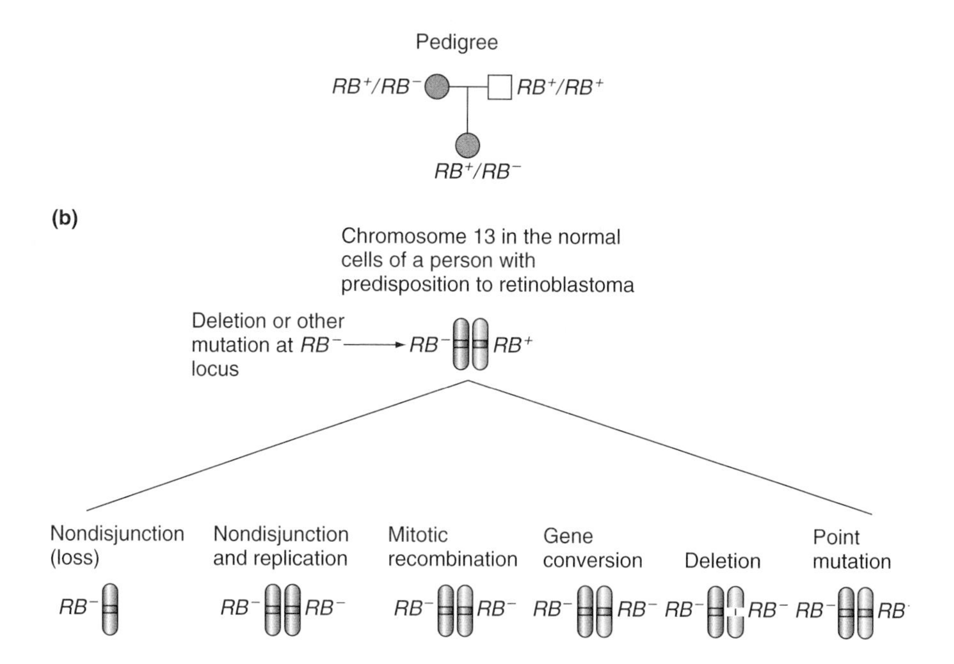

Pedigree

RB^+/RB^- ○—□ RB^+/RB^+

RB^+/RB^-

(b)

Chromosome 13 in the normal
cells of a person with
predisposition to retinoblastoma

Deletion or other
mutation at RB^- ⟶ RB^- ▯▯ RB^+
locus

| Nondisjunction (loss) | Nondisjunction and replication | Mitotic recombination | Gene conversion | Deletion | Point mutation |

RB^- ▯ RB^- ▯▯ RB^- RB^- ▯▯ RB^- RB^- ▯▯ RB^- RB^- ▯▯ RB^- RB^- ▯▯ RB^-

The retinoblastoma tumor-suppressor gene
Figure 18.24

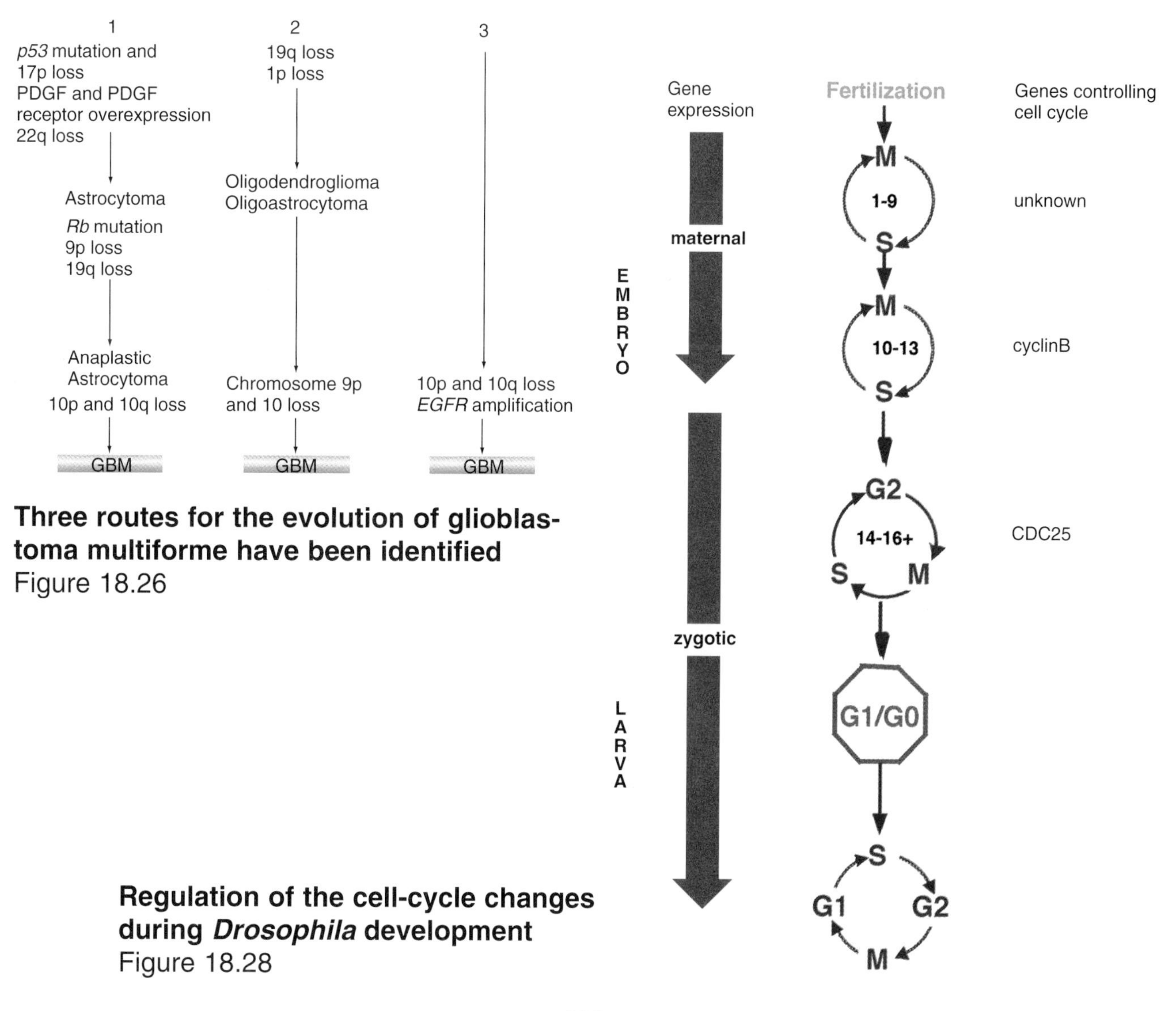

1
p53 mutation and
17p loss
PDGF and PDGF
receptor overexpression
22q loss

↓

Astrocytoma

Rb mutation
9p loss
19q loss

↓

Anaplastic
Astrocytoma
10p and 10q loss

↓

GBM

2
19q loss
1p loss

↓

Oligodendroglioma
Oligoastrocytoma

↓

Chromosome 9p
and 10 loss

↓

GBM

3

↓

10p and 10q loss
EGFR amplification

↓

GBM

Three routes for the evolution of glioblastoma multiforme have been identified
Figure 18.26

Gene
expression

Fertilization

Genes controlling
cell cycle

M
1-9
S   unknown

M
10-13
S   cyclinB

G2
14-16+
S   M   CDC25

G1/G0

S
G1   G2
M

maternal

EMBRYO

zygotic

LARVA

Regulation of the cell-cycle changes during *Drosophila* development
Figure 18.28

Alignment of histone H4 from humans and four model systems.

```
H. sapiens       MSGRGKGGKGLGKGGAKRHRKVLRDNIQGITKPAIRRLARRGGVKRISGLIYEETRGVLKVFLENVIRDAVTYTEHAKRKT
M. musculus      MSGRGKGGKGLGKGGAKRHRKVLRDNIQGITKPAIRRLARRGGVKRISGLIYEETRGVLKVFLENVIRDAVTYTEHAKRKT
C. elegans       MSGRGKGGKGLGKGGAKRHRKVLRDNIQGITKPAIRRLARRGGVKRISGLIYEETRGVLKVFLENVIRDAVTYCEHAKRKT
D. melanogaster  MTGRGKGGKGLGKGGAKRHRKVLRDNIQGITKPAIRRLARRGGVKRISGLIYEETRGVLKVFLENVIRDAVTYTEHAKRKT
A. thaliana      MSGRGKGGKGLGKGGAKRHRKVLRDNIQGITKPAIRRLARRGGVKRISGLIYEETRGVLK IFLENVIRDAVTYTEHARRKT
                 * • * * * * * * * * * * * * * * * * * * * * * * * * * * * * * * * * * * * * * * * * * * * * * * * * * * * * * * * * * * * * • * * * * * * * * * * • * * * • * * *
```

* Indicates identical amino acid
• Indicates similar amino acid

The amino-acid sequence of histone H4 is highly conserved in evolution
Figure 19.4

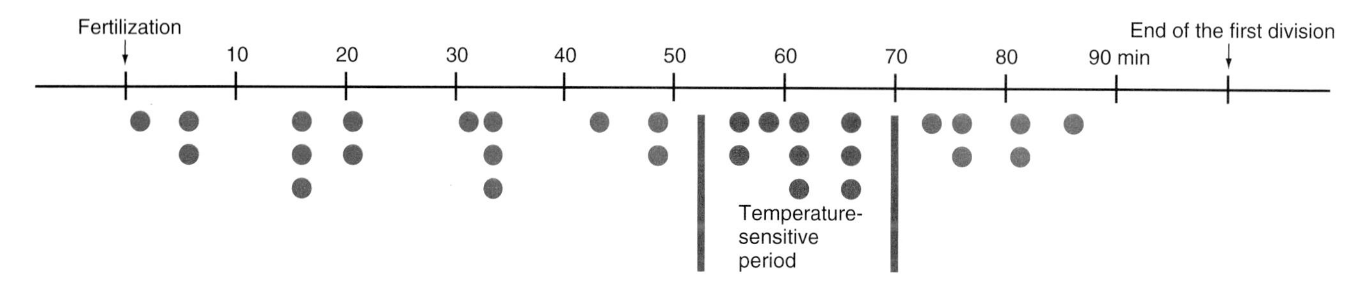

Time-of-function analysis
Figure 19.6

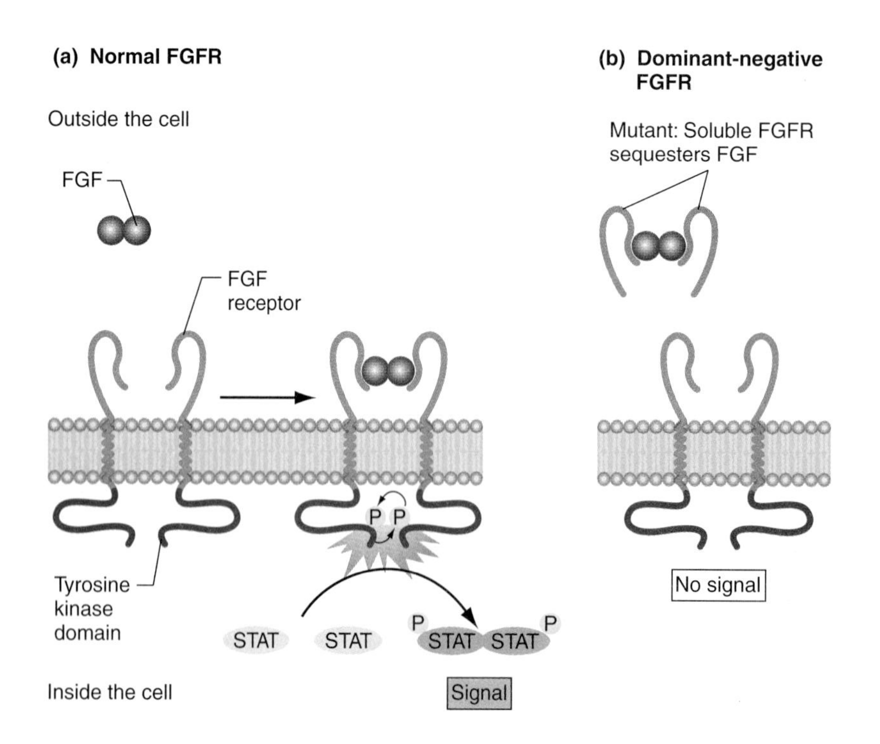

(a) Normal FGFR

(b) Dominant-negative FGFR

Engineering a dominant-negative mutation in a mouse fibroblast growth factor receptor (FGFR) gene
Figure 19.7

(a) Synthesis of dsRNA

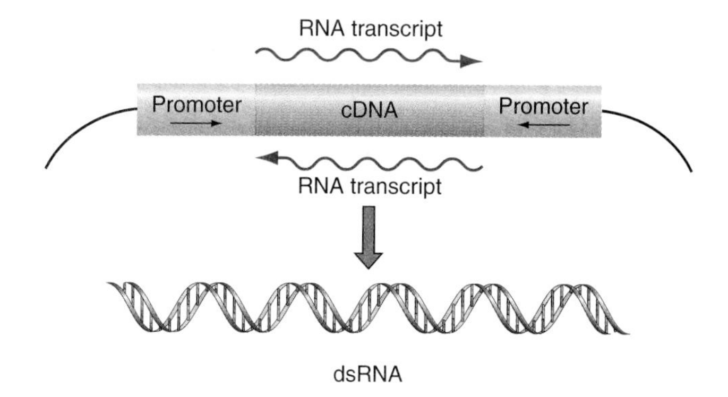

RNA interference (RNAi): A new tool for studying development
Figure 19.8

(a) Redundant gene function

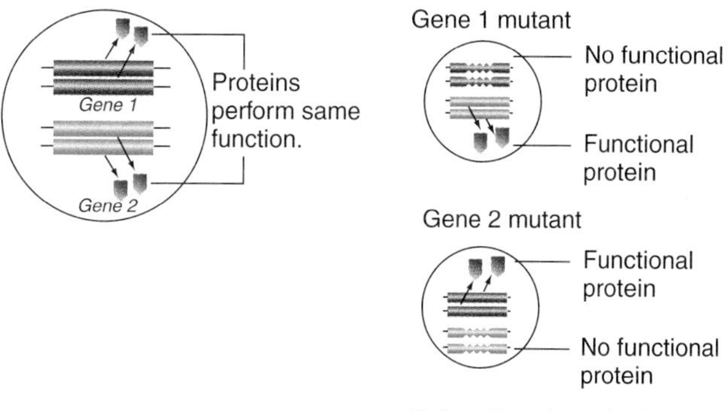

(b) Multiple roles in development

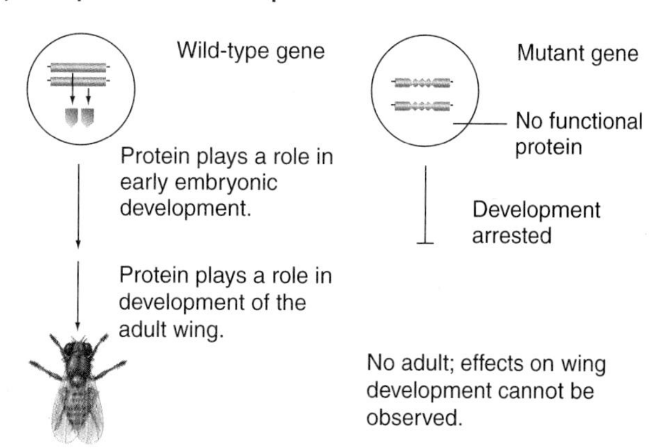

Why saturation mutagenesis may miss some genes
Figure 19.11

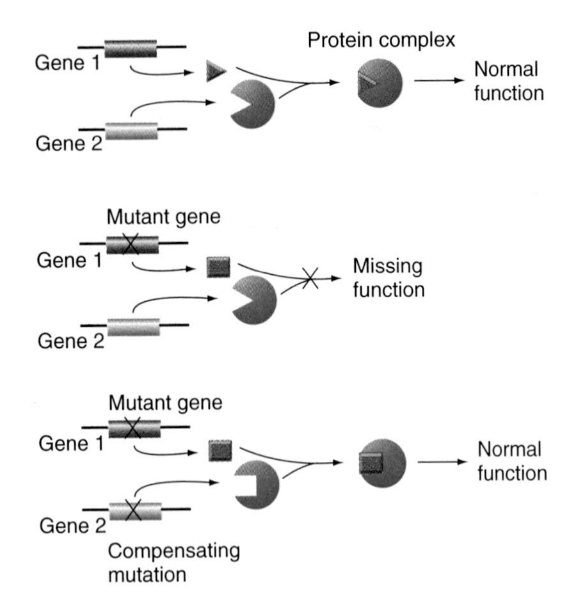

Suppressor mutations
Figure 19.12

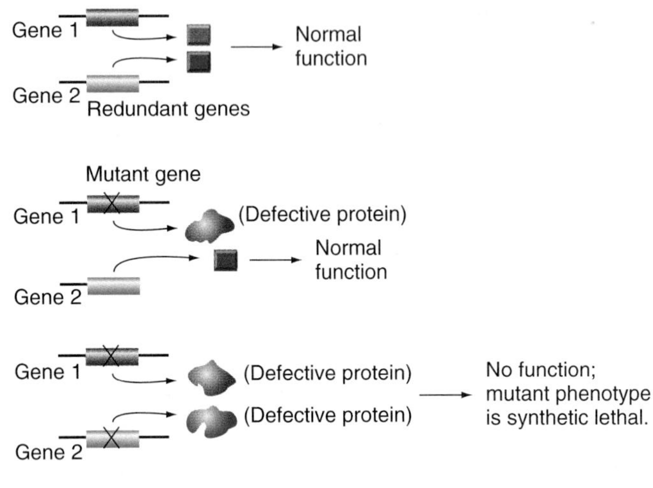

Synthetic lethality: Mutations in two genes together cause lethality, but a mutation in either gene alone does not
Figure 19.14

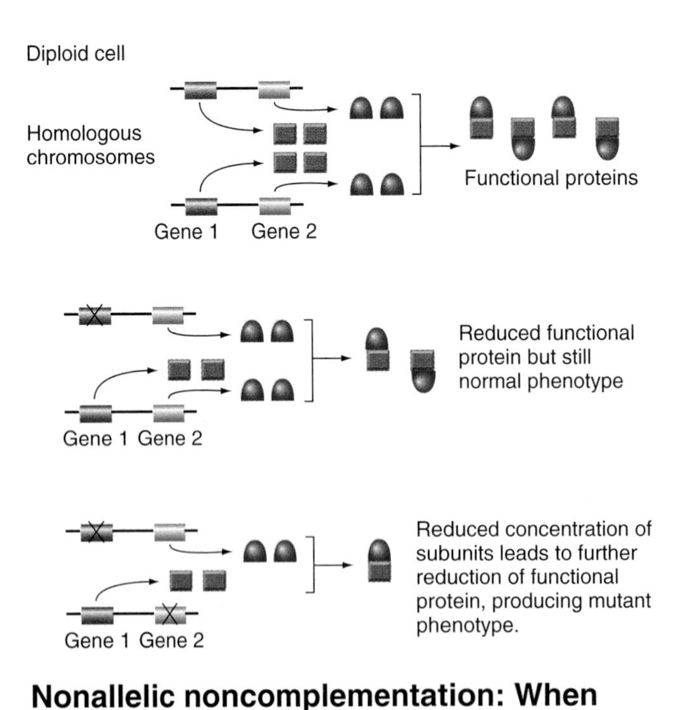

Nonallelic noncomplementation: When mutations in two different genes fail to complement each other
Figure 19.15

(a) Fusion protein gene in *E. coli*

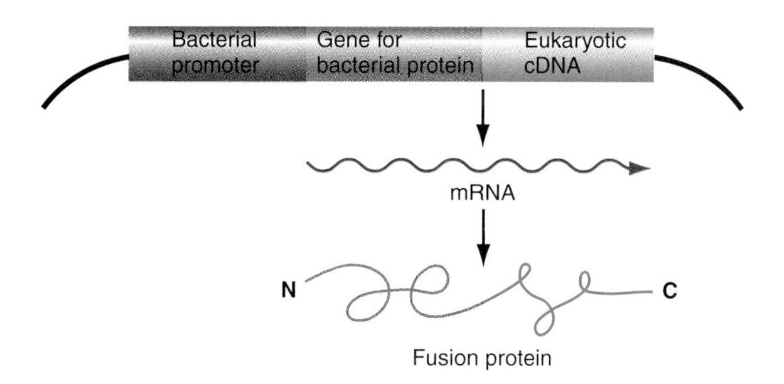

(c) Tagging a protein with GFP

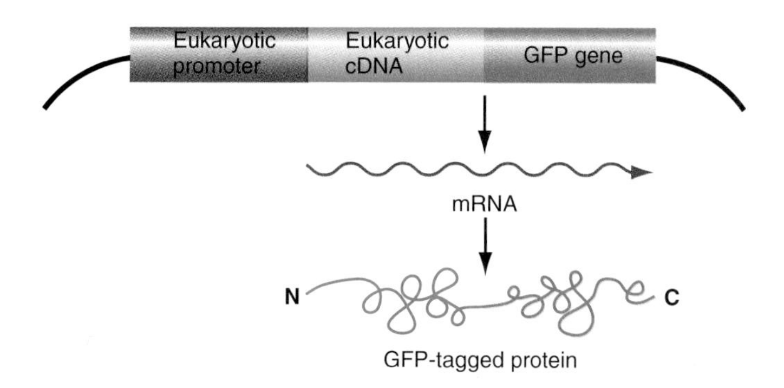

Using antibodies and GFP tagging to follow the localization of proteins
Figure 19.18

(b) Using mosaics to study cell signaling

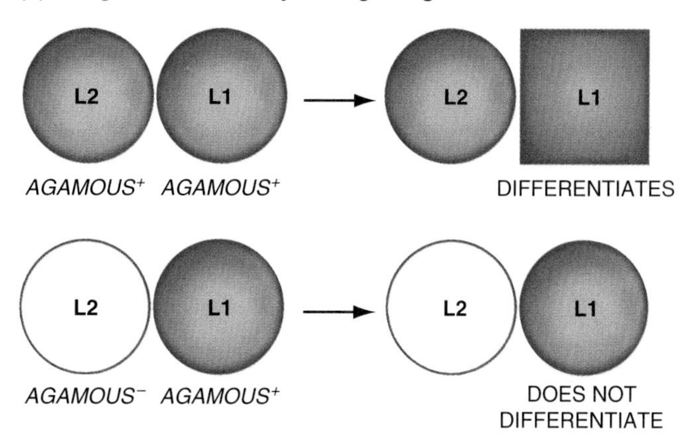

Mosaic analysis
Figure 19.19

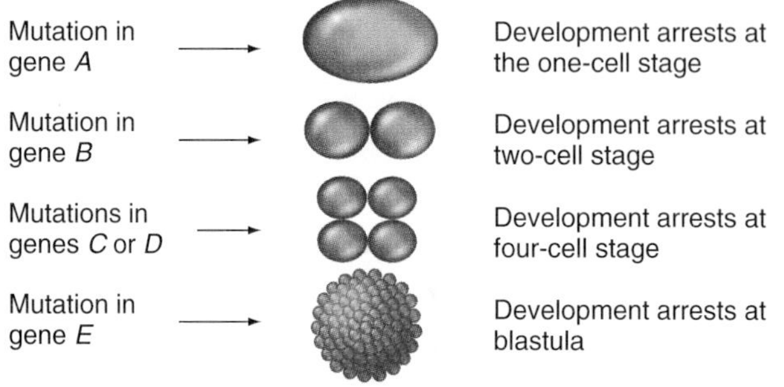

Mutation in gene A → Development arrests at the one-cell stage

Mutation in gene B → Development arrests at two-cell stage

Mutations in genes C or D → Development arrests at four-cell stage

Mutation in gene E → Development arrests at blastula

Suggests that gene A acts first, then gene B, then genes C and D, then gene E.

When do mutations arrest development?
Figure 19.20

(a) Epistasis in the secretion pathway

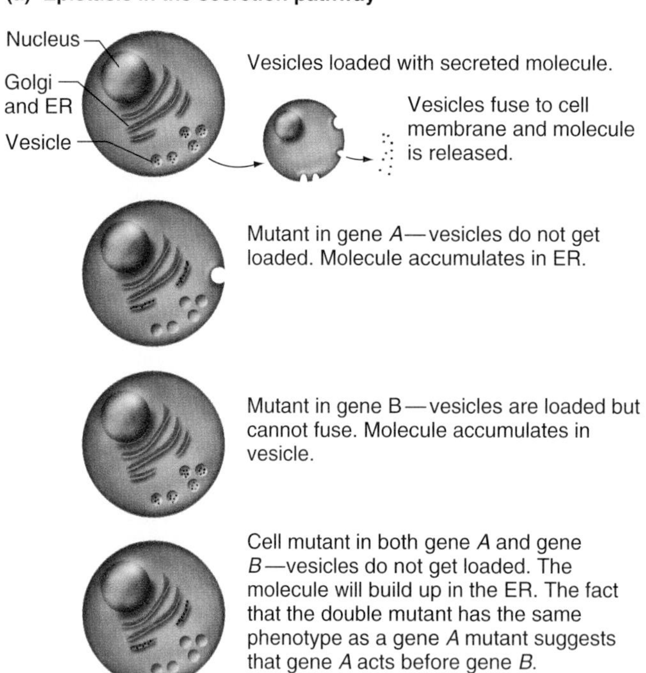

Nucleus
Golgi and ER
Vesicle

Vesicles loaded with secreted molecule.

Vesicles fuse to cell membrane and molecule is released.

Mutant in gene A—vesicles do not get loaded. Molecule accumulates in ER.

Mutant in gene B—vesicles are loaded but cannot fuse. Molecule accumulates in vesicle.

Cell mutant in both gene A and gene B—vesicles do not get loaded. The molecule will build up in the ER. The fact that the double mutant has the same phenotype as a gene A mutant suggests that gene A acts before gene B.

(b) Epistasis in the pathway for vulva formation

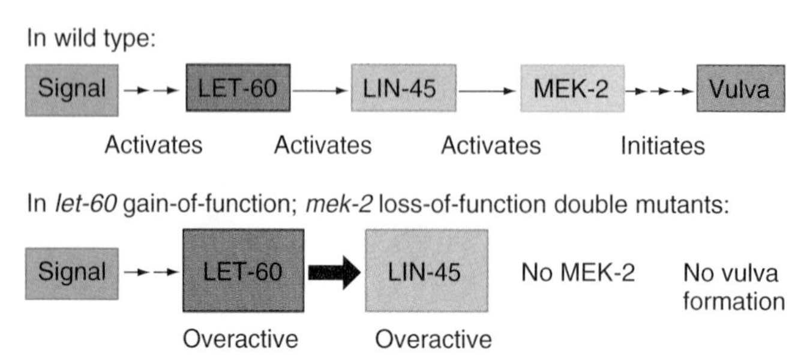

In wild type:

Signal ⇢ LET-60 → LIN-45 → MEK-2 ⇢ Vulva

Activates   Activates   Activates   Initiates

In let-60 gain-of-function; mek-2 loss-of-function double mutants:

Signal ⇢ LET-60 ⟹ LIN-45   No MEK-2   No vulva formation

Overactive   Overactive

Double mutant analysis
Figure 19.21

Bicoid mRNA localizes to the anterior cortex of the oocyte
Figure 19.26

(a) Diagram of a *Drosophila* egg chamber

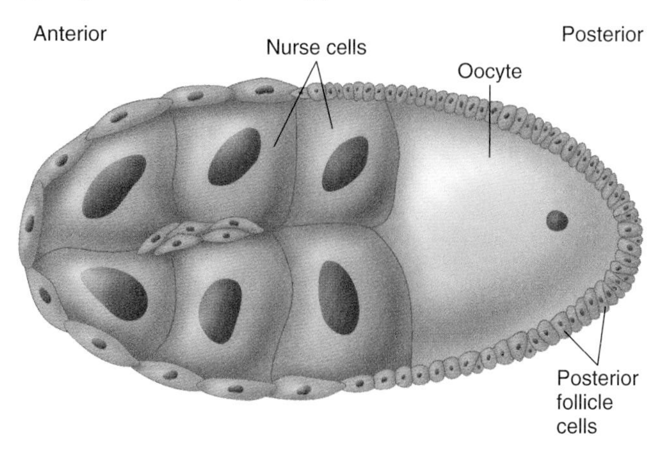

Anterior

Nurse cells

Oocyte

Posterior

Posterior follicle cells

(a) Asymmetric neuroblast stem cell divisions

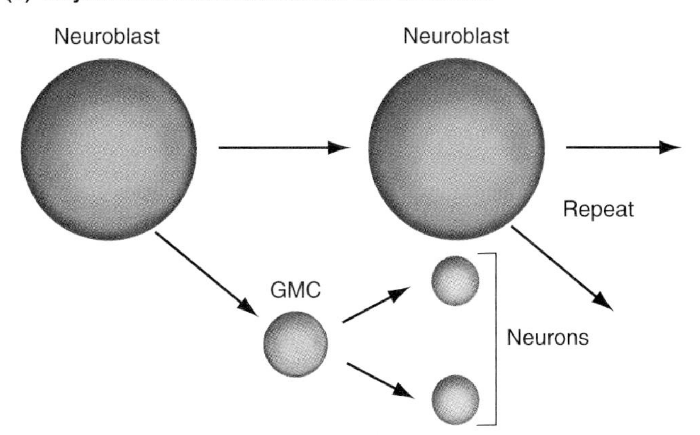

Neuroblast

Neuroblast

Repeat

GMC

Neurons

Asymmetric cell division
Figure 19.27

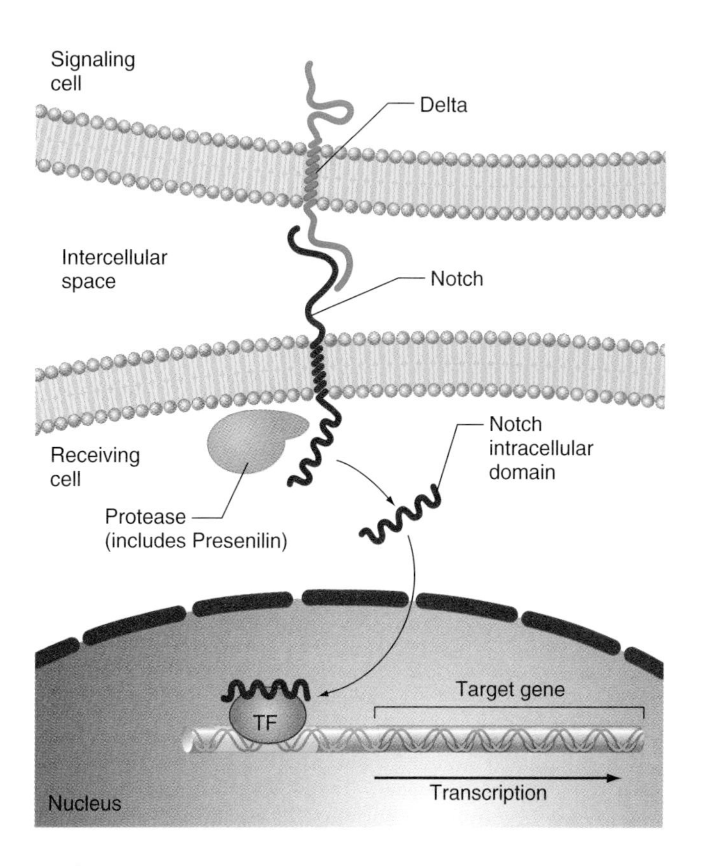

Signaling cell

Delta

Intercellular space

Notch

Receiving cell

Notch intracellular domain

Protease (includes Presenilin)

Target gene

TF

Nucleus

Transcription

Juxtacrine signaling
Figure 19.28

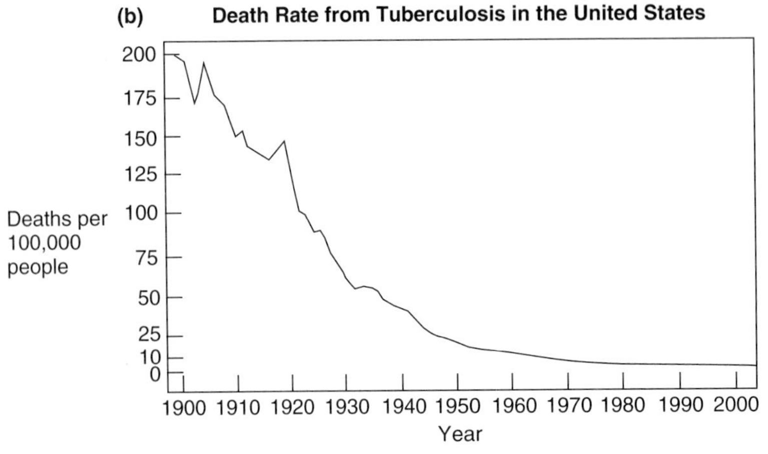

(b) **Death Rate from Tuberculosis in the United States**

Deaths per 100,000 people

Year

Prior to 1933 data are for only areas with death-registration; after 1933 data are for entire U.S.

Tuberculosis in human populations
Figure 20.1

| Genotypes in first generation | RR | Rr | rr |

Genotypes in first generation RR Rr rr

Number of individuals in first generation N_{RR} N_{Rr} N_{rr}

Allele types in first generation R r

Allele frequencies in first generation

$$p = \text{frequency of } R = \frac{2N_{RR} + N_{Rr}}{2N_{total}}$$

$$q = \text{frequency of } r = \frac{N_{Rr} + 2N_{rr}}{2N_{total}}$$

$N_{total} = N_{RR} + N_{Rr} + N_{rr}$

$2N_{total}$ = total chromosomes

Computing allele frequencies from genotype frequencies
Figure 20.2

	Allele R	r
	Frequency p	q
Allele R / Frequency p	RR / p^2	Rr / pq
Allele r / Frequency q	Rr / pq	rr / q^2

Eggs

Derivation of the Hardy-Weinberg equation
Figure 20.3

Gametes	R		r
Allele frequencies	p		q
Zygotes' genotype frequencies	$G_{RR} = p^2$	$G_{Rr} = 2pq$	$G_{rr} = q^2$
Relative fitness: basis of selection	w_{RR}	w_{Rr}	w_{rr}
Adults' genotype frequencies	$\dfrac{p^2 w_{RR}}{\overline{w}}$	$\dfrac{2pq w_{Rr}}{\overline{w}}$	$\dfrac{q^2 w_{rr}}{\overline{w}}$
Gametes' (next generation) allele frequencies	$p' = \dfrac{p^2 w_{RR} + pq w_{Rr}}{\overline{w}}$		$q' = \dfrac{q^2 w_{rr} + pq w_{Rr}}{\overline{w}}$

$\overline{w} = p^2 w_{RR} + 2pq w_{Rr} + q^2 w_{rr}$

Changes in allele and genotype frequencies when selection acts on genotype-dependent differences in fitness
Figure 20.5

217

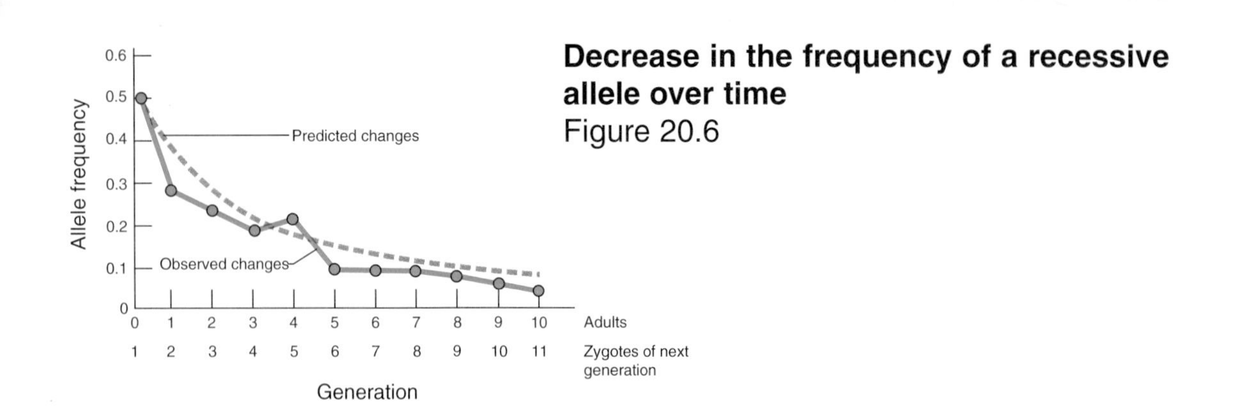

Decrease in the frequency of a recessive allele over time
Figure 20.6

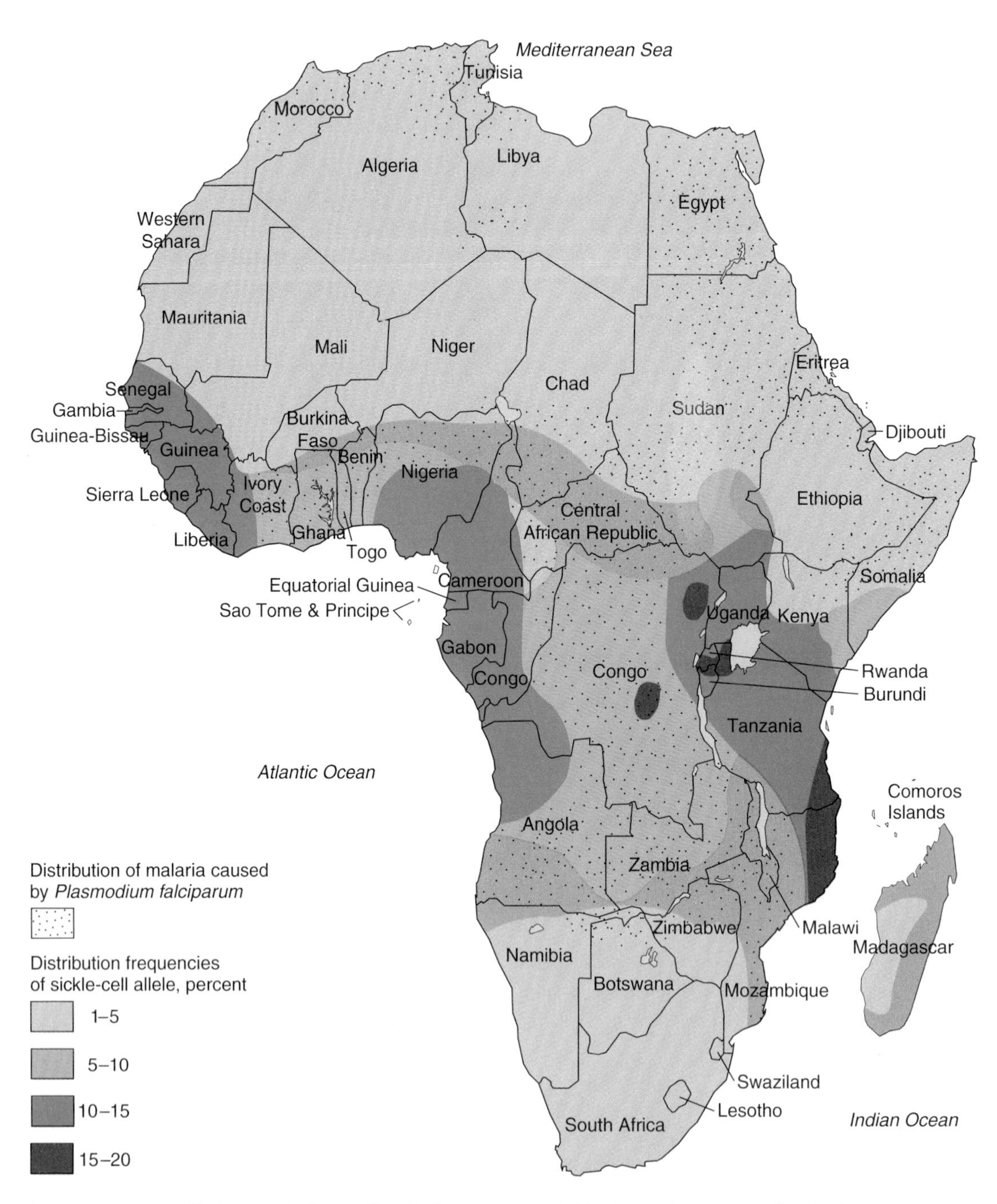

Distribution of malaria caused by *Plasmodium falciparum*

Distribution frequencies of sickle-cell allele, percent

- 1–5
- 5–10
- 10–15
- 15–20

Frequency of the sickle-cell allele across Africa where malaria is prevalent
Figure 20.7

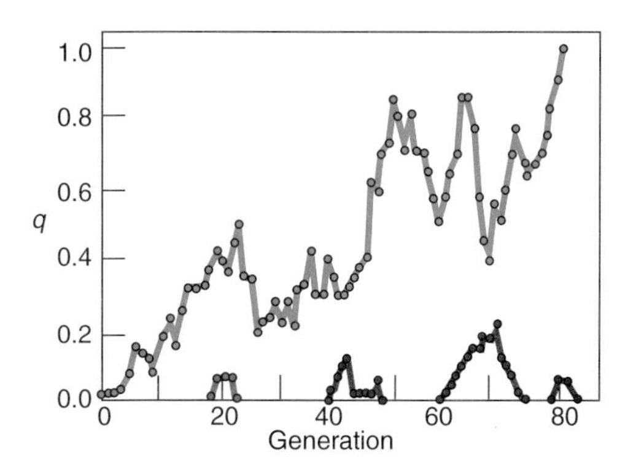

The effect of genetic drift
Figure 20.8

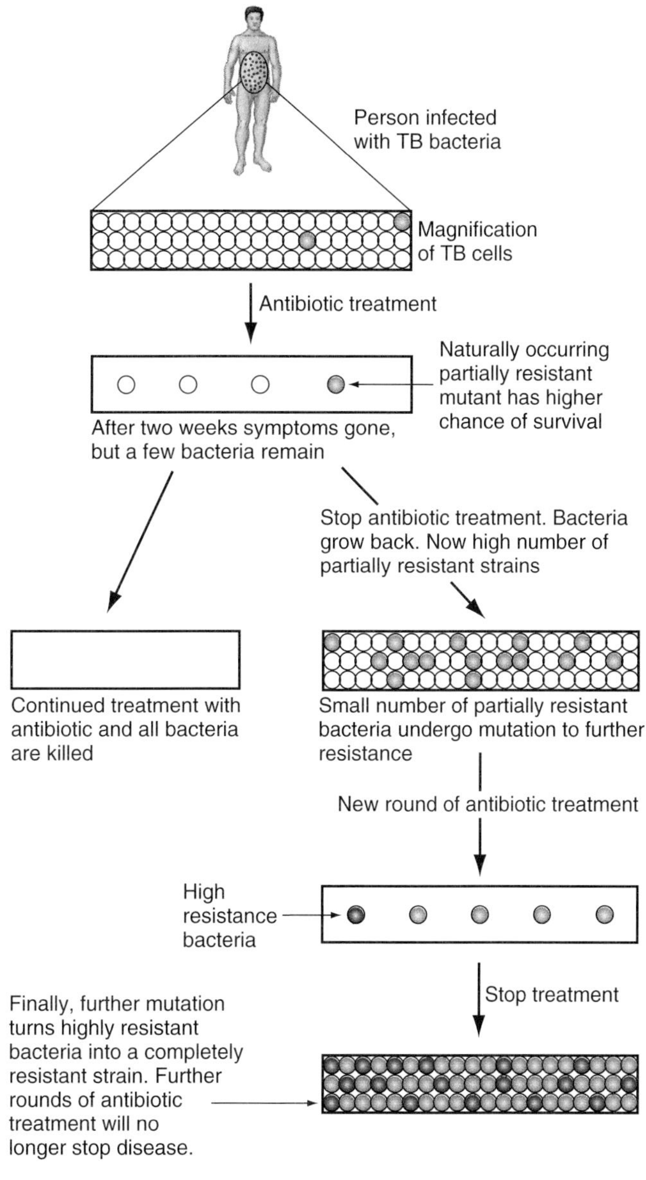

Person infected
with TB bacteria

Magnification
of TB cells

Antibiotic treatment

Naturally occurring
partially resistant
mutant has higher
chance of survival

After two weeks symptoms gone,
but a few bacteria remain

Stop antibiotic treatment. Bacteria
grow back. Now high number of
partially resistant strains

Continued treatment with
antibiotic and all bacteria
are killed

Small number of partially resistant
bacteria undergo mutation to further
resistance

New round of antibiotic treatment

High
resistance
bacteria

Stop treatment

Finally, further mutation
turns highly resistant
bacteria into a completely
resistant strain. Further
rounds of antibiotic
treatment will no
longer stop disease.

The evolution of resistance in TB bacteria
Figure 20.9

(b)

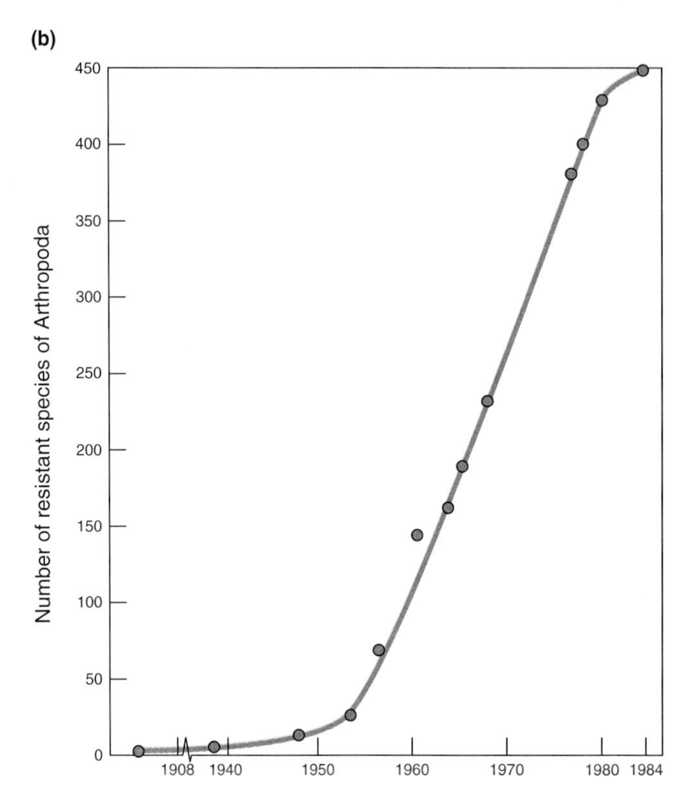

Increase of insecticide resistance from 1908–1984

Figure 20.10

(b)

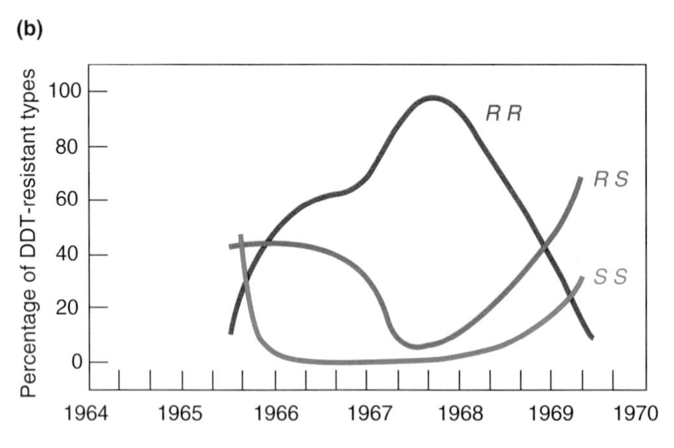

How genotype frequencies among populations of *A. aegypti* mosquito larvae change in response to insecticide

Figure 20.11

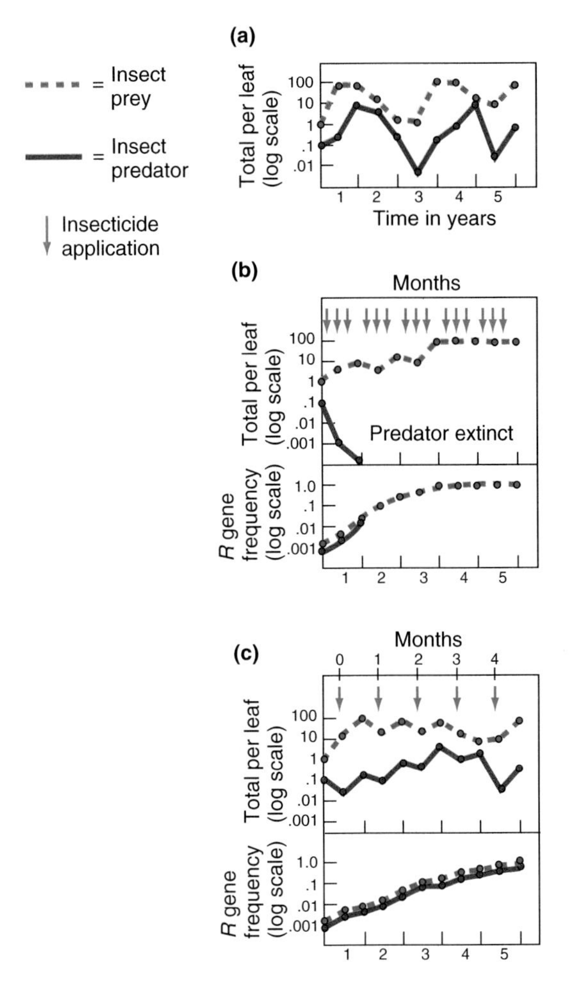

Changes in population density (insects per leaf) of prey and pest over time, with and without insecticide spraying
Figure 20.12

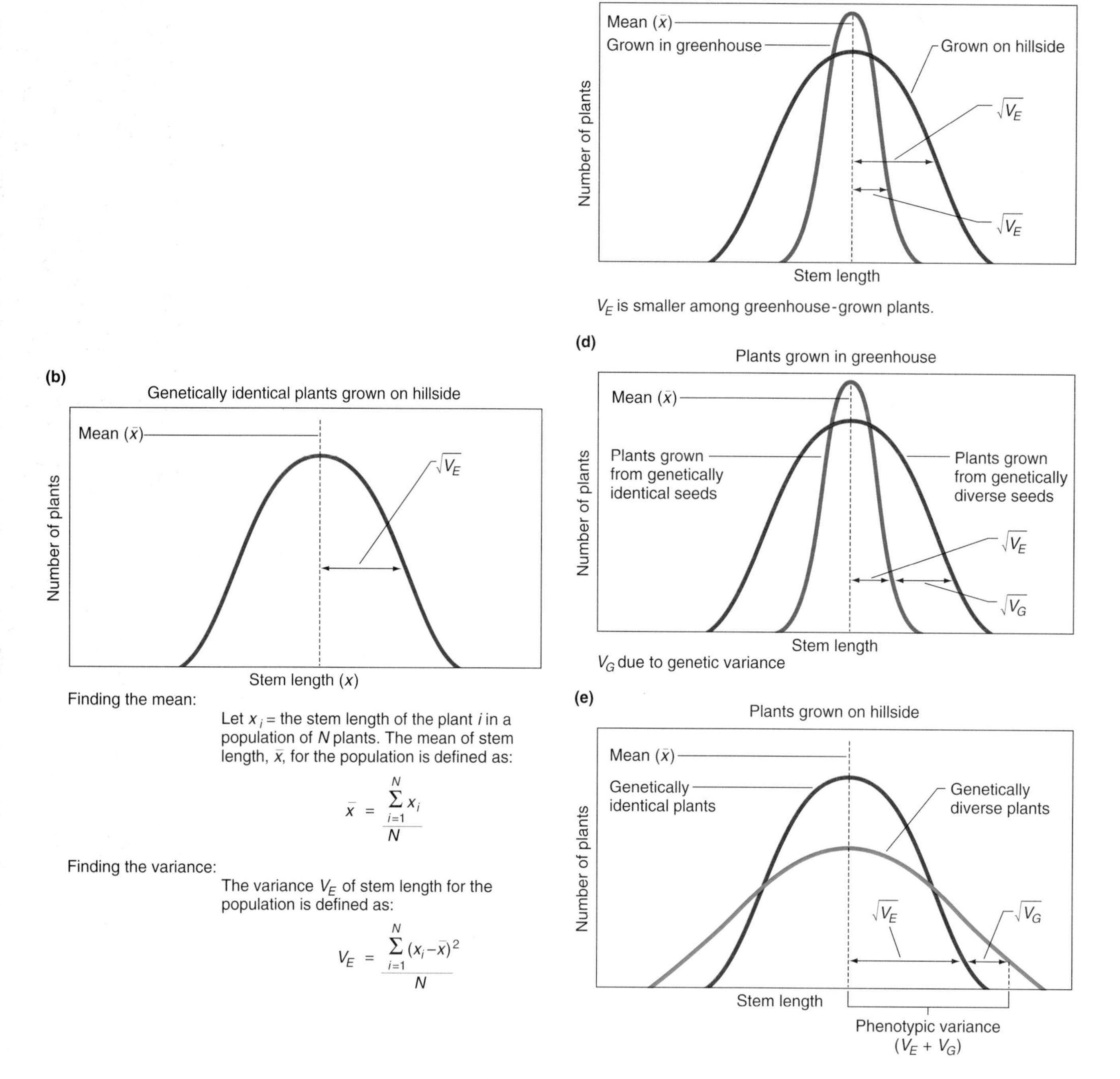

(c) Genetically identical seeds

Mean ($\bar{x}$)
Grown in greenhouse
Grown on hillside
$\sqrt{V_E}$
$\sqrt{V_E}$

Number of plants

Stem length

V_E is smaller among greenhouse-grown plants.

(b) Genetically identical plants grown on hillside

Mean ($\bar{x}$)
$\sqrt{V_E}$

Number of plants

Stem length (x)

Finding the mean:

Let x_i = the stem length of the plant i in a population of N plants. The mean of stem length, $\bar{x}$, for the population is defined as:

$$\bar{x} = \frac{\sum_{i=1}^{N} x_i}{N}$$

Finding the variance:

The variance V_E of stem length for the population is defined as:

$$V_E = \frac{\sum_{i=1}^{N} (x_i - \bar{x})^2}{N}$$

(d) Plants grown in greenhouse

Mean ($\bar{x}$)
Plants grown from genetically identical seeds
Plants grown from genetically diverse seeds
$\sqrt{V_E}$
$\sqrt{V_G}$

Number of plants

Stem length

V_G due to genetic variance

(e) Plants grown on hillside

Mean ($\bar{x}$)
Genetically identical plants
Genetically diverse plants
$\sqrt{V_E}$
$\sqrt{V_G}$

Number of plants

Stem length

Phenotypic variance ($V_E + V_G$)

Studies of dandelions can help sort out the effects of genes versus the environment
Figure 20.13

(b) Correlation between parents and offspring

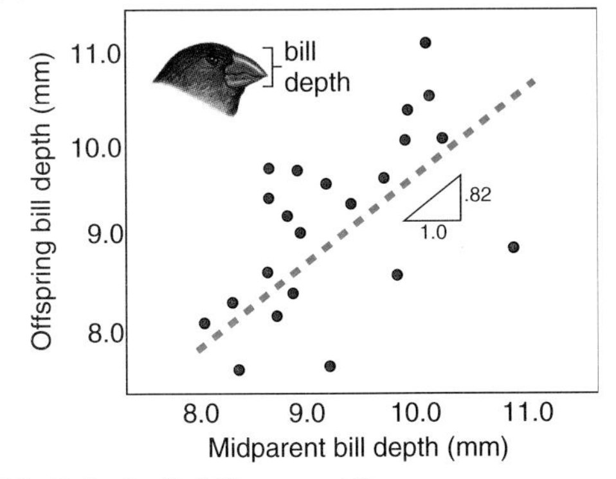

(c) If the heritability were 1.0

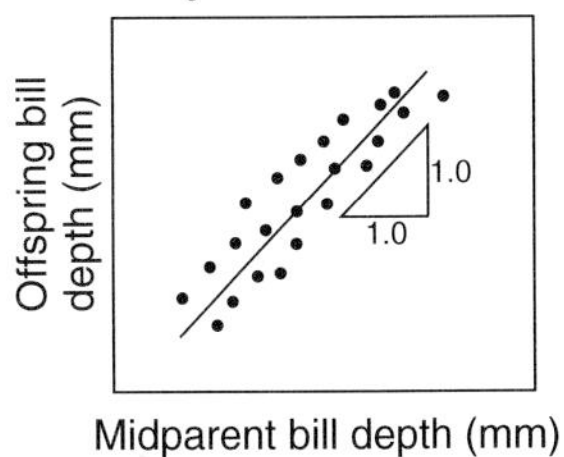

(d) If the heritability were 0.0

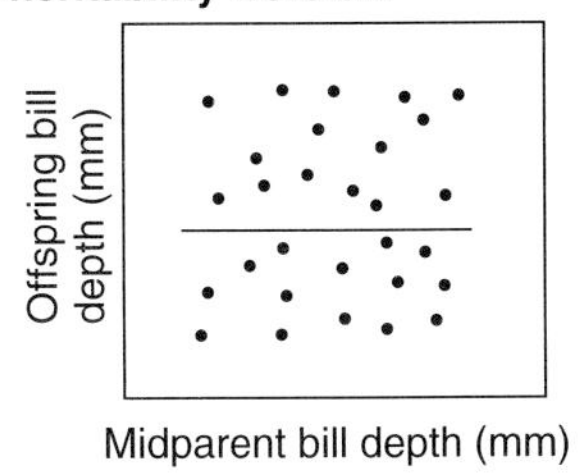

Measuring the heritability of bill depth in populations of Darwin's finches
Figure 20.14

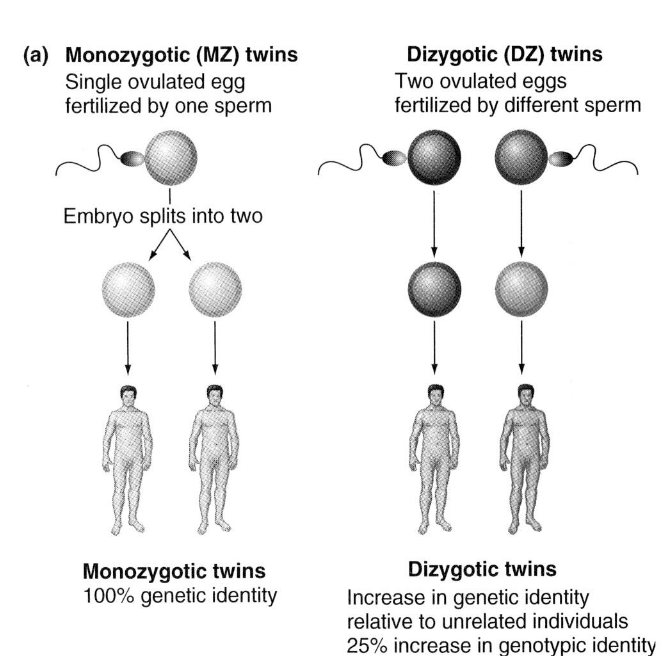

(a) Monozygotic (MZ) twins
Single ovulated egg fertilized by one sperm

Embryo splits into two

Monozygotic twins
100% genetic identity

Dizygotic (DZ) twins
Two ovulated eggs fertilized by different sperm

Dizygotic twins
Increase in genetic identity relative to unrelated individuals
25% increase in genotypic identity
50% increase in allele sharing

(b) Examples of results expected with traits of 0.0 heritability

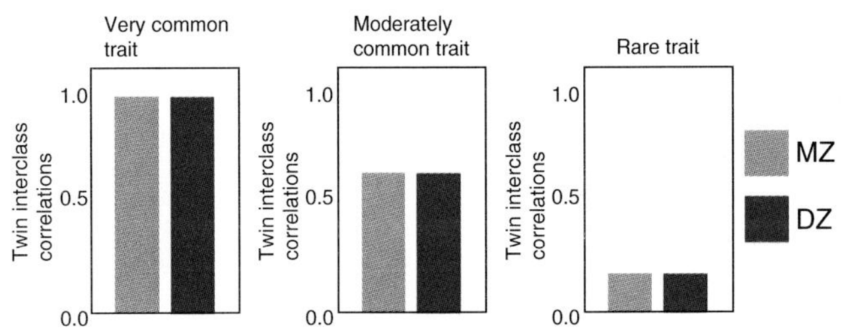

(No difference in correlation levels between MZ and DZ classes)

(c) Examples of results expected with traits of 1.0 heritability

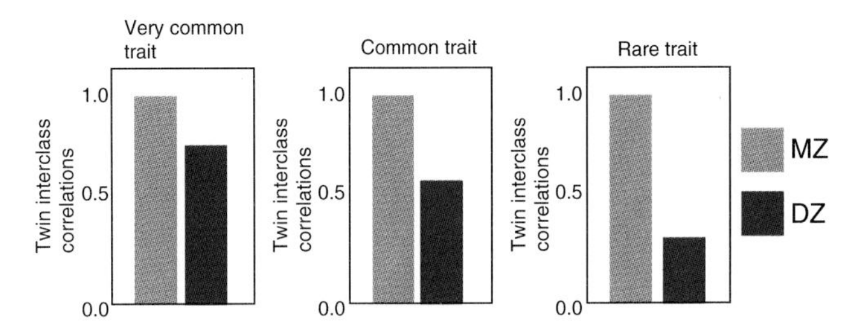

(Differences observed in correlation levels of genotypic traits between MZ and DZ classes)

Comparing phenotypic differences between sets of MZ and DZ twins to show the heritability of a trait
Figure 20.15

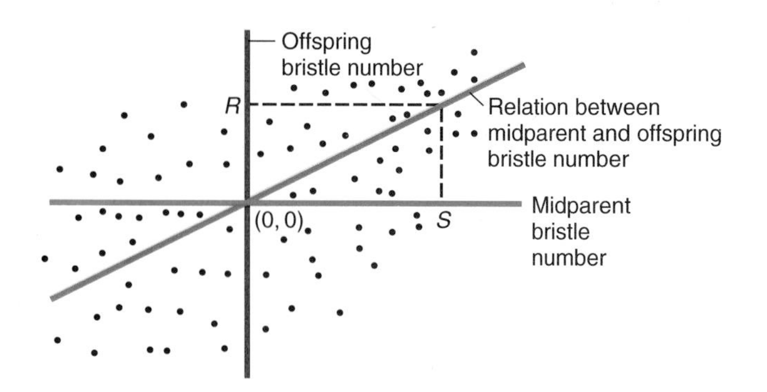

Relationship between midparent number of abdominal bristles and bristle number in offspring for a hypothetical laboratory population of *Drosophila*
Figure 20.16

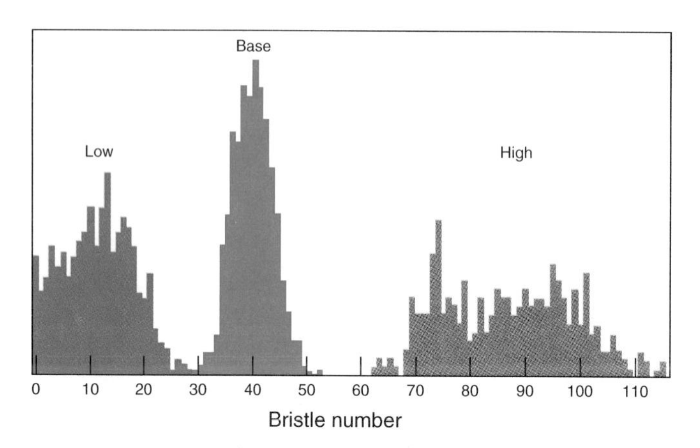

Evolution of abdominal bristle number in response to artificial selection in *Drosophila*
Figure 20.17

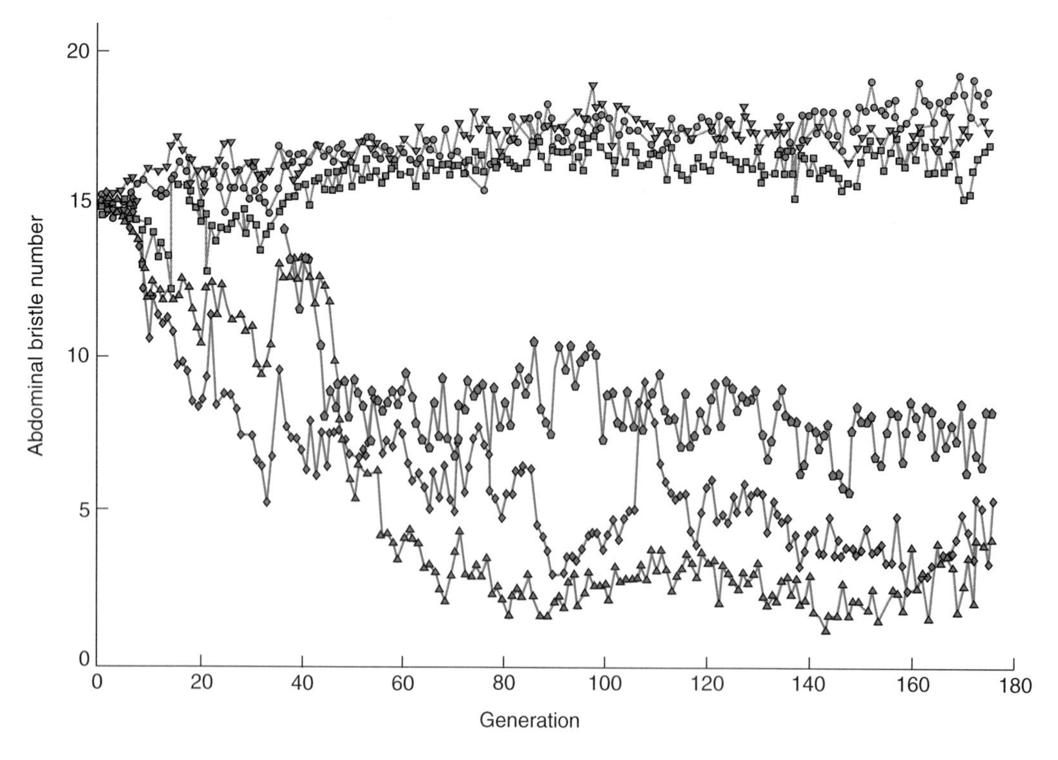

The effect of new mutations on mean bristle number in *Drosophila*
Figure 20.18

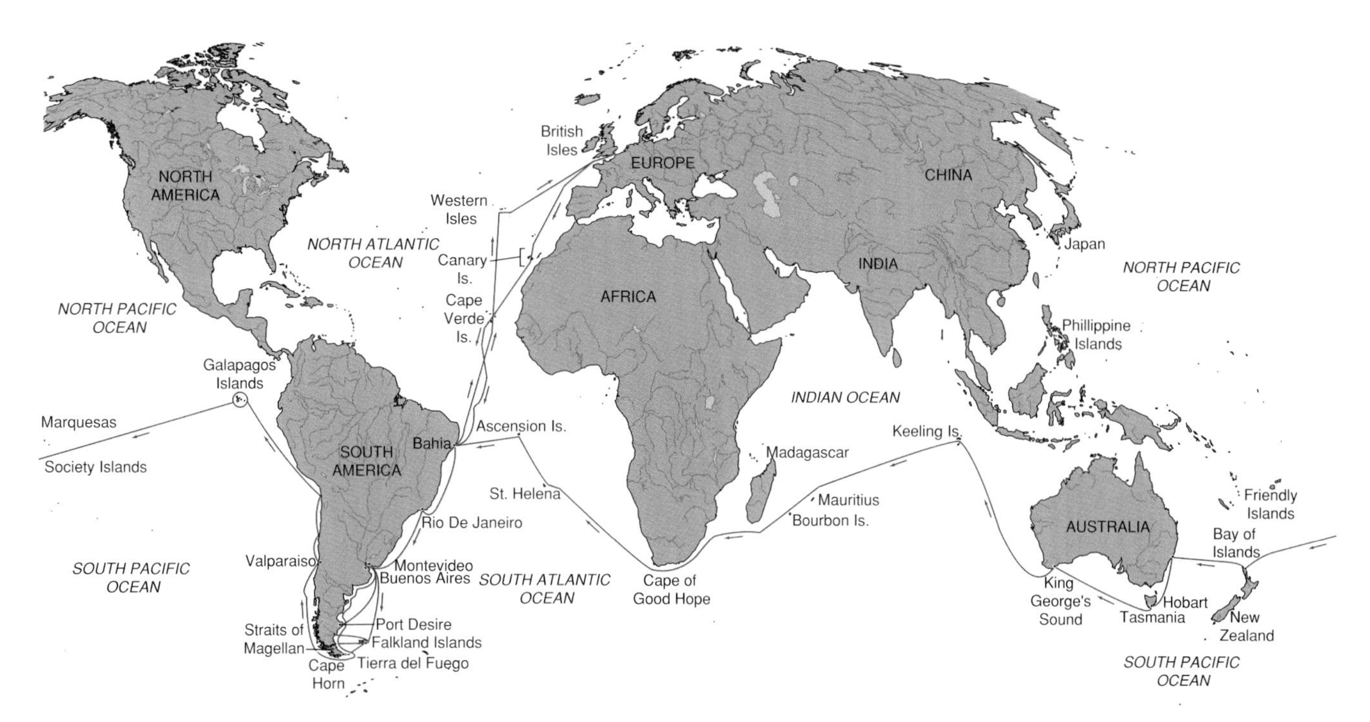

The voyage of the HMS *Beagle*
Figure 21.1

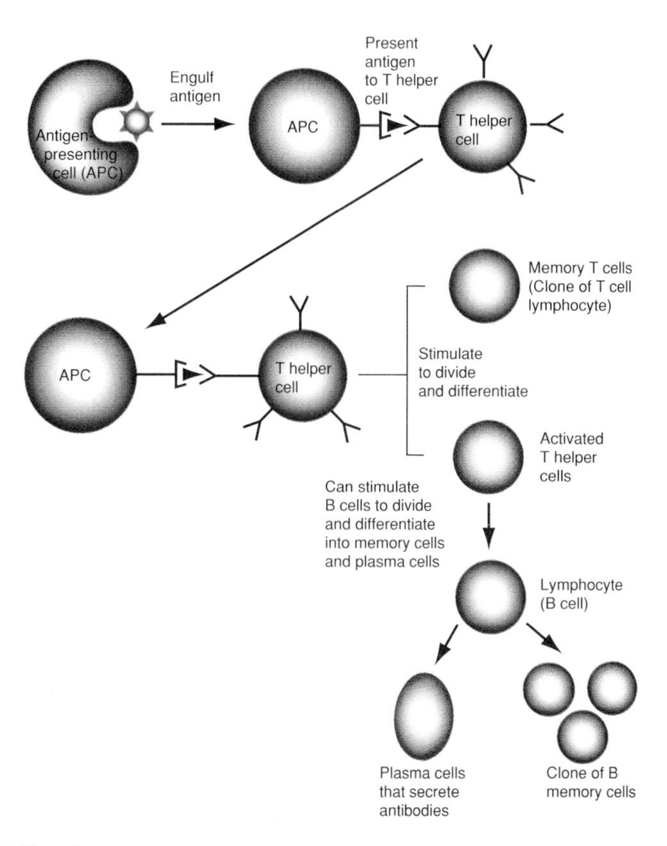

The immune response
Figure 21.2

225

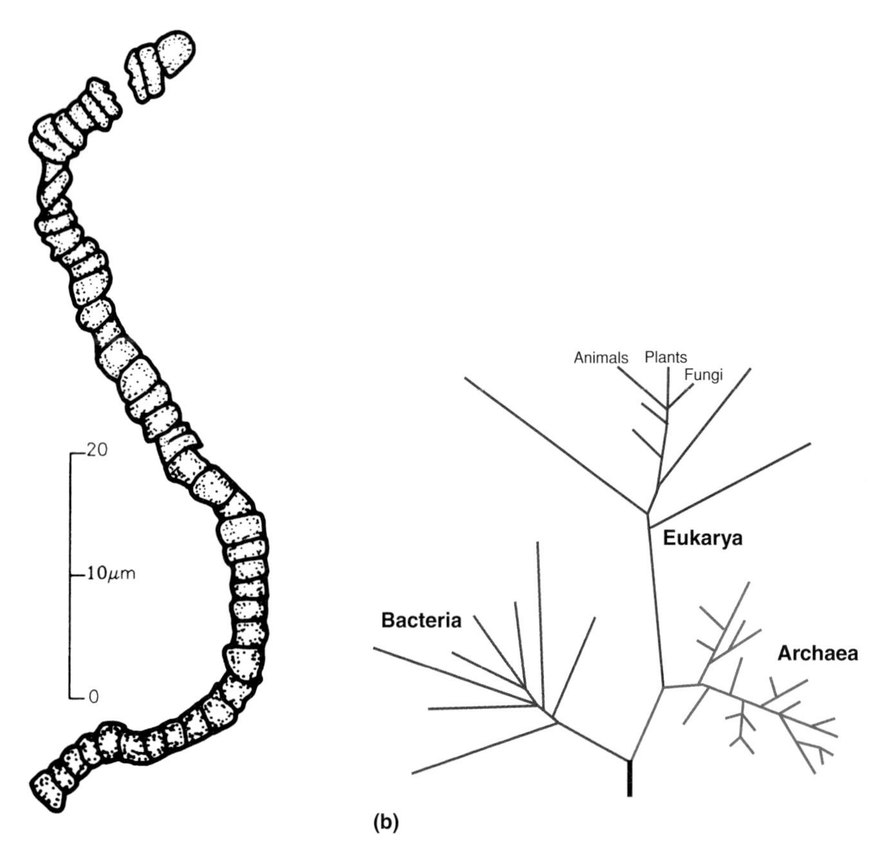

The earliest cells evolved into three kingdoms of living organisms
Figure 21.3

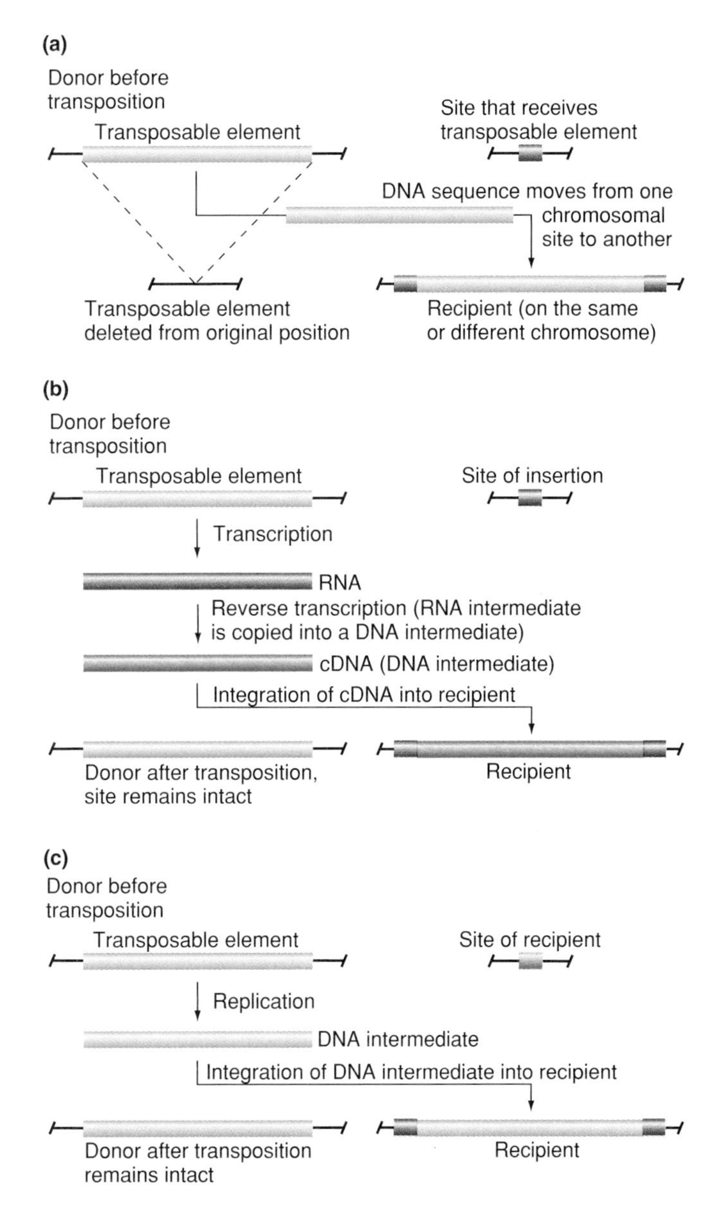

(a)

Donor before
transposition

Transposable element

Site that receives
transposable element

DNA sequence moves from one
chromosomal
site to another

Transposable element
deleted from original position

Recipient (on the same
or different chromosome)

(b)

Donor before
transposition

Transposable element

Site of insertion

Transcription

RNA

Reverse transcription (RNA intermediate
is copied into a DNA intermediate)

cDNA (DNA intermediate)

Integration of cDNA into recipient

Donor after transposition,
site remains intact

Recipient

(c)

Donor before
transposition

Transposable element

Site of recipient

Replication

DNA intermediate

Integration of DNA intermediate into recipient

Donor after transposition
remains intact

Recipient

**Duplication by transposition: Three
possibilities**

Figure 21.6

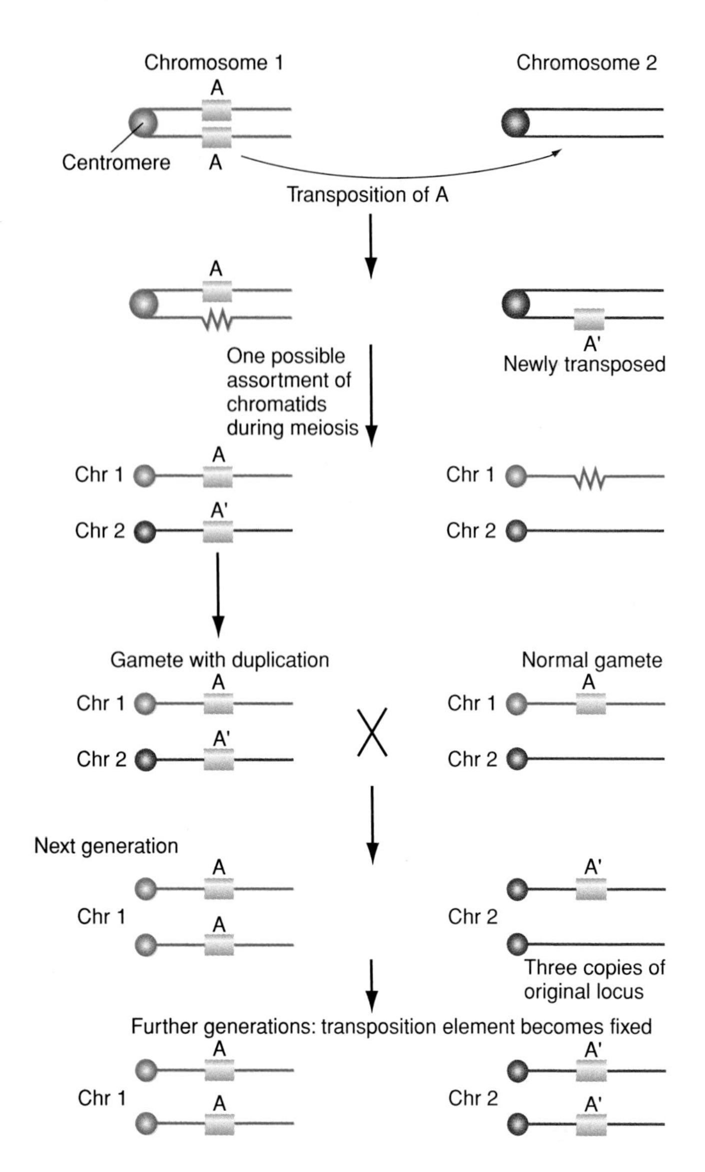

**Transposition through direct movement of
a DNA sequence**
Figure 21.7

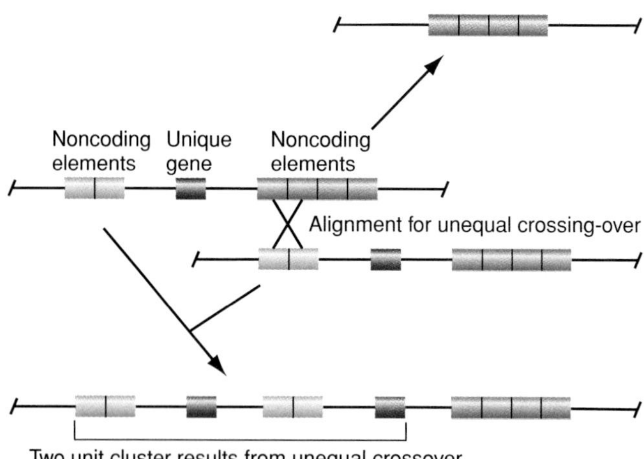

**Duplications arise from unequal
crossing-over**
Figure 21.8

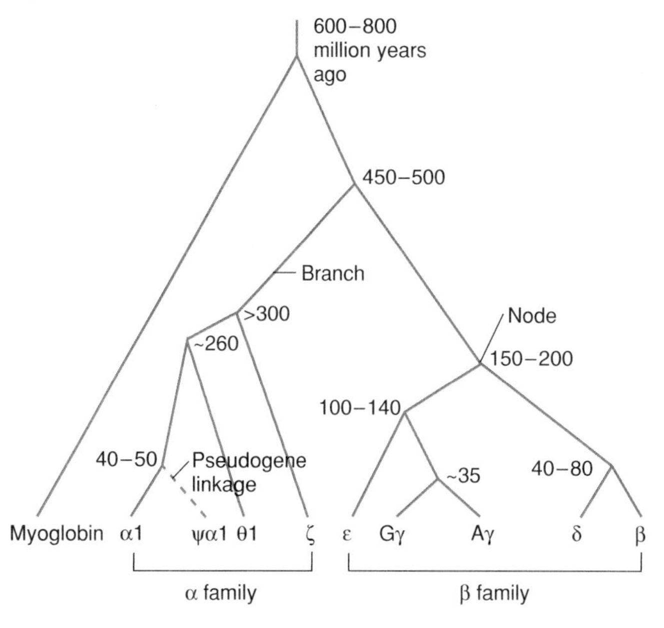

A phylogenetic tree
Figure 21.9

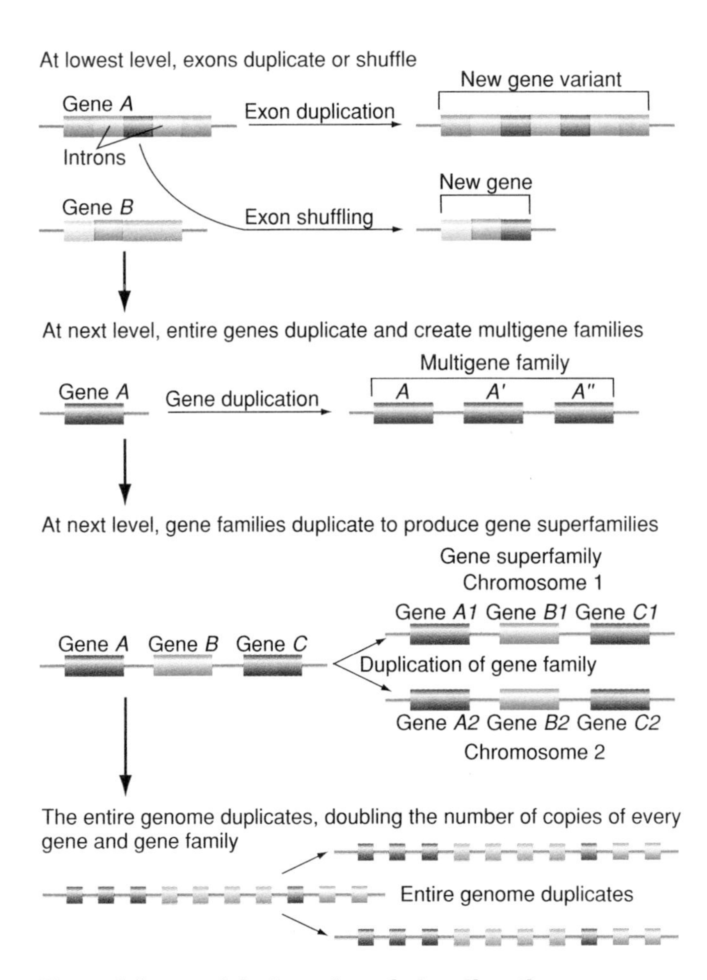

At lowest level, exons duplicate or shuffle

At next level, entire genes duplicate and create multigene families

At next level, gene families duplicate to produce gene superfamilies

The entire genome duplicates, doubling the number of copies of every gene and gene family

**Four hierarchic levels of duplications
increase genome size**
Figure 21.10

229

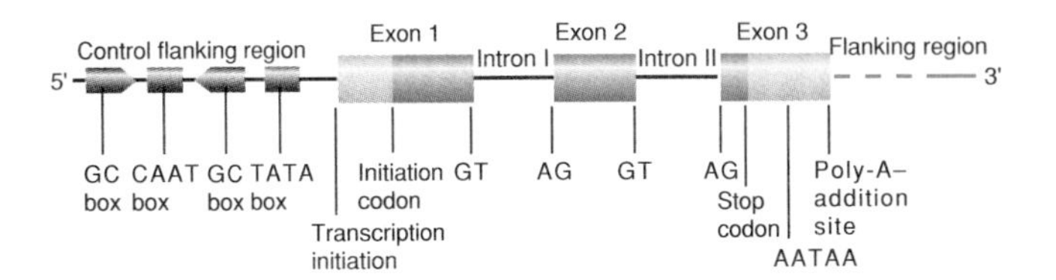

The basic structure of a gene
Figure 21.11

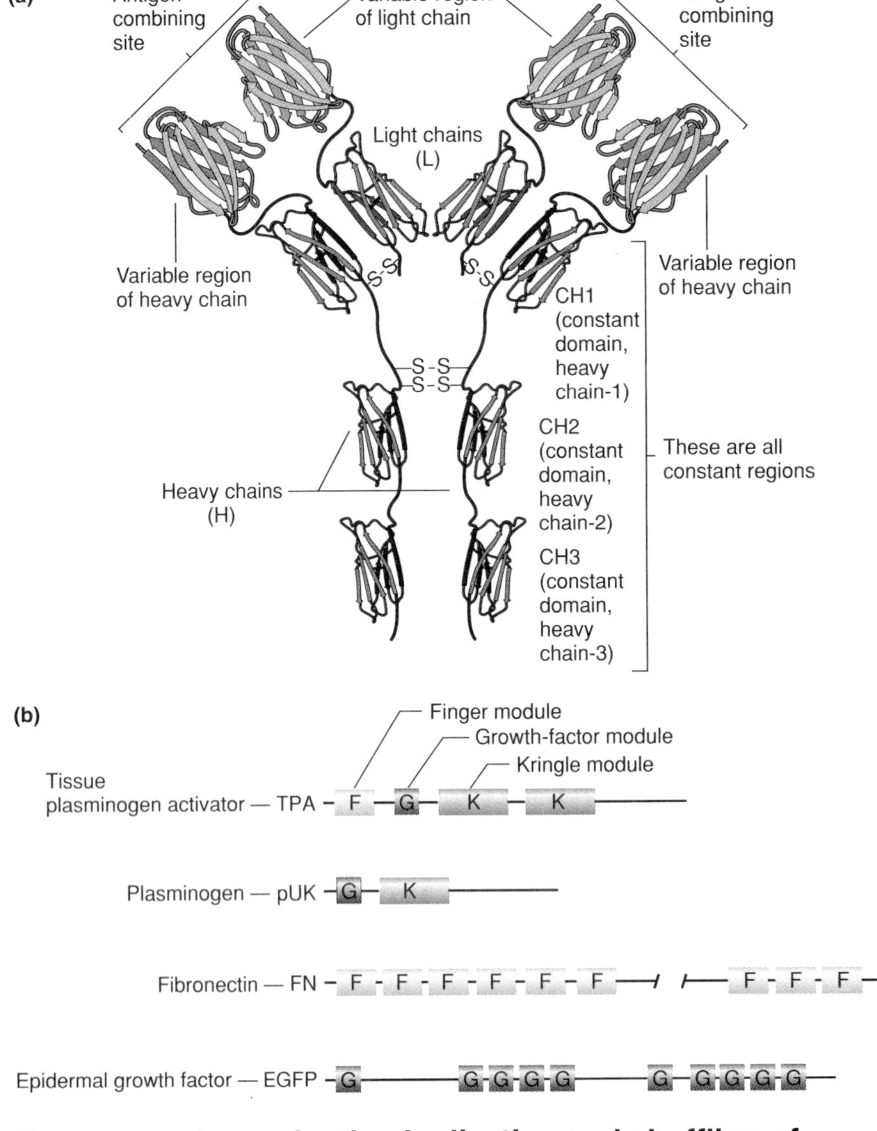

(a)

Antigen-combining site

Variable region of light chain

Antigen-combining site

Light chains (L)

Variable region of heavy chain

S-S

S-S

Variable region of heavy chain

CH1 (constant domain, heavy chain-1)

S-S
S-S

CH2 (constant domain, heavy chain-2)

These are all constant regions

Heavy chains (H)

CH3 (constant domain, heavy chain-3)

(b)

Finger module
Growth-factor module
Kringle module

Tissue plasminogen activator — TPA — F — G — K — K

Plasminogen — pUK — G — K

Fibronectin — FN — F - F - F - F - F - F - F — / — / — F - F - F —

Epidermal growth factor — EGFP — G — G-G-G-G — G - G-G-G-G —

Genes can change by the duplication and shuffling of exons
Figure 21.12

Multigene families
Figure 21.13

Tandem gene family: Members of the multigene family are clustered on the same chromosome

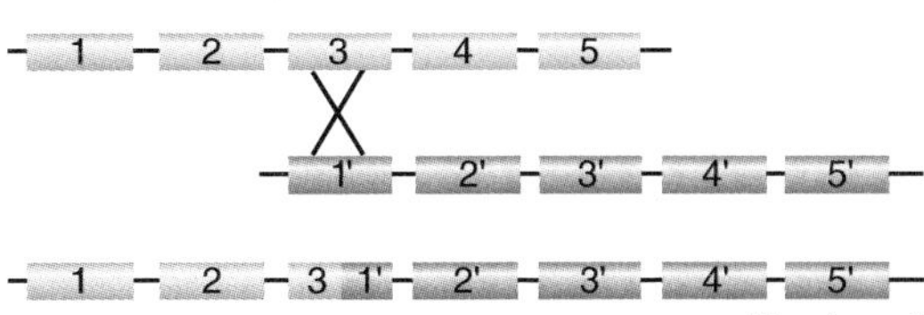

Dispersed gene family: Members of the multigene family are on different chromosomes

Chromosome 1
Chromosome 2
Chromosome 3
Chromosome 4

1 — 2 — 3 — 4 — 5

1' — 2' — 3' — 4' — 5'

1 — 2 — 3 1' — 2' — 3' — 4' — 5'

Number of members in this cluster increases

1' 3 — 4 — 5

Number of members in this cluster decreases

(a)

Gene family

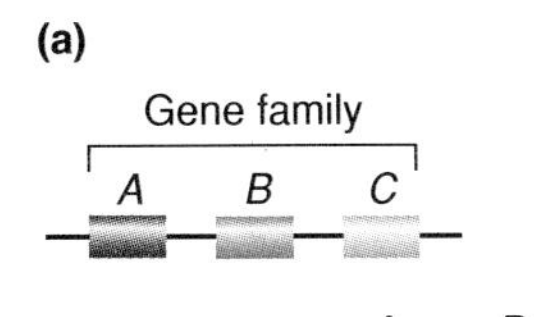

A *B* *C*

During synapsis gene *A* aligns with gene *C* because of homology between the two. Then a portion of gene *C* is changed to the sequence of gene *A*.

Note: Gene *A* remains unchanged, only gene *C* has a portion (dark) of the gene *A* sequence.

Converted region

Alternative resolution: unequal crossing-over

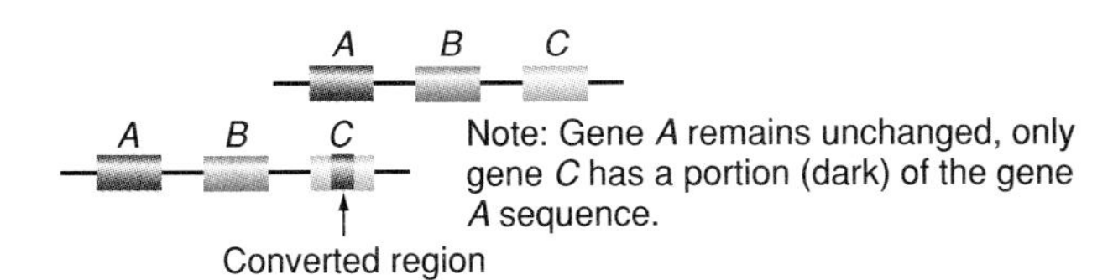

A *B* *C/A* *B* *C*

↓ Divergence

A *B* *C* *D* *E*

(b)

MHC (mice)

Small set of functional class I genes

Class I pseudogenes (25 – 38)
Nonfunctional pseudogenes

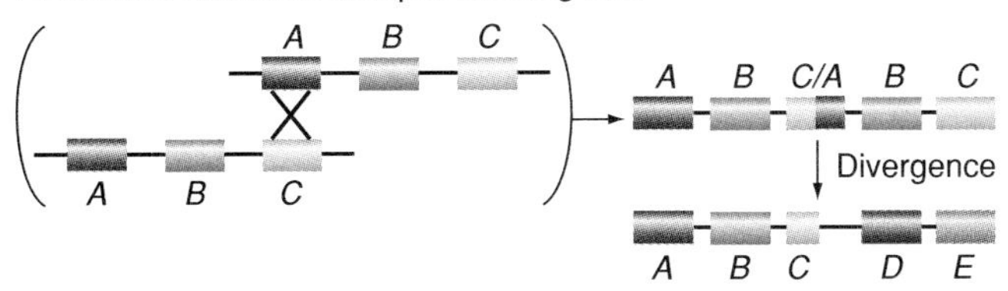

Information flow through gene conversion

Many alleles in population

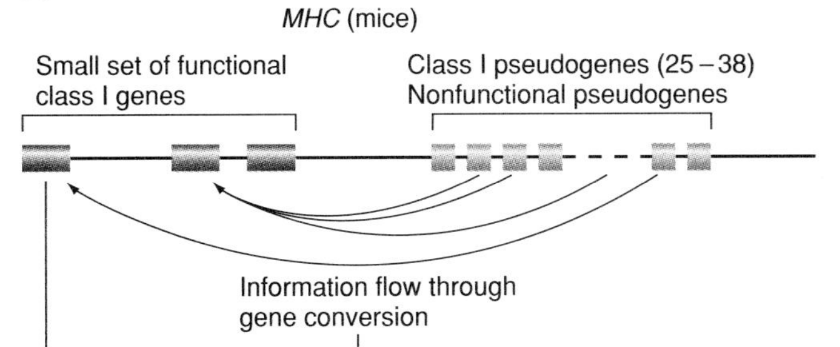

Intergenic gene conversion
Figure 21.14

(a)

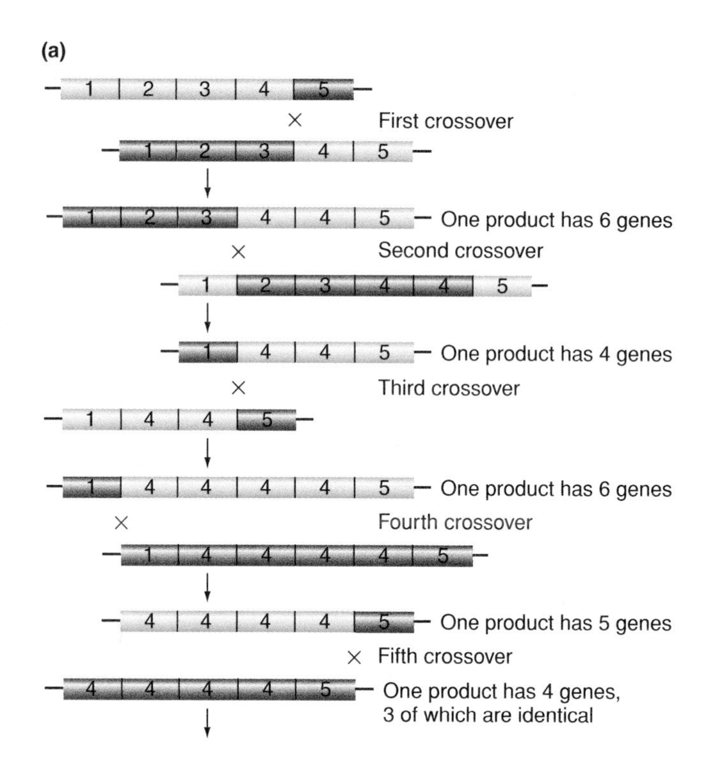

(b)

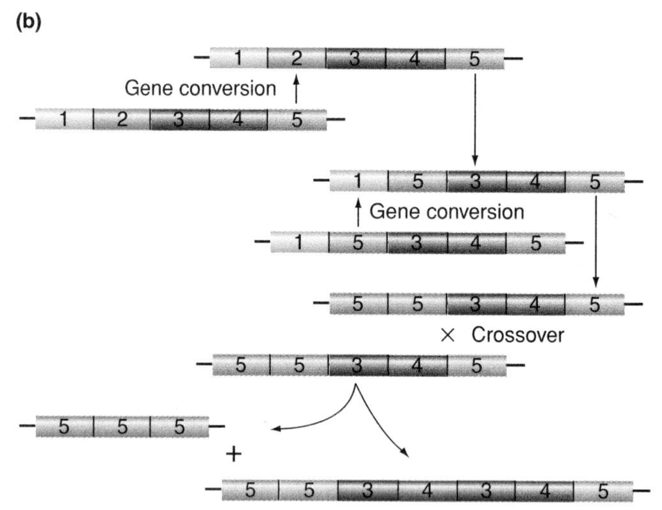

Concerted evolution can lead to gene homogeneity

Figure 21.15

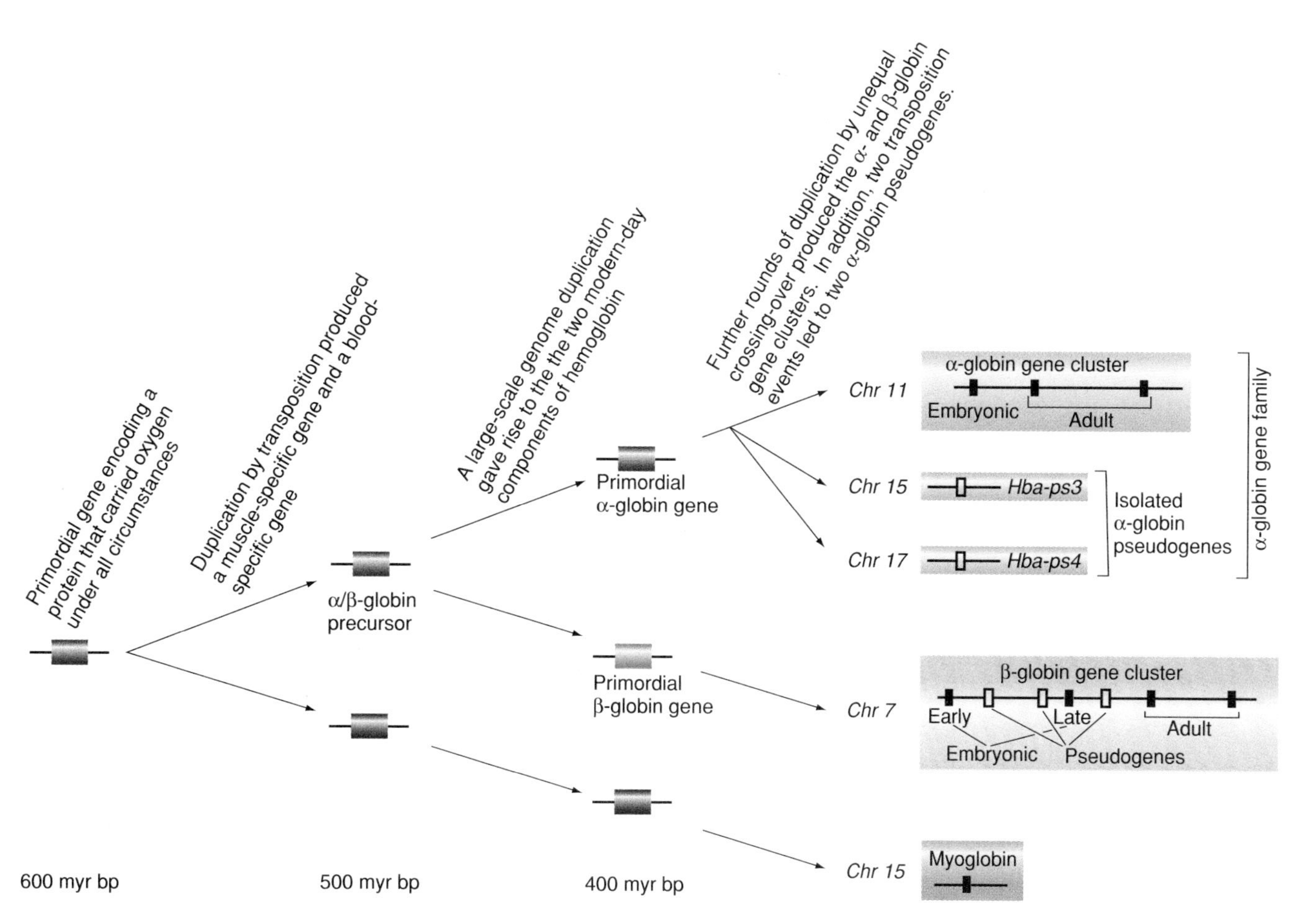

Evolution of the mouse globin superfamily
Figure 21.16

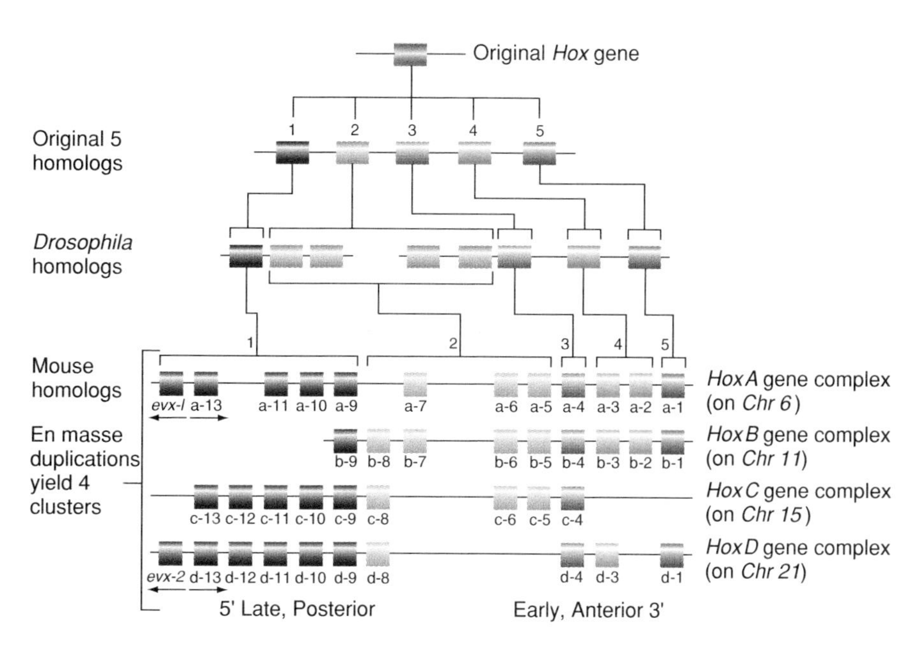

Evolution of the *Hox* gene superfamily of mouse and *Drosophila*
Figure 21.17

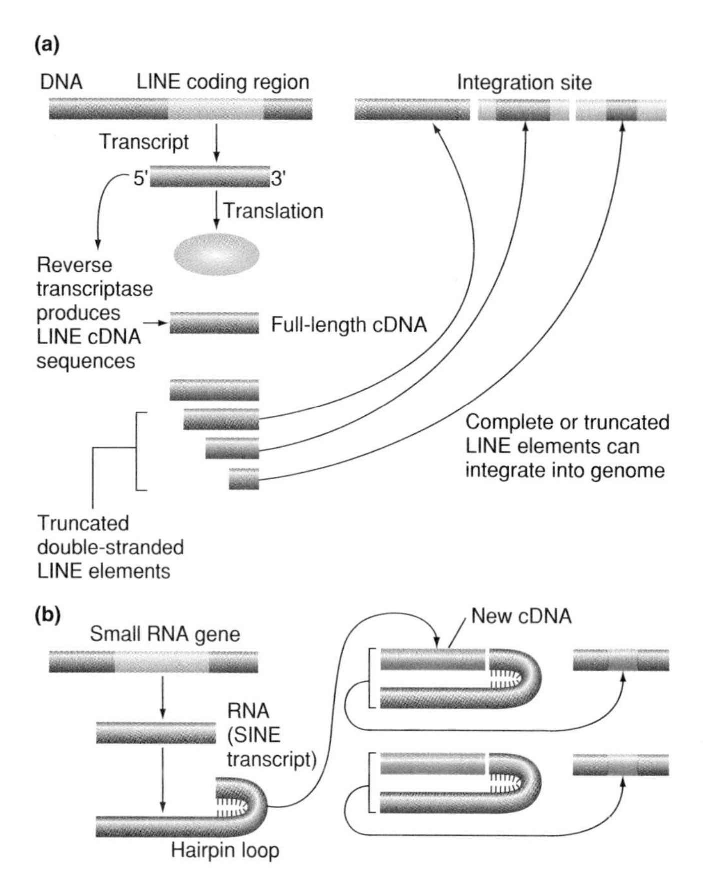

(a)

DNA   LINE coding region   Integration site

Transcript

5'   3'

Translation

Reverse transcriptase produces LINE cDNA sequences

Full-length cDNA

Complete or truncated LINE elements can integrate into genome

Truncated double-stranded LINE elements

(b)

Small RNA gene   New cDNA

RNA (SINE transcript)

Hairpin loop

The creation of LINE and SINE gene families
Figure 21.18

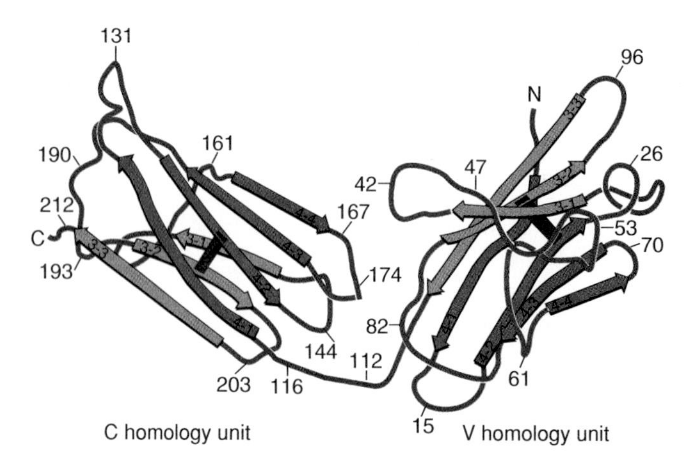

131

190

161

96

N

47   26

42

167

212

C

53

70

193

174

82

144   112

203   116

61

15

C homology unit   V homology unit

Three-dimensional structures encoded by immunoglobulin homology units
Figure 21.19

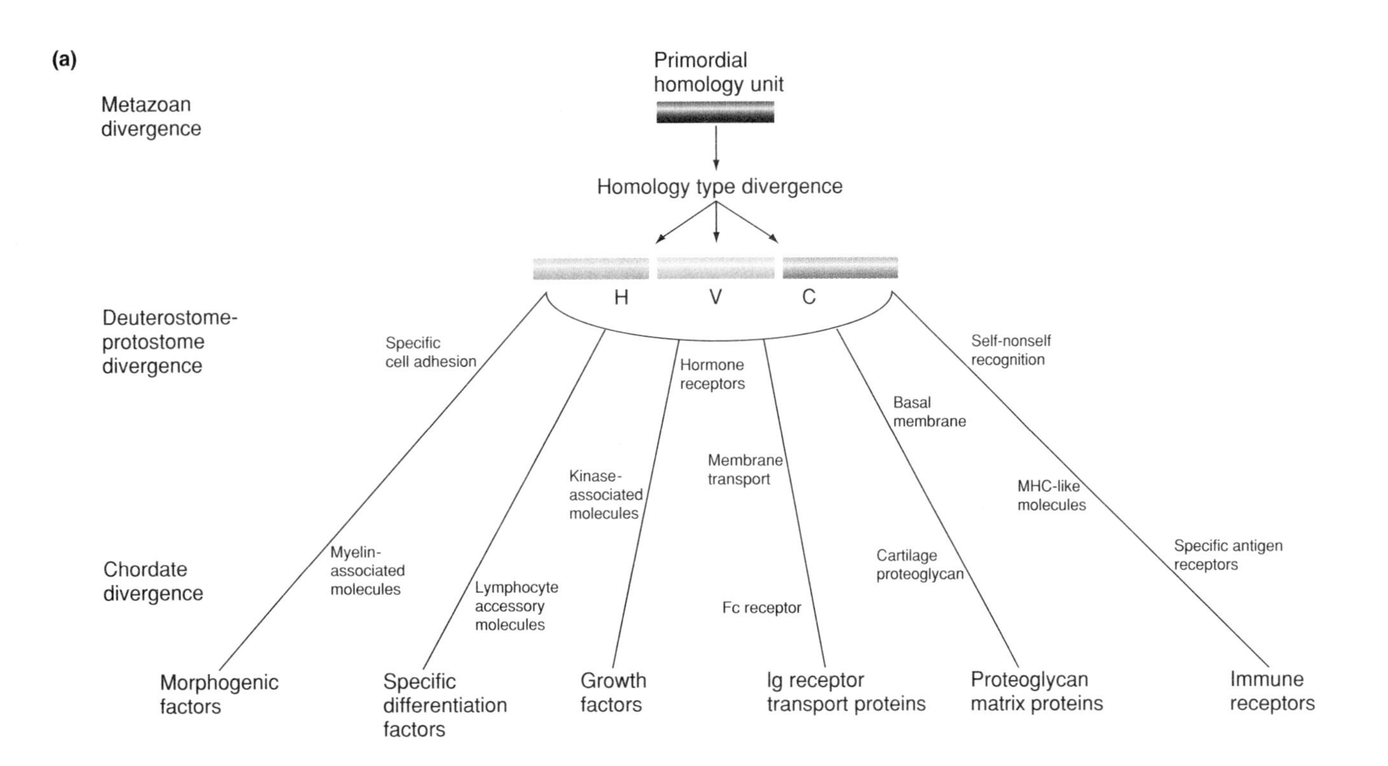

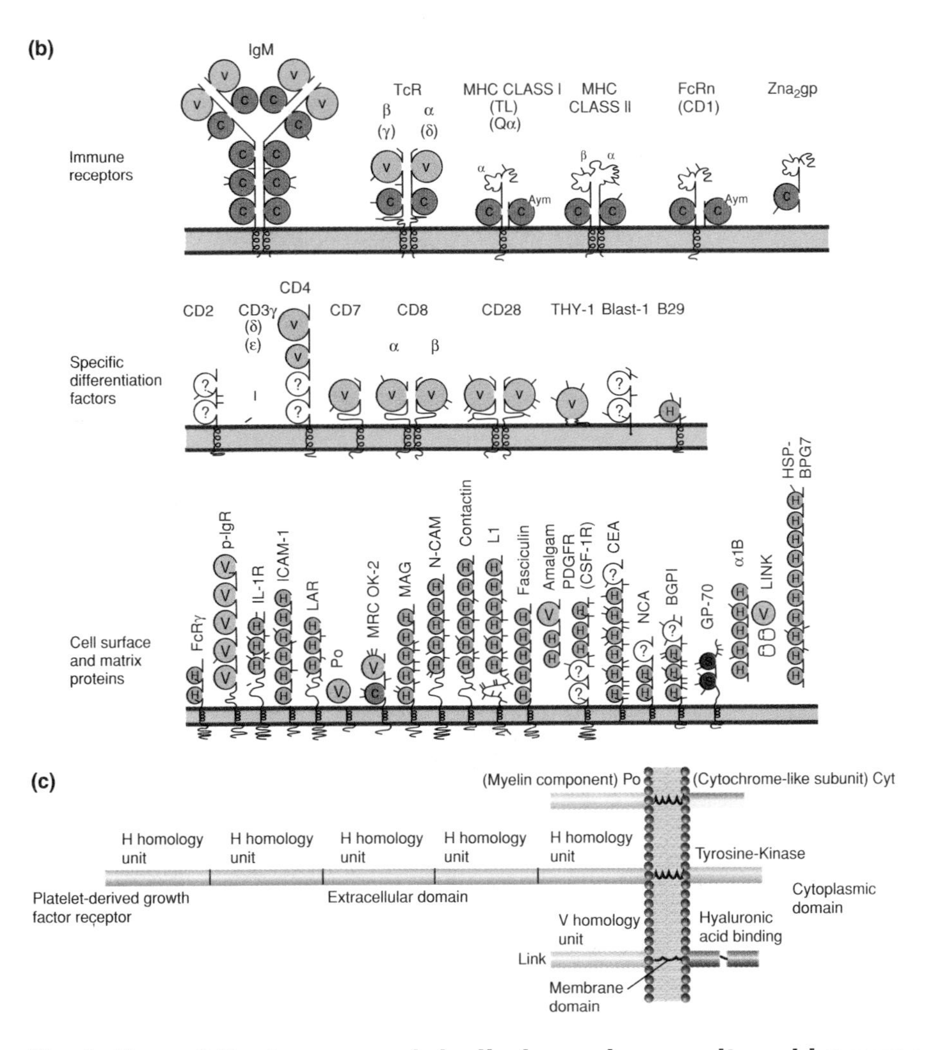

Evolution of the immunoglobulin homology unit and immunoglobulin gene superfamily
Figure 21.20

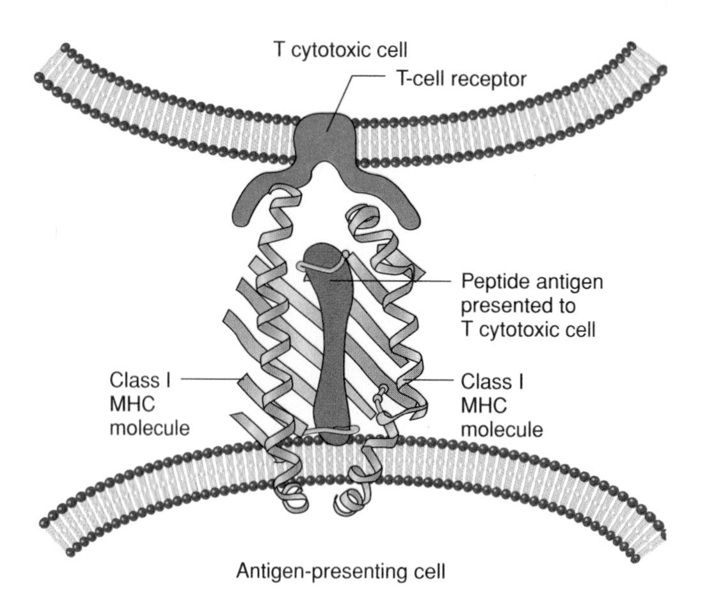

T cytotoxic cell

T-cell receptor

Peptide antigen
presented to
T cytotoxic cell

Class I
MHC
molecule

Class I
MHC
molecule

Antigen-presenting cell

Immune-cell receptors
Figure 21.21

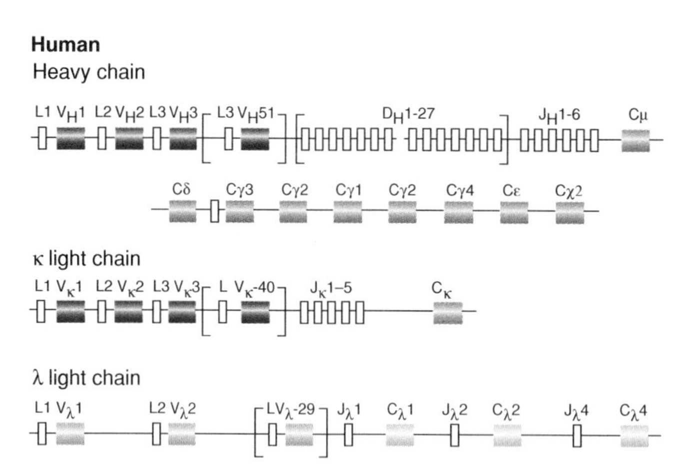

Human
Heavy chain

L1 V$_H$1 L2 V$_H$2 L3 V$_H$3 L3 V$_H$51   D$_H$1-27   J$_H$1-6   Cμ

Cδ   Cγ3   Cγ2   Cγ1   Cγ2   Cγ4   Cε   Cχ2

κ light chain

L1 V$_κ$1 L2 V$_κ$2 L3 V$_κ$3 L V$_κ$-40   J$_κ$1–5   C$_κ$

λ light chain

L1 V$_λ$1   L2 V$_λ$2   LV$_λ$-29   J$_λ$1   C$_λ$1   J$_λ$2   C$_λ$2   J$_λ$4   C$_λ$4

**Three families of antibody genes encode
three types of antibody chains: One heavy
and two light—λ and κ**
Figure 21.22

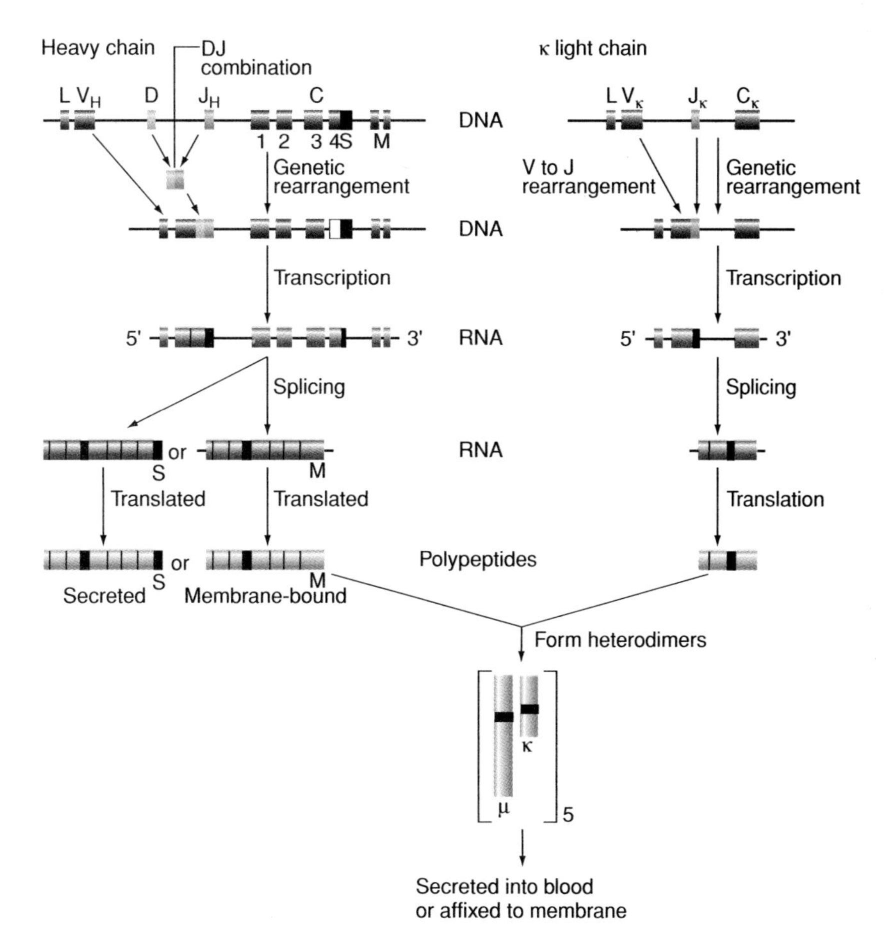

DNA rearrangements bring together segments of a gene for expression
Figure 21.23

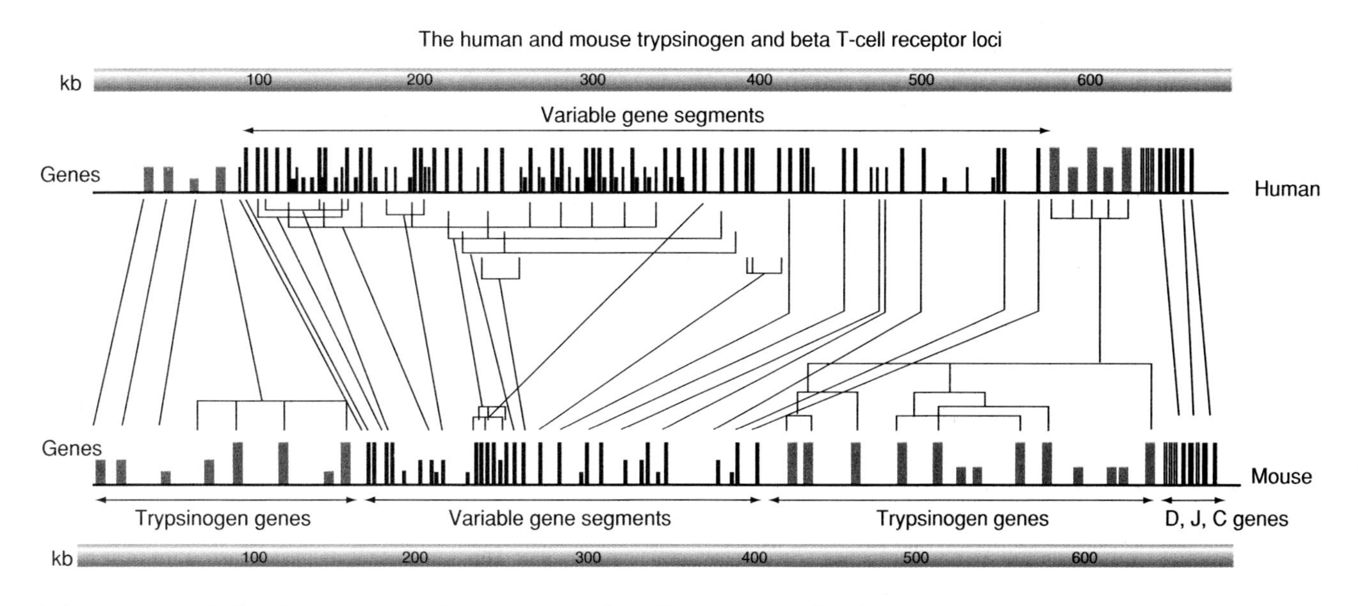

The human and mouse trypsinogen and beta T-cell receptor loci

Diagrams of the human and mouse β T-cell receptor loci
Figure 21.24